Driving Climate Change: Cutting Carbon from Transportation

Driving Climate Change: Cutting Carbon from Transportation

Edited by
**Daniel Sperling and
James S. Cannon**

AMSTERDAM • BOSTON • HEIDELBERG • LONDON
NEW YORK • OXFORD • PARIS • SAN DIEGO
SAN FRANCISCO • SINGAPORE • SYDNEY • TOKYO
Academic Press is an imprint of Elsevier

ELSEVIER

Academic Press is an imprint of Elsevier
30 Corporate Drive, Suite 400, Burlington, MA 01803, USA
525 B Street, Suite 1900, San Diego, California 92101-4495, USA
84 Theobald's Road, London WC1X 8RR, UK

This book is printed on acid-free paper.

Copyright © 2007, Elsevier Inc. All rights reserved.

No part of this publication may be reproduced or transmitted in any form or by any means, electronic or mechanical, including photocopy, recording, or any information storage and retrieval system, without permission in writing from the publisher.

Permissions may be sought directly from Elsevier's Science & Technology Rights Department in Oxford, UK: phone: (+44) 1865 843830, fax: (+44) 1865 853333, E-mail: permissions@elsevier.com. You may also complete your request on-line via the Elsevier homepage (http://elsevier.com), by selecting "Support & Contact" then "Copyright and Permission" and then "Obtaining Permissions."

Library of Congress Cataloging-in-Publication Data
Application Submitted

British Library Cataloguing-in-Publication Data
A catalogue record for this book is available from the British Library.

ISBN 13: 978-0-12-369495-9
ISBN 10: 0-12-369495-7

For information on all Academic Press publications
visit our Web site at www.books.elsevier.com

Printed in the United States of America
07 08 09 10 9 8 7 6 5 4 3 2 1

Working together to grow
libraries in developing countries

www.elsevier.com | www.bookaid.org | www.sabre.org

ELSEVIER   BOOK AID International   Sabre Foundation

Table of Contents

Acknowledgments	vii
Preface	ix
1. Introduction and Overview by Dan Sperling and James S. Cannon	1
2. Peaking of World Oil Production and Its Mitigation by Robert L. Hirsch, Roger Bezdek, and Robert Wendling	9
3. Toward a Policy Agenda for Climate Change: Changing Technologies and Fuels and the Changing Value of Energy by Duncan Eggar	29
4. Coordinated Policy Measures for Reducing the Fuel Use of the U.S. Light Duty Vehicle Fleet by Anup P. Bandivadekar and John B. Heywood	41
5. Carbon Burdens from New Car Sales in the United States by John DeCicco, Freda Fung, and Feng An	73
6. Reducing Vehicle Emissions Through Cap-and-Trade Schemes by John German	89
7. North American Feebate Analysis Model by Alexandre Dumas, David L. Greene, and André Bourbeau	107
8. Reducing Growth in Vehicle Miles Traveled: Can We Really Pull It Off by Gary Toth	129
9. International Comparison of Policies to Reduce Greenhouse Gas Emissions from Passenger Vehicles by Feng An	143
10. Reducing Transport-Related Greenhouse Gas Emissions in Developing Countries: The Role of the Global Environmental Facility by Walter Hook	165

11. What Multilateral Banks (and Other Donors) Can Do to Reduce Greenhouse Gas Emissions: A Case Study of Latin America and the Caribbean by Deborah Bleviss ... 189

12. From Public Understanding to Public Policy: Public Views on Energy, Technology, and Climate Science in the United States by David M. Reiner ... 201

13. Narrative Self-Identity and Societal Goals: Automotive Fuel Economy and Global Warming Policy by Kenneth S. Kurani, Thomas S. Turrentine, and Reid R. Heffner ... 217

14. Lost in Option Space: Risk Partitioning to Guide Climate and Energy Policy by David L. Bodde ... 239

15. Toward a Transportation Policy Agenda for Climate Change by David Burwell and Daniel Sperling ... 253

Appendix A About the Editors and Authors ... 269

Appendix B Asilomar Attendee List: 2005 ... 279

Index ... 285

Acknowledgments

This book is the product of many people and much time, money, and talent. The book is an outgrowth of the Tenth Biennial Asilomar Conference on Transportation and Energy, held August 23–26, 2005, in Pacific Grove, California. The chapters evolved from presentations and discussions at the conference. The opinions presented are those of the chapter authors.

The conference was hosted and organized by the Institute of Transportation Studies at the University of California, Davis (ITS-Davis), under the auspices of the U.S. National Research Council's Transportation Research Board—in particular, the standing committees on Energy, Alternative Fuels, and Transportation and Sustainability.

The conference would not have been possible without the generous support of the following organizations: American Honda Motor Company, California Air Resources Board (California Environmental Protection Agency), California Department of Transportation, Energy Foundation, Natural Resources Canada, Surdna Foundation, U.S. Department of Transportation Center for Climate Change, U.S. Environmental Protection Agency Offices of Research and Transport and Air Quality, U.S. National Oceanic and Atmospheric Administration, University of California Transportation Center, WestStart-CALSTART, and William and Flora Hewlett Foundation.

Plus we want to acknowledge the Corporate Affiliate Members of ITS-Davis who provide valuable support that allows the Institute the flexibility to initiate new activities and events such as the conference upon which this book is based. Those companies are ExxonMobil, Nissan, Toyota, Aramco, Chevron, Subaru, Nippon Oil, and Pacific Gas & Electric.

The conference program was directed by Daniel Sperling, along with David Burwell, John DeCicco, Lew Fulton, David Greene, Judi Greenwald, Jack Johnston, Robert Larson, Marianne Mintz, Peter Reilly-Roe, Farideh Ramjerdi, Mike Savonis, Chris Sloane, Lee Schipper, and Steve Winkelman. This committee worked closely in crafting a set of speakers and topics that was engaging and insightful.

In addition to the many authors, we want to acknowledge the generous assistance of a large set of peer reviewers, drawn from attendees at the conference, who provided valuable feedback and suggestions to the authors. Book production was assisted by Jeff Georgeson, who provided copyediting assistance. We especially appreciate the efforts of Christine Minihane of Elsevier, who enthusiastically and capably shepherded the entire book project from its inception.

Most of all, we want to acknowledge the many attendees of the conference listed in Appendix B. These invited leaders and experts, coming from many parts of the world and many segments of society, enriched the conference with their deep insights and rich experiences.

Preface

Climate change is creeping into the public consciousness. Arcane scientific debates are front-page news. Novels and movies feature climate change. Presidents and prime ministers are becoming conversant in climate change science and policy. With greater attention, though, comes greater controversy and conflict. The public debate is more cacophonous than ever. The public is confused, but so are experts and leaders in government and industry, for good reason. The debate over climate policy is riddled with huge uncertainties and knowledge gaps.

Frustrated by simplistic public discourse and overwhelmed by the enormity of the challenge, a group of individuals organized a high-level meeting on one aspect of the debate: transportation energy policy and investments. Two hundred leaders and experts were assembled from the automotive and energy industries, start-up technology companies, public interest groups, academia, U.S. energy laboratories, and governments from around the world. Three broad strategies for reducing greenhouse gas emissions were investigated: reducing motorized travel, shifting travel to less-energy-intensive modes, and changing fuel and propulsion technologies. This book is an outgrowth of that conference.

The conference was not a one-time event. It was the latest in a series of conferences held (almost) every two years on some aspect of transportation and energy policy, always at the same Asilomar Conference Center near Monterey on the California coast. The first conference in 1988 addressed alternative transportation fuels. The full list appears below:

I. Alternative Transportation Fuels in the '90s and Beyond (July 1988)
II. Roads to Alternative Fuels (July 1990)
III. Global Climate Change (August 1991)
IV. Strategies for a Sustainable Transportation System (August 1993)
V. Is Technology Enough? Sustainable Transportation-Energy Strategies (July 1995)
VI. Policies for Fostering Sustainable Transportation Technologies (August 1997)

VII. Transportation Energy and Environmental Policies into the 21st Century (August 1999)
VIII. Managing Transitions in the Transport Sector: How Fast and How Far? (September 2001)
IX. The Hydrogen Transition (July 2003)
X. Toward a Policy Agenda for Climate Change (August 2005)

Except for the 1991 climate change conference, all were organized and hosted by the Institute of Transportation Studies at the University of California, Davis, under the auspices of several committees of the United States Transportation Research Board. Sponsorship was provided by a variety of government agencies, companies, and foundations. Sponsors for this latest conference on climate policy are thanked in the Acknowledgments.

This book contributes understandings that support the development of better research and development programs, better investments, and better policy. But the real contribution to the contentious climate debate may be more basic: to conceptualize what we know about climate change, to articulate the problem with greater clarity, and to identify key questions and a range of possible answers.

These are not modest ambitions. The environmental community has been struggling with these same challenges for some time, highlighted by *Death of Environmentalism*, a controversial 2004 report, authored by two insiders. The treatise suggests that environmental leaders have largely failed to engage society in addressing climate change because they have not been able to conceptualize and articulate what is important in a way that resonates broadly or to devise compelling responses. But the failure to meet climate challenges is not theirs alone. The challenges and the consequences face us all.

This brings us back to knowledge and expertise. The culture of the academic world is built around the search for knowledge. Academics speak in terms of metrics, analytical frameworks, and statistics. But as Henry Kissinger once said, "Most foreign policies that history has marked highly, in whatever country, have been originated by leaders who were opposed by experts." He went on to say, "It is, after all, the responsibility of the expert to operate the familiar and that of the leader to transcend it." We agree. Consider Rachel Carson on environmental awareness (*Silent Spring*), Jane Jacobs on urban planning (*Death and Life of American Cities*), and Betty Freidan on the role of women (*The Feminine Mystique*). None were experts. All were leaders.

More knowledge and more experts are certainly needed in the energy area. But generating more knowledge is not the problem right now. Humans have tremendous intellectual capacity. What is needed is a framework—a reconceptualization of the problem—that will allow human society to create the mechanisms and incentives that will channel creativity productively and efficiently. Channeled creativity is needed soon and on a massive scale.

New ways of thinking and new leaders are needed that reach beyond stovepipe expertise, reductionist approaches, and narrow self interests. The boundaries of knowledge and leadership need to be pushed. Much is at stake. This book, we hope, is one step forward.

CHAPTER 1

Introduction and Overview

Dan Sperling, Institute of Transportation Studies, University of California, Davis, and James S. Cannon, Energy Futures, Inc.

Climate policy was center stage as government officials from more than 150 countries converged on Montreal in December 2005 for the eleventh meeting of the United Nations Framework Convention on Climate Change (UNFCCC). It was the largest intergovernmental climate conference to date, with over 10,000 participants in attendance. More than 40 actions strengthening global efforts to fight climate change were endorsed at the conference (UNFCCC, 2005).

Although a significant event, the Montreal meeting was neither the first nor the most important international meeting on climate change, and it certainly won't be the last. As far back as 1992, voluntary reductions in emissions of heat-trapping greenhouse gases (GHG) were endorsed by delegates from 189 countries at an international conference held in Rio de Janeiro, Brazil. Then in 1997, as part of the Kyoto Protocol, these voluntary reductions were replaced by a set of mandatory emission reduction targets for developed nations.

The Kyoto Protocol formally went into effect in February 2005 after countries contributing 55 percent of all GHG emissions had finally approved the Protocol (with Russia's approval pushing it over the threshold). Every rich country in the world adopted the Protocol except the United States and Australia. While opposition by the United States, the world's largest GHG emitter, was hugely unpopular in Europe, it was indicative of withering enthusiasm for the Protocol itself.

The program of international diplomacy and mandatory GHG reductions created by the Kyoto Protocol is the most comprehensive response to date, but not the only one. Overall, the global political commitment to GHG reduction is clearly growing. Within the United States, growing numbers of cities, states, and corporations are embracing strategies to reduce GHGs. Nearly every developed nation is now making an effort to address GHG emissions.

Emissions continue to grow, however, not only from the United States and developing countries but also Kyoto signatories. Indeed, the Kyoto Protocol imposes no penalties for noncompliance and provides many opportunities to buy compliance without making any effort to actually reduce emissions. Russia, for instance, with the collapse of its economy since 1990 and corresponding drop in energy use, is allowed to sell these unearned credits to others. And giant developing countries such as China and India have no responsibilities under the Protocol.

As the political wheels spin, scientific evidence continues to mount, suggesting that impacts on Earth's climate from GHG emissions, mainly from the burning of fossil fuels, is likely causing significant shifts in climate. As the Montreal meeting was winding down in December 2005, new independent scientific assessments by the U.S. National Oceanic and Atmospheric Administration and the United Kingdom Meteorological Office concurred that the eight hottest years in more than a century of record keeping have occurred in the last decade. Analyses at the Center for Atmospheric Research in Colorado concluded that 75 percent of the 4 million square miles of permafrost in Arctic regions could melt in the next century, and a multinational assessment predicted an almost complete melting of the Arctic ice cap each summer in this century.

The Third Assessment Report of the Intergovernmental Panel on Climate Change (IPCC 2001), representing the consensus opinion of 1,500 scientists, in its most recent report concludes that Earth's climate system has demonstrably changed on both global and regional scales since the preindustrial era, and that there is new and stronger evidence that most of the warming observed over the last 50 years is attributable to human activities. Scenarios based on a range of climate models point to an increase in globally averaged surface temperatures of 1.4° to 5.8°C over the period 1990 to 2100.

Despite the mounting scientific evidence of warming and melting ice, the exact scientific connections between increased GHG emissions and climate change remain uncertain. It is not clear how much, how fast, and where the climate will change, but it is becoming increasingly certain that there is a connection—and that transportation is a major source of those GHG emissions.

Transportation is the largest source of GHG emissions in the United States and is the sector where GHG emissions are growing the fastest. Nearly all transportation emissions stem from the burning of petroleum-derived gasoline and diesel fuel.

Against this backdrop, 200 climate change leaders assembled at the Tenth Biennial Conference on Transportation Energy and Environmental Policy at the Asilomar Conference Center in August 2005 to address what could or should be done to reduce emissions from the transport sector. This book, which emerged from that conference, addresses strategies and policies to reduce GHG emissions from transportation.

Almost all programs currently under way to reduce GHG emissions from transportation are incremental. They are mostly aimed at reducing fuel consumption by vehicles. Their modest goal is to stem the tide of growing GHG emissions. They are not aimed at transforming our transport and energy systems or altering travel behavior and land use development. But if climate models are right and carbon dioxide concentrations must be stabilized, even at twice preindustrial levels, then emissions will need to be reduced by one third from projected growth by 2050, and by 90 percent by the end of this century.

The long-term solution to climate change will most likely involve a complete transformation of the energy sources used to propel human society. This will include elimination of most carbon-containing fossil fuels, capturing and sequestering carbon from the remaining fossil fuels, and a more efficient use of energy. A global shift to hydrogen is one possible long-term solution to climate change. It was the subject of the previous book in this series, *The Hydrogen Energy Transition* (Sperling and Cannon, 2004).

In this book, and at the conference, we take a first step toward developing a strategy for the transport sector. What is the role of technology versus behavioral changes? Are entirely new technologies needed? What type of research is needed, and by whom? What is the role of transportation vis-à-vis other sectors? Which policy instruments might be most effective, and which might be most acceptable? Definitive answers are not possible at this time. They may never be—but we make a strong beginning.

The authors in this book identify and discuss promising programs and policies. They address the many opportunities to reduce emissions. They address new and improved vehicle technologies and low-carbon fuels; international programs that refocus mobility on efficient mass transit and walking and bicycling; innovative urban planning that leads to less fuel consumptive lifestyles; and the role of public involvement.

GHG Emissions Headed in Wrong Direction

Despite a variety of political commitments around the world to reduce CO_2 emissions, including the Kyoto Protocol, emissions continue to increase. The majority of GHG emissions occur in the form of carbon dioxide (CO_2), and most CO_2 emissions are the result of the combustion of fossil fuels. According to official U.S. government sources, global CO_2 emissions grew from 21.4 billion metric tons in 1990 to roughly 26 billion tons in 2004, and they are expected to increase another 50 percent by 2025, an annual increase of 2 percent per year (EIA, 2005). Table 1-1 charts this real and projected global growth in CO_2 emissions, including those of the world's five leading emitters.

4 *Driving Climate Change*

TABLE 1-1. World Carbon Dioxide Emissions: 1990–2025 (Billion Metric Tons Per Year)

	1990	2002	2010	2015	2020	2025	Annual % Change 2002–2025
Total World	21.4	24.4	30.0	33.0	35.6	38.4	2.0
United States	5.0	5.7	6.4	6.7	7.1	7.6	1.2
China	2.3	3.3	5.5	6.5	7.4	8.1	4.0
Former Soviet Union	3.8	2.4	2.8	3.0	3.2	3.4	1.5
India	0.6	1.0	1.4	1.6	1.8	2.0	2.9
Japan	1.0	1.2	1.2	1.2	1.2	1.2	0.2

Source: EIA, 2005.

TABLE 1-2. U.S. Carbon Dioxide Emissions by Energy Sector: 1990–2004 (Million Metric Tons Per Year)

	1990	1998	2002	2004
Total U.S.	5,002	5,598	5,809	5,973
Transportation	1,570	1,758	1,865	1,934
Industrial	1,692	1,791	1,671	1,730
Residential	954	1,089	1,190	1,212
Commercial	781	934	1,020	1,024

Source: EIA, 2005.

Table 1-1 shows that the United States is by far the world's leading CO_2 emitter, accounting for about 27 percent of the total. It also shows great projected emission growth among the developing nations in the next two decades, with China projected to eclipse the United States before 2020. The United States, however, will remain far ahead of all others in emissions per capita into the foreseeable future.

Transportation is the largest and fastest-growing source of CO_2 in the United States among all energy sectors. No approach to climate change prevention can be comprehensive without a major focus on transportation. In the United States, transportation accounts for about one third of all emissions (see Table 1-2). Since 1990, transportation emissions have grown at an annual average rate of 1.5 percent, and that rate is not diminishing.

Most of the emissions come from cars and trucks burning petroleum fuels. Sixty percent of transportation CO_2 emissions results from gasoline combustion in cars, and 22 percent from diesel fuel combustion in trucks and buses.

The Asilomar Declaration

In summary, GHG emissions are growing, the scientific evidence linking GHG emissions to troubling climate changes is gathering momentum, and the global political response, though strengthening, remains largely ineffective. Transportation is a particularly difficult challenge. Against this backdrop, roughly 200 climate change leaders and experts were invited to focus on the transportation GHG challenge at the tenth Biennial Conference on Transportation Energy and Environmental Policy convened at the Asilomar Conference Center in Pacific Grove, California, August 23–26, 2005.

The three-day meeting featured more than 25 presentations by international leaders and experts from industry, government, academia, and non-governmental organizations. From the presentations and discussions, 14 chapters were prepared for this book, 12 by presenters and 2 by other participants. Specific session topics at Asilomar included climate change trends and research, CO_2 reduction through new technologies and alternative fuels, options to restrain vehicle travel growth, responses in developing nations, GHG policy instruments, and U.S. GHG reduction initiatives.

The lively discussions highlighted the lack of a clear consensus, even among experts, about what steps should be taken and when to prevent global climate change. Nonetheless, several threads of agreement surfaced among the experts. These were put into writing and endorsed by participants near the end of the conference. Called the Asilomar Declaration, the agreement states the following three commonly held beliefs.

Declaration 1: It is the consensus of the Tenth Biennial Conference on Transportation Energy and Environmental Policy that climate change is real. Transportation-related GHG emissions are a major part of this global problem, and they must be reduced.

This initial assertion indicated agreement among the various representatives of the national and international transportation community—practitioners, suppliers, consumers, researchers, policymakers, and advocates—that the time has come for the transportation sector to squarely confront the challenges for reducing GHG emissions.

Declaration 2: U.S. national policy has so far failed to adequately address the role of transportation in climate change. This must be remedied.

Transportation is a principal contributor to climate change, and transportation infrastructure is threatened by changes in climate. More and better planning is needed to anticipate and respond to changes in climate. That is a challenge for the traditional transportation infrastructure community. But of greater concern is how to reduce GHG emissions—and, for additional reasons, oil use. Sometime in the near future, most likely between 2010 and 2020, world conventional oil production in non-OPEC countries will peak, even as demand for oil, especially for transportation, continues to climb. New transportation fuels and new fuel technologies are needed to deal with the resulting shortfall. These new technologies require long lead times, often in excess of 20 years. While there was not detailed agreement about how and when to proceed, there was agreement that actions to reduce GHG emissions and oil use need to accelerate.

Declaration 3: *By judiciously crafting a portfolio of solutions, it is possible to reduce transportation-related GHG emissions, while creating an efficient and effective transportation system for current and future generations.*

Opportunities abound to reduce transportation-related GHG emissions. Reductions can be realized even while increasing people's (and firms') access to goods and services. Opportunities include improved fuel efficiency, improved fuel and vehicle technologies, a more robust mix of transportation fuels, and demand-side strategies that improve the efficiency of the transportation system. These latter strategies include improved land-use planning, tolls and other pricing schemes, greater public investment in alternative modes of travel, consumer incentives, improved system integration, and mobility management. These many strategies for improving the transportation and energy systems are pursued in isolation from each other and are relatively ineffective, especially in the United States. There is an increasing urgency to pursue those that are most effective and beneficial.

The underlying issues and insights that led to the Asilomar Declaration are addressed within this book. The book is organized into five groups of chapters:

- Global oil and climate change
- Policies to reduce transportation GHG emissions
- International GHG reduction programs
- Public opinion and climate change issues
- Conclusions

The first two chapters that follow this introduction address the global trends in oil and climate change. World oil demand is expected to grow more than 40 percent by 2025. It is highly questionable today whether global production can expand even to meet this relatively short-term increase in demand before it reaches its peak output.

There exists no silver bullet for reducing petroleum fuel use in transportation, or for the resulting GHG emissions. There are, however, policy measures available to encourage the production, purchase, and use of more fuel efficient vehicles, as well as reduce driving. Speakers at Asilomar presented an encouraging array of fiscal and regulatory strategies that could be applied in the United States and elsewhere to displace petroleum consumption in transportation and flatten the current GHG emissions trajectory. Three chapters discuss policies to reduce transportation GHG emissions.

While oil use in industrialized countries is growing by only 1 percent per year, it is growing by 6 percent annually in Africa, Asia, and Latin America. Climate change is not a binding concern in those regions, but all face escalating oil imports, air pollution, road construction costs, and traffic congestion. Efforts to address these other challenges also address climate change. Three chapters in this section of the book examine the cobenefits for climate change of creating better transportation systems.

The next section of the book includes three chapters discussing public opinion and climate change issues. Consumer behavior and public attitudes can inhibit or accelerate the use of energy and emissions of GHGs. But the public remains startlingly uninformed about climate change policy choices, and researchers and policymakers have only a weak understanding of consumer behavior. It is becoming increasingly clear that transportation has symbolic meaning to consumers beyond its utility in providing access to goods and services. How might behaviors and attitudes shift in ways that influence climate change, and what is the role of policy?

Finally, the book concludes with a summary chapter. It builds on the Asilomar Declaration in which the participants agreed that global climate change is real and that it is possible to reduce transportation-related GHG emissions while creating an efficient and effective transportation system. It notes that most transportation innovations have come about in response to policy objectives unrelated to climate change. That is not surprising. Large uncertainties continue to surround climate change, and politicians are accountable to local and national constituencies. But circumstances are changing. The science of climate change is improving. Climate change is becoming a public issue. And politicians and the general public are beginning to appreciate the links between oil imports, global tensions, and climate change. This book provides insights into the messy process of reducing GHG emissions from transportation without sacrificing quality of life.

References

Greene, David, and Danilo Santini, eds. *Transportation and Global Climate Change.* Washington, D.C.: American Council for an Energy-Efficient Economy, 1993.

Intergovernment Panel on Climate Change (IPCC). *Climate Change 2001: The Scientific Basis.* Cambridge, U.K.: Cambridge University Press, 2001.

Sperling, Daniel, and James Cannon, eds. *The Hydrogen Transition.* Burlington, MA: Elsevier Academic Press, 2004.

United Nations Framework Convention on Climate Change (UNFCCC). Website www.unfccc.int.

U.S. Energy Information Administration (EIA). *Emissions of Greenhouse Gases in the United States, 2004.* Washington, D.C.: EIA, 2005.

CHAPTER 2

Peaking of World Oil Production and Its Mitigation

Robert L. Hirsch, Roger Bezdek, and Robert Wendling

Oil is the lifeblood of modern civilization. It fuels the vast majority of the world's mechanized transportation equipment, including its automobiles, trucks, airplanes, trains, ships, and farm equipment. Oil is also the primary feedstock for many of the chemicals that are essential to modern life.

The Earth's endowment of oil is finite, and demand for oil continues to increase at an significant rate. Accordingly, geologists know that at some future date the conventional oil supply will no longer be capable of satisfying world demand. At that point, world conventional oil production will have peaked and will begin to decline.

Under business-as-usual conditions, world oil demand will continue to grow, increasing approximately 2 percent per year, driven primarily by the transportation sector. The economic and physical lifetimes of existing transportation equipment are measured on decade time scales. Since turnover rates are low, rapid changeover in transportation end-use equipment is inherently time consuming.

Oil peaking represents a liquid fuels problem, not an "energy crisis." Motor vehicles, aircraft, trains, and ships simply have no ready alternative to liquid fuels. Nonhydrocarbon energy sources, such as solar, wind, photovoltaics, nuclear power, geothermal, and fusion technology, produce electricity, not liquid fuels, so their widespread use in transportation is certainly decades away. Accordingly, mitigation of declining world oil production must be narrowly focused.

The world has never faced a problem like the peaking of oil. Previous energy transitions—such as wood to coal or coal to oil—were gradual and evolutionary; oil peaking will be abrupt and revolutionary. Without massive mitigation more than a decade before the fact, the problem will be pervasive and devastating.

Driving Climate Change   Copyright © 2006 by Academic Press.
All rights of reproduction in any form reserved.

The inevitable peaking of world oil production presents the world with an unprecedented risk management problem. The rapid rise in world oil prices from 2004 through 2005 may appear modest in comparison to the price escalations and oil shortages that are almost certain to accompany the peaking of world conventional oil production. As peaking is approached, oil prices and price volatility will increase dramatically, and, without timely mitigation, the economic, social, and political costs will be unprecedented. Viable mitigation options exist on both the supply and demand sides, but to have substantial impact on a world scale, they must be initiated more than a decade in advance of peaking.

Peaking of World Conventional Oil Production

Oil was formed by geological processes hundreds of millions of years ago and is not found everywhere. It is typically found in underground reservoirs of dramatically different sizes, at varying depths, and with widely varying characteristics. The largest oil reservoirs are called "super giants," many of which were discovered in the Middle East. Because of their size and other characteristics, super-giant reservoirs are generally the easiest to find, the most economic to develop, and the longest lived. The last super-giant oil reservoirs discovered worldwide were found in 1967 and 1968. Since then, smaller reservoirs of varying sizes have been discovered in what are called "oil prone" locations worldwide. Many smaller reservoirs must be discovered to replace one of the world's super giants.

Future world oil production must also include the output from all the yet-to-be discovered oil fields in their various states of development. This is an extremely complex summation problem because of the variability and large possible biases in publicly available data. The remarkable complexity of the problem can easily lead to incorrect conclusions, either positive or negative.

Reaching a global oil production peak does not mean that the world is "running out" of oil. Each tapped reservoir reaches a maximum oil production rate, which typically occurs after roughly half of the recoverable oil in the field has been produced. Satisfying increasing oil demand, therefore, not only requires continuing to produce older oil fields with their declining production, it also requires finding new ones, capable of producing sufficient quantities of oil to both compensate for shrinking production from older fields and to provide the increases demanded by the market. As discovery rates fail to replace diminishing production rates from known fields in the face of increased demand, a global oil shortage emerges.

A pattern of declining oil discovery has already emerged. Extensive exploration has occurred worldwide for the last 30 years, but results have been disappointing. If recent trends hold, there is little reason to expect that exploration success will dramatically improve in the future. This situation is evident in Figure 2-1, which shows the difference between annual world

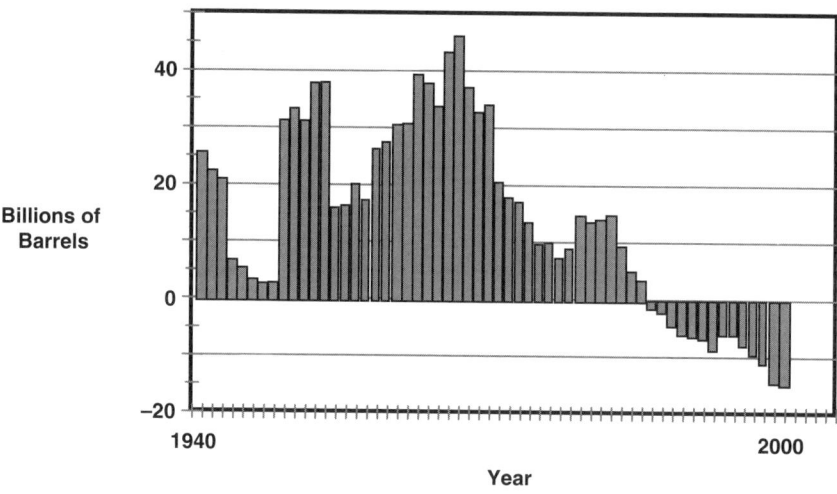

FIGURE 2-1. Difference between annual world oil reserves additions and annual consumption; 1940–2000.

oil reserves additions minus annual consumption (Aleklett and Campbell, 2003). It clearly shows a world moving from a long period in which reserve additions were much greater than consumption into an era where annual additions are falling increasingly short of annual consumption.

Oil Reserves

Once oil has been discovered via an exploratory well, full-scale production requires many more wells across the reservoir to provide multiple paths that facilitate the flow of oil to the surface. This multitude of wells also helps to define the total recoverable oil in a reservoir—its so-called "reserves."

The concept of reserves is often misunderstood. Reserves is an estimate of the amount of oil in an oil field that can be extracted at an assumed cost. Thus, a higher oil price outlook often means that more oil can be produced, but geology places an upper limit on price-dependent reserves growth. In well-managed oil fields, the upper limit is often only 10 to 20 percent more than what is available at lower prices. Reserves and production should not be confused. An oil field can have large estimated reserves, but if the field is past its maximum production, the remaining reserves will still be produced at a declining rate.

Specialists who estimate reserves use an array of methodologies and a great deal of judgment. Thus, different estimators might calculate different reserves from the same data. Sometimes politics or self-interest influences reserves estimates. An oil reservoir owner, for example, may want a higher estimate in order to attract outside investment or to influence other producers.

Projections of the Peaking of World Oil Production

World oil demand is expected to grow more than 40 percent by 2025 (U.S. Department of Energy, 2005). It is questionable whether global pro-duction can expand to meet this increase in demand before production reaches its peak. Recently, many credible analysts have become much more pessimistic about the possibility of finding the huge new reserves needed to meet growing world demand. Even the optimistic forecasts frequently suggest that world oil peaking will occur in less than 25 years. If this occurs, enormous economic disruption, as only glimpsed during the 1973 oil embargo and the 1979 Iranian oil cutoff, is likely to result.

Various individuals and groups have used available information and geological estimates to develop projections for when world oil production might peak. A sampling of recent projections is shown in Table 2-1.

Previous Oil Supply Shortfalls and Disruptions

There have been over a dozen global oil supply disruptions over the past half-century (U.S. Department of Energy 2000). Disruptions ranged in dura-

TABLE 2-1. Projections of the Peaking of World Oil Production

Projected Date	Source of Projection	Background & Reference
2006–2007	Bakhitari, A. M. S.	Oil executive (Iran) (Bakhitari, 2004)
2007–2009	Simmons, M. R.	Investment banker (U.S.) (Simmons, 2003)
After 2007	Skrebowski, C.	Petroleum journal editor (U.K.) (Skrebowski, 2004)
Before 2009	Deffeyes, K. S.	Oil company geologist (ret., U.S.) (Deffeyes, 2003)
Before 2010	Goodstein, D.	Vice Provost, Cal Tech (U.S.) (Goodstein, 2004)
Around 2010	Campbell, C. J.	Oil company geologist (ret., Ireland) (Campbell, 2003)
After 2010	World Energy Council	World Non-Government Org. (World Energy Council, 2003)
2012	Pang Xiongqi	Petroleum engineer (China) (Xiongqi, 2005)
2010–2020	Laherrere, J.	Oil geologist (ret., France) (Laherrere, 2003)
2016	EIA nominal case	DOE analysis/information (U.S.) (U.S. Department of Energy, 2000)
After 2020	CERA	Energy consultants (U.S.) (Jackson et al., 2004)
2025 or later	Shell	Major oil company (U.K.) (Davis, 2003)

tion from 1 to 44 months. Percentage supply shortfalls varied from roughly 1 percent to nearly 14 percent of world production. The most traumatic disruption, in 1973–1974, was not the most severe, but it nevertheless led to greatly increased oil prices and caused significant worldwide economic damage. A second major disruption, in 1979, was also neither the longest nor the most severe, although it, too, led to significant global economic problems. The 1973 and the 1979 disruptions are frequently assumed to be the most relevant in predicting what might happen at world oil peaking.

Higher oil prices during these disruptions resulted in increased costs for the production and delivery of goods and services. High prices boosted inflation and unemployment, reduced demand for products other than oil, lowered capital investment, and undercut consumer and business confidence. Tax revenues declined and budget deficits increased, driving up interest rates. These effects were magnified during periods when oil price increases were the most abrupt and severe. Government policies could not eliminate the adverse impacts of sudden, severe oil disruptions. Some policies reduced the damage, while contradictory monetary and fiscal policies to control inflation exacerbated recessionary income and unemployment effects.

Estimates of the damage caused by past oil price disruptions vary substantially, but without a doubt, the effects were significant. Economic growth decreased in most oil importing countries following the disruptions of 1973–1974 and 1979. The impact of the first oil shock was accentuated by inappropriate policy responses (Lee and Ratti, 1995; Hamilton and Herrera, 2003). Despite a decline in the ratio of oil consumption to the gross domestic product (GDP) in many developed countries over the past three decades, oil remains vital, and there is considerable empirical evidence regarding the effects of oil price shocks:

- The loss suffered by the countries belonging to the Organization of Economic Cooperation and Development (OECD) in the recession from 1974 to 1975 that immediately followed the first oil disruption amounted to $350 billion in 1974 dollars, or $1.1 trillion in 2003 dollars, although part of this loss was related to factors other than oil (Bird, 2003).
- The loss resulting from the 1979 oil disruption was about 3 percent of GDP, or $350 billion (1980 dollars) in 1980, rising to 4.25 percent or $570 billion in 1981, and accounted for much of the decline in economic growth and the increase in inflation and unemployment in the OECD in the recession of 1981 and 1982. These losses totaled about $700 billion and $1.1 trillion, respectively, in 2003 dollars.
- The effect of the oil price upsurge in 1990–1991 was more modest because price increases were smaller, they did not persist, and oil intensity in OECD countries had declined, thereby decreasing their economic vulnerability to oil price jumps.

- Although oil intensity and the share of oil in total imports have declined in recent years, OECD economies remain vulnerable to higher oil prices because of the "life blood" nature of liquid fuel use.

The impact of sustained, significantly increased oil prices associated with oil peaking will be severe. Virtually certain are increases in inflation and unemployment, declines in the output of goods and services, and a decline in living standards. Without timely mitigation, the impact on the developed economies will almost certainly be extremely damaging, while many developing nations will likely be much worse off.

Mitigation Options and Issues

Fortunately, a number of options exist that can be applied to lessen oil dependence and reduce economic vulnerability to oil price increases. They include reduction in energy demand, methods to increase oil production, and the introduction of alternative fuels that can substitute for oil in key market applications.

Energy Conservation

Practical mitigation of the problems associated with world oil peaking must include fuel efficiency technologies that could have an impact on a large scale over time. It is clear that automobiles and light trucks, together termed light-duty vehicles (LDVs), represent the largest targets for consumption reduction worldwide.

Government-mandated vehicle fuel efficiency requirements are certain to be an element in the mitigation of world oil peaking. In addition to major fuel efficiency improvements in conventional vehicles, one result would almost certainly be the more rapid deployment of hybrid electric vehicles. Market penetration of these technologies cannot happen rapidly because of the time and effort required for manufacturers to retool their factories for large-scale production and because of the slow turnover of existing stock. In addition, a shift from gasoline to diesel fuel would require a major refitting of refineries, which would take time.

It is difficult to project what the fuel economy benefits of hybrid electric or diesel powered LDVs might be on an international scale because consumer preferences will likely change once the public understands the potential impacts of the peaking of world oil production. The fuel efficiency benefits that hybrid electric drivetrains might provide for heavy-duty trucks and buses are likely smaller than for LDVs for a number of reasons, including the fact that there has long been a commercial demand for higher-efficiency technologies in order to minimize fuel costs in these professional fleets.

Hybrid electric technology can also impact the medium duty truck fleet, which is now heavily populated with diesel engines. For example, road testing of diesel hybrid electric drivetrains in FedEx trucks began recently, with fuel economy benefits claimed to be 33 percent (Eaton Corporation, 2004). On the other hand, there appear to be limits to the fuel economy benefits of hybrid drivetrains in large vehicles; for example, the fuel savings in hybrid buses might only be in the 10 percent range (National Renewable Energy Laboratory, 2002).

Improved Oil Recovery

Improved oil recovery (IOR) is used to varying degrees in all oil fields. A particularly notable opportunity to increase production from existing oil fields is to use enhanced oil recovery technology (EOR), also known as tertiary recovery. EOR is usually initiated after primary and secondary recovery techniques have maximized their productivity. Primary production is the process by which oil naturally flows to the surface because oil is under pressure underground. Secondary recovery involves the injection of water into a reservoir to force additional oil to the surface.

EOR has been practiced since the 1950s in various conventional oil fields, primarily in the United States. The process that likely has the largest worldwide potential is miscible flooding wherein CO_2 or light hydrocarbons are injected into oil reservoirs, where they act as solvents to move residual oil.

Heavy Oil and Oil Sands

This category of unconventional oil includes a variety of viscous oils that are called heavy oil, bitumen, oil sands, and tar sands. These oils have the potential to play a much larger role in satisfying the world's needs for liquid fuels in the future.

The largest deposits of unconventional oils exist in Canada and Venezuela, with smaller resources in Russia, Europe, and the United States. While the sizes of the Canadian and Venezuela resources are enormous, 3 to 4 trillion barrels in total, the amount of oil estimated to be economically recoverable is of the order of 600 billion barrels. This relatively low fraction is in large part due to the extreme difficulty in extracting these oils (Williams, 2003). While recovery may increase with higher world oil prices, estimation of the increased reserves would be highly speculation.

Here are some of the reasons why the production of unconventional oils has not been more extensive:

- Production costs for unconventional oils are typically much higher than for conventional oil.
- Significant quantities of energy are required to recover and transport unconventional oils.
- Unconventional oils are of lower quality and, therefore, are more expensive to refine into clean transportation fuels than conventional oils.

- There can be significant environmental problems associated with the production of these unconventional oils.

Gas-to-Liquids (GTL)

Very large reservoirs of natural gas exist around the world, many in locations that are isolated from natural gas–consuming markets. Significant quantities of this "stranded gas" are being liquefied and transported to markets in refrigerated, pressurized ships in the form of liquefied natural gas (LNG). Another method of bringing stranded natural gas to world markets is to disassociate the methane molecules, add steam, and convert the resultant mixture to high-quality liquid fuels via the Fisher-Tropsch (F-T) process. F-T based GTL results in clean, finished fuels, ready for use in existing end-use equipment with only modest finishing and blending. GTL processes have undergone significant development over the past decade.

Coal Liquefaction

To derive liquid fuels from coal, the leading process involves gasification of the coal, removal of impurities from the resultant gas, and then synthesis of liquid fuels, using the F-T process. Gas cleanup technologies are well developed and deployed in refineries worldwide. F-T synthesis is also well developed and commercially practiced. A number of coal liquefaction plants were built and operated during World War II, and the Sasol Company in South Africa subsequently built several larger and more modern facilities (Kruger, 1983). Modern gasification technologies have been dramatically improved over the years, with the result that over 200 gasifiers are in commercial operation around the world, most using petroleum coke or coal as their feedstock. Coal liquids from gasification followed by F-T synthesis are of such high quality that they do not need to be refined. Coal liquefaction is believed capable of providing clean substitute fuels at between $35 and $45 per barrel (Gray et al., 2001; Gray, 2005).

Biomass

Biomass can be grown, collected, and converted to substitute liquid fuels by a number of processes. Currently, biomass-to-ethanol is produced on a large scale to provide a gasoline additive in the United States and Brazil, among other places. The market for ethanol derived from biomass is influenced by government requirements and facilitated by generous tax subsidies. Research holds promise of more economical ethanol production from cellulosic, or woody, biomass, but related processes are far from economic. Reducing the cost of growing, harvesting, and converting biomass crops will be necessary (Smith et al., 2004).

Hydrogen

Hydrogen has potential as an alternative to petroleum-based liquid fuels over the long term in some transportation applications. Like electricity, hydrogen is an energy carrier, not a primary fuel; hydrogen production requires an energy source for its production. Energy sources for hydrogen production include natural gas, coal, nuclear power, and renewable resources.

Recently, the U.S. National Research Council (NRC), the operating arm of the U.S. National Academies, completed a study that included an evaluation of the technical, economic, and societal challenges associated with the development of a hydrogen economy (National Research Council, 2004). The NRC concluded that fuel cells must improve by a factor of 10 to 20 in cost, a factor of 5 in lifetime, and roughly a factor of 2 in efficiency in order to become commercial. The NRC did not believe that such improvements could be achieved by technology development alone. It called for new concepts and technical breakthroughs. In other words, today's technologies do not appear practically viable, and the advent of commercial hydrogen vehicles cannot be predicted.

Three Mitigation Scenarios

Issues related to the peaking of world oil production are extremely complex, involve literally trillions of dollars, and are very time sensitive. To explore these matters, three mitigation scenarios with differing starting times (Hirsch, Bezdek, and Wendling, 2005):

- Scenario I: Action is not initiated until oil peaking occurs.
- Scenario II: Action is initiated 10 years before peaking.
- Scenario III: Action is initiated 20 years before peaking.

The analysis was simplified, and the estimates were approximate. Nevertheless, the mitigation envelope that resulted is believed to be indicative of the realities of such an enormous undertaking. The focus was on large-scale, physical mitigation, mainly energy efficiency improvements and the introduction of new fuels, as opposed to analysis of policy actions, such as tax credits, rationing, and automobile speed restrictions. Physical mitigation included implementation of technologies that can substantially reduce the consumption of liquid fuels or increase production, while still delivering comparable services.

The pace that governments and industry choose to mitigate the impacts of the peaking of world oil production is yet to be determined. As a limiting case, the analysis assumed the implementation of crash programs mandated by governments worldwide can be implemented very quickly.

18   *Driving Climate Change*

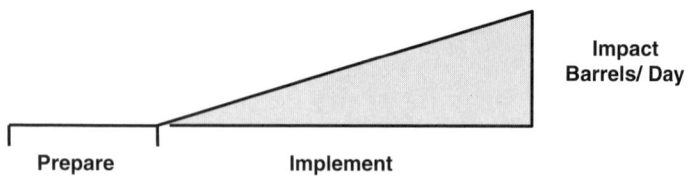

FIGURE 2-2. The delayed wedge approximation for illustrating the impact of the introduction of an energy saving or liquid fuel production technology.

This is obviously the most optimistic situation because government and corporate decision-making is rarely instantaneous.

The model chosen to illustrate the possible effects of likely mitigation actions involves the use of "delayed wedges" to approximate the scale and pace of each action, shown in Figure 2-2. Delayed wedgegraphs are composed of two parts. The first is the preparation time needed prior to tangible market impacts. In the case of efficient transportation, this time is required to redesign vehicles and retool factories to produce more efficient vehicles. In the case of the production of substitute fuels, the delay is associated with planning and construction of relevant facilities.

After the preparation phase, delayed wedges then approximate the penetration of mitigation effects into the marketplace. This might be the growing sales of more fuel-efficient vehicles or the growing production of substitute fuels. The wedges were assumed to continue to expand for a few decades, which simplifies the analysis, but is increasingly less realistic over time because markets will adjust and impact rates will change.

The criteria for selecting candidates for energy efficiency improvements and substitute oil production were as follows:

- The option must produce liquid fuels that can, as produced or as refined, substitute for liquid fuels currently in widespread use—for example, gasoline, diesel, and jet fuel. The end products will thus be compatible with existing distribution systems and end-use equipment.
- The option must be capable of reducing demand for liquid fuels or being implemented on a massive scale, ultimately millions to tens of millions of barrels per day worldwide.
- The option must include technology that is commercial or near commercial, which requires that the process has at least been demonstrated on a commercial scale.
- Substitute fuel production technologies must be inherently energy efficient, assumed to mean that greater than 50 percent of process energy input is contained in the clean liquid fuels product.

- Energy sources or energy efficiency technologies that produce or save electricity were not considered in this context because the focus was on liquid fuels.

Candidates Selected and Rejected

In the end-use efficiency category, a dramatic increase in the efficiency of petroleum-based fuel equipment is one attractive option. The imposition of Corporate Average Fuel Economy (CAFE) requirements for U.S. automobiles in 1975 was one of the most effective of the mandates initiated in response to the oil embargo of 1973 and 1974. In recent years, fuel economy for automobiles has not been a high national priority in the United States.

Nevertheless, fuel efficient hybrid electric drivetrain technology has been penetrating the automobile and truck markets in the United States and elsewhere since the late 1990s. In a period of national oil emergency, hybrid electric and other vehicle fuel efficiency technologies could be massively implemented in new vehicles. A variety of currently available technologies offer fuel economy improvements of 40 percent or more for automobiles and for light and medium trucks.

The fuel production options selected in this analysis were heavy oil and tar sands, coal liquefaction, improved oil recovery, and GTL systems. The rationale was as follows:

- Improved oil recovery is being applied worldwide.
- Oil sands production is currently commercial in Canada and heavy oil is produced in Venezuela and elsewhere.
- Coal liquefaction is a near-commercial technology.
- GTL is a viable commercial technology, which is economic where natural gas is remote from markets.

A number of options were excluded for different reasons. Shale oil represents a huge resource, particularly in the United States. However, practical recovery technologies are still in the development stage and are not yet ready for commercial deployment. Biomass options capable of producing liquid fuels were also not included due to high costs. Ethanol from biomass is currently produced in the United States and Brazilian transportation markets. It is mandated and subsidized in the U.S., but it is not yet economically viable without government support. Biodiesel fuel is a subject of considerable current interest but it, too, is not yet commercially viable. A major research and development effort might change the biomass outlook, if initiated in the near future.

Over 45 percent of world oil consumption is for nontransportation uses. Fuel switching away from some nontransportation uses of liquid fuels is already taking place worldwide. For significant world-scale impact, large substitute energy facilities would have to be constructed to provide the substitute energy. Building these plants would require decades.

20 *Driving Climate Change*

Nuclear power, wind, and photovoltaics produce electric power, which is not a near-term substitute fuel in transportation equipment that requires liquid fuels. Many decades after oil peaking, it is conceivable that a massive shift from liquid fuels to electricity might occur in some applications. However, consideration of such changes is speculative at this time.

Modeling World Oil Supply and Demand

It is not possible to predict with certainty when a global peak in conventional oil production will occur or how rapidly production will decline after the peak. Therefore, this analysis did not stipulate a date for peaking. Rather, peaking was assumed at year zero, and the analysis considered effects 20 years before to 20 years after peaking. A shape for world oil peaking was also required, and the production pattern for the U.S. lower 48 states was used as a model of what can occur in a large, complex oil province over the course of over five decades. As shown in Figure 2-3, U.S. lower 48 states production pattern is reasonably approximated by a simple triangular pattern with a roughly 2 percent annual rise before peaking, followed by a 2 percent annual decline after.

For this analysis, world production at peaking was assumed to be 100 million barrels per day (MMbpd), which is 16MMbpd above the current 84 MMbpd world production. The selection of 100MMbpd is not intended as a prediction of magnitude or timing; its use is for illustrative purposes only. Since the wedge estimates are rough estimates, a 100MMbpd peak represents a credible assumption for this kind of analysis. If global peaking were to occur in the next year or two, 100MMbpd might be high by 10 to

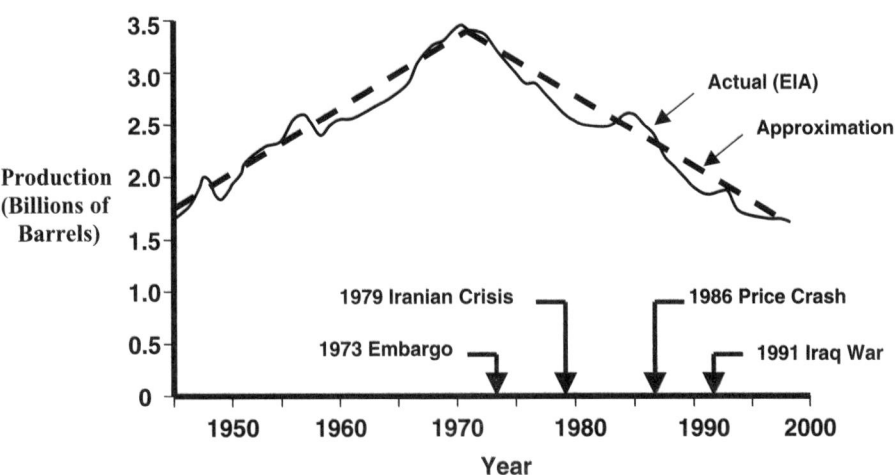

FIGURE 2-3. U.S. lower 48 oil production, 1945–2000.

15 percent. If peaking occurs at 125MMbpd at some future date, the 100MMbpd assumption would be low by 25 percent.

Another important variable is future world oil demand growth. The World Energy Council has made the following prediction: "Oil demand is projected to increase at about 1.9 percent per year, rising from an actual level of about 75.7MMbpd in 2000 to between 113 and 115MMbpd in 2020, an increase of between 37.5 and 39.5MMbpd" (World Energy Council, 2001). Recent trends indicate a world oil annual demand growth in excess of 3 percent, driven in part by rapidly increasing oil consumption in China and India. However, sustaining such a high growth rate on a continuing basis seems unlikely. With these considerations in mind, a 2 percent growth in demand after peaking was assumed. This extrapolation of demand after peaking provides a reference that facilitates calculation of supply shortfalls. The assumption has the benefit of simplicity, but it ignores the real-world feedback of oil price escalation on demand, which is sure to happen but will be extremely difficult to forecast.

Some other analysts have projected world oil production decline rates of 3 to 8 percent, well above the 2 percent assumed in this analysis. Such higher decline rates would make the mitigation problem much more difficult (Al-Husseini, 2004).

Results of Crash Program Mitigation

The estimated contributions over 15 years of a worldwide crash program to expand production of the energy resources, including efficiency, selected as viable options in this analysis are shown in Figure 2-4. The results of the analyses of future oil supply, demand, and supply shortfall, including the

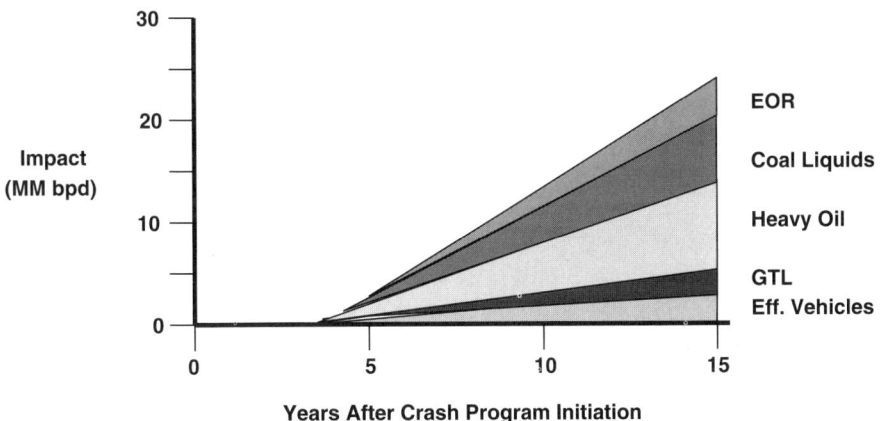

FIGURE 2-4. Estimated contributions of the worldwide crash program implementation of various oil-peaking mitigation options.

22 *Driving Climate Change*

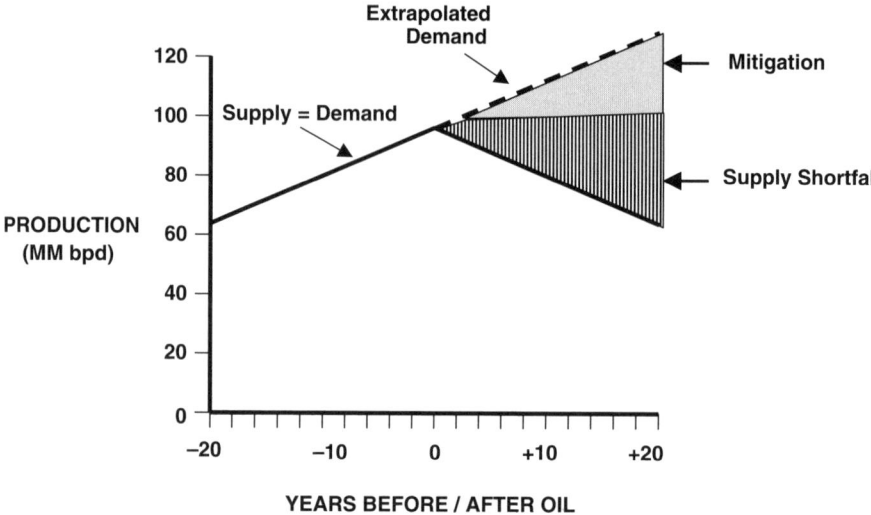

FIGURE 2-5. Mitigation crash programs started at the time of world oil peaking: A significant supply shortfall occurs over the forecast period.

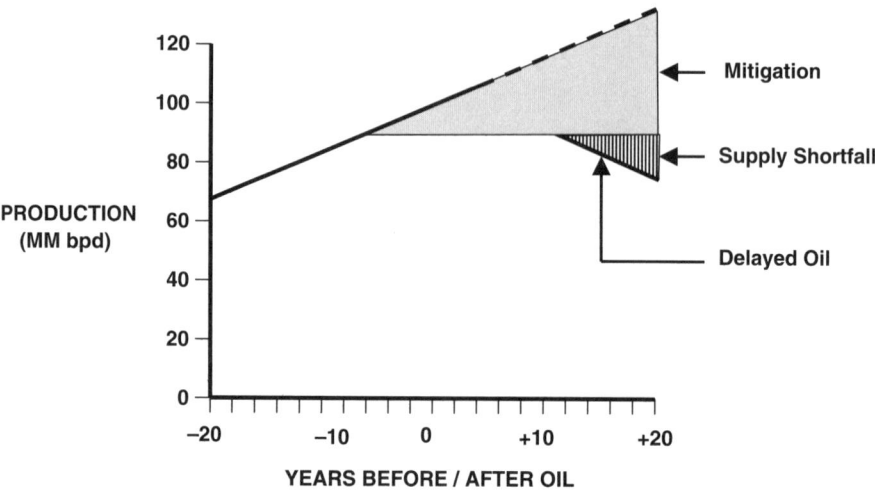

FIGURE 2-6. Mitigation crash programs started 10 years before world oil peaking: A moderate supply shortfall occurs after roughly 10 years.

impact of mitigation options on the assumed world oil peaking pattern, appear in Figures 2-5, 2-6, and 2-7. The major findings are as follows:

- Waiting until world oil production peaks before taking crash program action (Figure 2-5) leaves the world with a significant liquid fuel deficit for more than two decades.

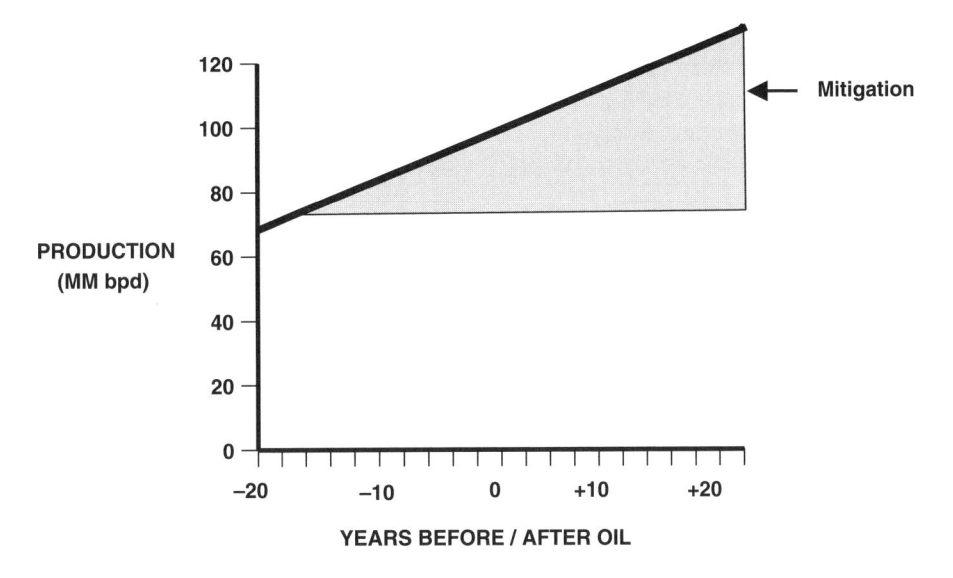

FIGURE 2-7. Mitigation crash programs started 20 years before world oil peaking: No supply shortfall occurs during the forecast period.

- Initiating a mitigation crash program 10 years before world oil peaking (Figure 2-6) helps considerably but still leaves a liquid fuels shortfall roughly a decade after the time that oil would have peaked.
- Initiating a mitigation crash program 20 years before peaking (Figure 2-7) appears to offer the possibility of avoiding a world liquid fuels shortfall for the forecast period.

The obvious conclusion from this analysis is that with adequate, timely mitigation, the worldwide economic costs of oil peaking can be minimized. If mitigation were to be too little or too late, a balance between oil supply and healthy world economic growth can only be achieved after a period of massive shortages, which would translate to significant economic hardship worldwide.

Risk Management

It is possible that peaking may not occur for several decades, but it is also possible that peaking may occur in the very near future. The world is thus faced with a daunting risk management problem. On the one hand, mitigation initiated too early would be premature if peaking is still several decades away. On the other hand, if peaking is imminent, failure to initiate mitigation quickly will have significant economic and social costs to the United States and the world.

The two risks are asymmetric. Mitigation actions initiated prematurely will be costly and could result in a suboptimal use of resources. Late initiation

24 *Driving Climate Change*

of mitigation may result in dire economic consequences. The world has never confronted a problem like this, and the failure to act on a timely basis is almost certain to have debilitating impacts. Risk minimization requires the implementation of mitigation measures well prior to peaking. Since it is uncertain when peaking will occur, the challenge is indeed significant.

Wildcards in Oil Peak Predictions

There are a number of factors that could conceivably impact the peaking of world oil production. Among the upsides, or factors that might ease the problems of world oil peaking, is the possibility that the pessimists are wrong again and peaking does not occur for many decades. Alternatively, Middle East oil reserves turn out to be much larger than publicly stated or a number of new super-giant oil fields are found and brought into production well before oil peaking might otherwise have occurred. Here are some other possible upsides:

- High world oil prices over a decade or more induce a higher level of energy conservation and efficiency.
- The United States and other nations decide to institute significantly more stringent fuel efficiency standards well before world oil peaking.
- World economic and population growth slows, and future demand is much less than anticipated.
- China and India decide to initiate or strengthen various fuel efficiency programs and other energy efficiency requirements, reducing the rate of growth of their oil requirements.
- Oil prices stay at a high enough level on a sustained basis so that industry begins construction of substitute fuels plants well before oil peaking.
- Huge new reserves of natural gas are discovered, a portion of which is converted to liquid fuels.
- Some kind of scientific breakthrough comes into commercial use, mitigating oil demand well before oil production peaks.

On the other hand, there are also several factors that might exacerbate the problem presented by a peaking of world oil. First of all, the advent of peak oil could come earlier than projected even by the most pessimistic forecasts. This could result if Middle East reserves turn out to be much less than stated or if extreme terrorism inflicts major damage to oil production, transportation, refining, and/or distribution facilities. Other concerns are that political instability in major oil producing countries, especially those in the volatile Middle East, results in unexpected, sustained world-scale oil shortages. The following are other possible downsides:

- Market signals mask the onset of peaking, delaying the initiation of mitigation.

- Consumers continue to demand larger, less-fuel-efficient vehicles.
- Expansion of energy production is hindered by increasing environmental, administrative, and institutional challenges, creating shortages beyond just liquid fuels.

Conclusion

Over the past century the development of economies and lifestyles has been fundamentally shaped by the availability of abundant, low-cost oil. A number of competent forecasters now project peaking within a decade; others contend it will occur later. Prediction of peaking is extremely difficult because of geological complexities, measurement problems, pricing variations, demand elasticity, and political distortions. Peaking will happen, but the timing is uncertain.

The analysis of the impact of mitigation strategies on future oil supplies performed in this study leads to several conclusions. First, oil scarcity and severalfold oil price increases due to world oil production peaking will almost certainly have dramatic negative economic impacts. The decade after the onset of world oil peaking may resemble the period after the 1973 to 1974 oil embargo but last for a decade or more. World economic losses could be measured on a $10 trillion scale.

Mitigation strategies will require an intense effort to implement over decades. This inescapable conclusion is based on the time required to replace vast numbers of liquid fuel–consuming vehicles and the time required to build a substantial number of substitute fuel production facilities. Scenario analysis suggests that waiting until world oil production peaks before taking crash program action would leave the world with a significant liquid fuel deficit for more than two decades. Initiating a crash mitigation program 10 years before world oil peaking helps considerably but still leaves a liquid fuels shortfall roughly a decade after the time that oil would have peaked.

Only by initiating a mitigation crash program 20 years before peaking might the world avoid a significant world liquid fuels shortfall. The conclusion is that the global disruption from oil peaking can be minimized through adequate, timely mitigation. If mitigation is too little or too late, however, a global balance of oil supply and healthy economic demand will be achieved only after a prolonged period of massive shortages and economic hardships.

Sustained high oil prices in the future are likely to stimulate some level of forced demand reduction. Stricter end-use efficiency requirements can further reduce embedded demand, but substantial, world-scale change will require more than a decade. Production of large amounts of substitute liquid fuels can and must be provided. A number of commercial or near-commercial substitute fuel production technologies are currently available, so the production of large amounts of substitute liquid fuels is technically and economically feasible, albeit time-consuming and expensive.

The peaking of world conventional oil production presents a classic risk management problem. Mitigation efforts initiated earlier than required may turn out to be premature, if peaking is long delayed. On the other hand, if peaking is imminent, failure to initiate timely mitigation could be extremely damaging. Prudent risk management requires the planning and implementation of mitigation well before peaking. Early mitigation will almost certainly be less expensive and less damaging to the world's economies than delayed mitigation.

Intervention by governments will be required in order to implement early mitigation strategies. The experiences of the 1970s and 1980s offer important lessons and guidance as to government actions that might be more or less desirable. But the process will not be easy. Expediency may require major changes to existing administrative and regulatory procedures for new facility development, such as lengthy environmental reviews and extensive public involvement.

Without mitigation, however, the peaking of world oil production will almost certainly cause major worldwide economic upheaval. Given enough lead time, the problems are solvable with existing technologies. New technologies are certain to help but on a longer time scale. Appropriately executed risk management could dramatically minimize the damages that might otherwise occur.

References

Al-Husseini, S. I. "A Producer's Perspective on the Oil Industry." Oil and Money Conference. London. October 26, 2004.
Aleklett, K., and C. J. Campbell. "The Peak and Decline of World Oil and Gas Production." Uppsala, Sweden: Uppsala University, ASPO website, 2003.
Bakhtiari, A. M. S. "World Oil Production Capacity Model Suggests Output Peak by 2006–07." *Oil and Gas Journal.* April 26, 2004.
Bird, Faith. "Analysis of the Impact of High Oil Prices on the Global Economy." Paris, France: International Energy Agency, 2003.
Campbell, C. J. "Industry Urged to Watch for Regular Oil Production Peaks, Depletion Signals." *Oil and Gas Journal.* July 14, 2003.
Davis, G. "Meeting Future Energy Needs." *The Bridge.* Washington, DC: National Academies Press. Summer 2003.
Deffeyes, K. S. *Hubbert's Peak—The Impending World Oil Shortage.* Princeton, NJ: Princeton University Press, 2003.
Eaton Corporation. Press release. March 30, 2004.
Goodstein, D. *Out of Gas—The End of the Age of Oil.* New York: W.W. Norton, 2004.
Gray, D. "Producing Liquid Fuels From Coal." Presentation at the National Academy of Sciences Systems Workshop on Trends in Oil Supply and Demand, Washington, D.C., October 20–21, 2005.
Gray, D. et al. "Coproduction of Ultra Clean Transportation Fuels, Hydrogen, and Electric Power from Coal." Mitretek Systems Technical Report MTR 2001-43, July 2001.
Hamilton, J. D., and A. M. Herrera. "Oil Shocks and Aggregate Macroeconomic Behavior: The Role of Monetary Policy." *Journal of Money, Credit, and Banking.* April 2004.

Hirsch, R. L., R. H. Bezdek, and R. M. Wendling. *Peaking of World Oil Production: Impacts, Mitigation, and Risk Management*. Pittsburgh, Pennsylvania: U.S. Department of Energy National Energy Technology Laboratory, February 2005.

Jackson, P., et al. "Triple Witching Hour for Oil Arrives Early in 2004—But, As Yet, No Real Witches." *CERA Alert*. April 7, 2004.

Kruger, P du P. "Startup Experience at Sasol's Two and Three." South Africa: Sasol, 1983.

Laherrere, J. Seminar at the Center of Energy Conversion. Zurich, Switzerland. May 7, 2003.

Lee, K., S. Ni, and R. Ratti. "Oil Shocks and the Macroeconomy: The Role of Price Variability." *Energy Journal*, Vol. 16, No. 4, 1995.

National Renewable Energy Technology Laboratory. Press release. February 8, 2002.

National Research Council. *The Hydrogen Economy: Opportunities, Costs, Barriers and R & D Needs*. Washington, D.C.: National Academies Press, 2004.

Simmons, M. R. ASPO Workshop. May 26, 2003.

Skrebowski, C. "Oil Field Mega Projects—2004." *Petroleum Review*. January 2004.

Smith, S. J. et al. "Near-Term US Biomass Potential." PNWD-3285. Columbus, Ohio: Battelle Memorial Institute, January 2004.

U.S. Department of Energy, Energy Information Administration. "Long Term World Oil Supply." Appendix I. April 18, 2000.

———. "Global Oil Supply Disruptions Since 1951." 2001.

———. Annual Energy Review. 2002.

———. "Latest Oil Supply Disruption Information." 2004.

———. "World Oil Market and Oil Price Chronologies: 1970–2003." March 2004.

———. International Petroleum Monthly. April 2004.

———. International Energy Outlook—2005. July 2005.

Williams, B. "Heavy Hydrocarbons Playing Key Role in Peak Oil Debate, Future Supply." *Oil and Gas Journal*. July 28, 2003.

World Energy Council. "Hydrocarbon Resources: Future Supply and Demand." Buenos Aires: World Energy Council—18th Congress, October 2001.

———. Drivers of the Energy Scene. London, England: World Energy Council, 2003.

Xiongqi, P. "The Challenges Brought by Shortages of Oil and Gas in China and Their Countermeasures." ASPO Lisbon Conference. May 19–20, 2005.

CHAPTER 3

Toward a Policy Agenda for Climate Change: Changing Technologies and Fuels and the Changing Value of Energy

Duncan Eggar

The business as usual case for global affairs suggests that some very disturbing trends are apparent today and could lead to significant changes by 2050. By then, the global population will grow by around 40 percent. The number of megacities will grow considerably. The number of vehicles in the world will increase from 700 million to 2 billion. The global demand for energy will increase between two- and threefold, and energy security will become an increasingly significant issue, with the cost of energy remaining high. These trends will be particularly marked in the developing world.

One of the greatest challenges posed by the current trends is to address the increasing buildup in the atmosphere of carbon dioxide (CO_2) and other greenhouse gases (GHGs) associated with increased use of fossil fuels as the world's primary energy source. CO_2 and other GHG emissions have been implicated, almost without doubt, as a major factor in global climate change.

In 2004, policy analysts proposed a "wedge and slices" approach to addressing the need to reduce CO_2 emissions (Browne, 2004; Pacala and Socolow, 2004). In essence, this approach involved reducing the seemingly inexorable growth of emissions to retain emissions in 2050 at 2000 levels. This would involve taking 7 gigatonnes (Gte) of carbon out of the atmosphere. Estimates showed that, by breaking this into seven or more slices of up to 1 Gte each, the problem becomes manageable. The good news is that technologies available today are capable of making sizeable reductions in carbon emitted into the atmosphere. For example, it has been estimated that, on a global basis, the universal introduction of hybrid electric vehicles (HEVs) powered by biofuels alone would contribute a 1 Gte reduction.

Another practical example of a technology that will lead to reduced CO_2 emissions is the proposed first power station to be fueled by hydrogen derived from natural gas. The CO_2 captured during hydrogen production will be reinjected into an oil reservoir, where it will be stored safely and also used to increase oil recovery. This is a significant step forward in developing "carbon free" energy technologies and will reduce the CO_2 emitted into the atmosphere from this 350 megawatt (MW) power plant by 90 percent. Nonetheless, hydrogen alone is not the universal panacea to the world's global climate change threat. The energy and environmental costs of a hydrogen economy will remain high until there are significant technical breakthroughs in fuel cells and hydrogen production, transportation, and storage.

History teaches us that improved technical performance leads to increased consumer expectation. If the world is to fully address the challenge of climate change through the stabilization of GHG concentrations in the atmosphere, it will also need to embrace demand management. To achieve this, the public must believe that reduced demand will not lead to a loss of lifestyle. Moreover, government policies must be developed that reward a sustainable level of energy demand. Technology alone will not be sufficient.

This chapter discusses global societal trends and future energy challenges that, on a business as usual basis, indicate a major increase in CO_2 and other GHG emissions by midcentury. It then discusses transport energy policy trends and technology trends in transportation offering alternative energy futures that could significantly reduce the contribution of transportation emissions to climate change.

Global Societal Trends

The United Nations (UN) forecasts under its medium case scenario that the world's population will rise from 6.5 billion today to 9.1 billion in 2050 (UN, 2004). Today 95 percent of all population growth occurs in the developing world, where a widespread pattern of increased urbanization complicates the problems, including that of energy demand. For example, in China between 2005 and 2030, the UN forecasts that the urban proportion of the population will rise from 36 to 60 percent (UN, 2004). This translates to 10 million new urban dwellers per year or about 27,400 per day for the next 25 years. Figure 3-1 shows the projected growth in urban and rural populations in China.

Worldwide, the UN forecasts that by 2015 there will be at least 368 cities with populations of more than 1 million (Prahalad, 2005). The growth of megacities, or urban agglomerations of more than 10 million inhabitants,

China is undergoing a transformation that many western nations experienced in the mid to late nineteenth century during the industrial revolution. The result is a large-scale urbanization, driven mainly by the policy of economic growth.

Forecast Population Change 2005–2030

Overall: +10%
Rural: −27%
Urban: +64%

10 million new urban dwellers per year

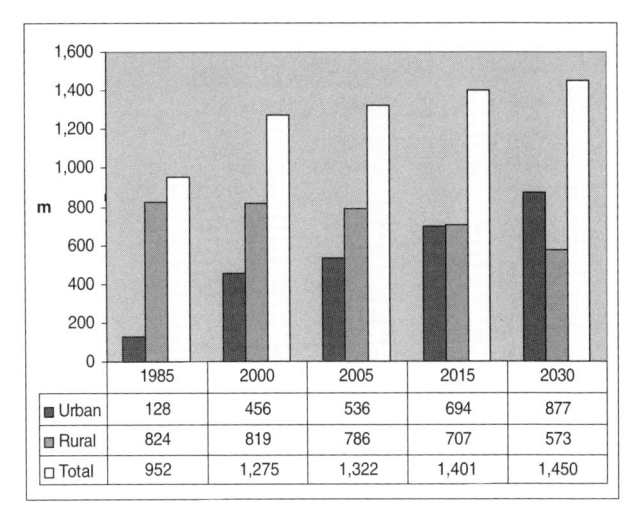

	1985	2000	2005	2015	2030
Urban	128	456	536	694	877
Rural	824	819	786	707	573
Total	952	1,275	1,322	1,401	1,450

FIGURE 3-1. China: Urban/rural population trend. *Source:* UN Department of Economic & Social Affairs.

has gone from five in 1975 to 14 in 1995 and is projected to reach 26 by 2015 (Times Atlas, 2000). Most of these megacities are in the developing world.

The trend toward urbanization is accompanied by economic development and the increased earnings potential of urban dwellers. Again using a Chinese example, it is understood that the per capita income ratio between urban and rural incomes is in excess of 3.5. A frequent demonstration of newfound wealth is in personal transportation, leading to more and often bigger cars in cities, despite the traffic jams and the 2.5 billion gallons of fuel that are wasted by them as they sit stalled in traffic every year (Glenn, 2005).

Most energy use in the world today is obtained from the burning of fossil fuels. While fossil fuels are far from running out, they are a finite resource and, increasingly, confined to remote locations or in areas of political sensitivity. This is especially true for petroleum, virtually the sole transportation fuel used in the world today. There is also a distinct regional imbalance between consumption and reserves, as shown in Figure 3-2.

It is reasonable to forecast that a huge increase in energy demand in all sectors will occur as developing countries pursue their aspirations for economic development, leading to growing personal affluence to be enjoyed by a rapidly growing urban population. Figure 3-3 shows the tripling of world oil consumption predicted by the International Energy Agency between 1971 and 2030 (IEA, 2004).

During the past century, the developed world has increasingly taken access to virtually unlimited supplies of energy for granted. Despite the

32 *Driving Climate Change*

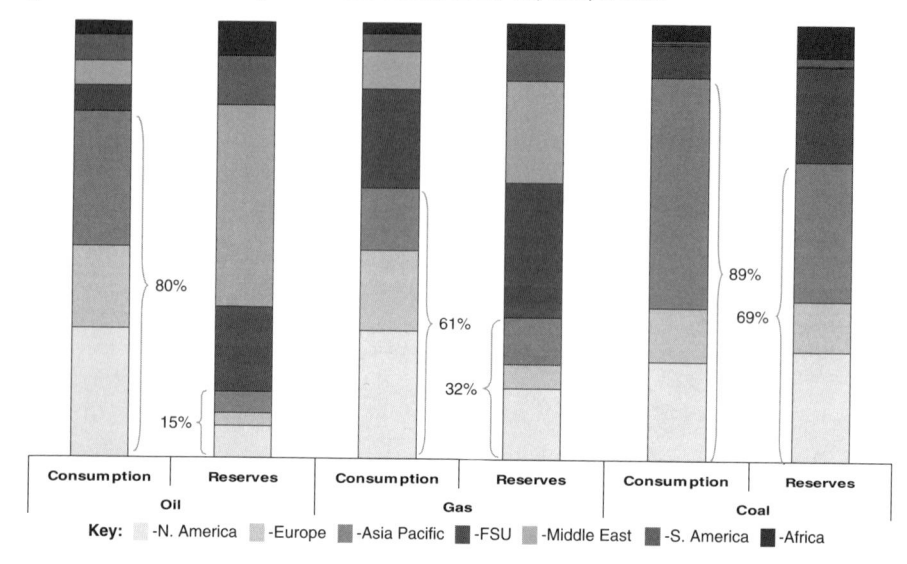

Regional Share of Consumption versus Reserves for Oil, Gas, & Coal

FIGURE 3-2. Growing dislocation of supply and demand. *Source:* BP Data, IEA WEO 2004. *Note:* Oil reserve figures do not include unconventional reserves estimates.

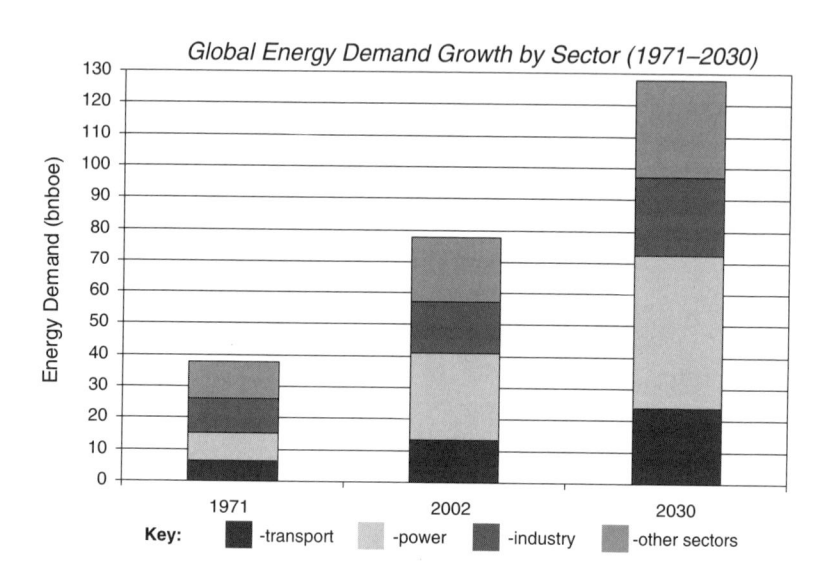

FIGURE 3-3. Rapid demand growth projected for all sectors. *Source:* IEA WEO 2004. *Notes:* 1. Power includes heat generated at power plants. 2. Other sectors include residential, agricultural, and service.

TABLE 3-1. Per Capita Energy Demand

	Primary Energy (million toe equivalent)	Population (million)	Per capita usage (toe/cap)
OECD—Europe	1,849.8	530.6	3.49
USA	2,331.6	293.6	7.94
Japan	514.6	127.6	4.03
Brazil	187.7	179.1	1.05
Russia	668.6	144.1	4.64
India	375.8	1,086.6	0.35
China	1,386.2	1,300.1	1.07

Sources: Primary energy from BP, 2005. Population from Population Reference Bureau, 2004.

burgeoning growth in energy consumption in the developing world, an imbalance between the developed and the less-developed worlds continues to exist. The imbalance of per capita energy demand between selected countries is shown in Table 3-1.

Future Energy Challenges

Figure 3-4 shows how the energy demand of the developing world may reasonably be expected to increase as those countries achieve higher GDP per capita. The interesting question is by what degree, for example, will they follow a European, an Asia Pacific, or an American trajectory? China has declared an aspiration to increase GDP by a factor of four, while only increasing energy demand by a factor of two.

The World Business Council for Sustainable Development (WBCSD) has forecast that if the UN goal to eliminate extreme poverty is to be met, global energy demand will increase over the period 2000 to 2050 by a factor of two to three, with the greater demand associated with the greater reduction in poverty (WBCSD, 2004).

Future transportation energy demand will be driven by an increasing expectation of the developing world for the access and mobility that the developed world takes for granted. Figure 3-5 illustrates global energy supply and demand in 2002 based on work by the IEA. It shows that in 2002 transportation accounted for 27 percent of world energy demand, met almost entirely by oil (IEA, 2004).

Separately, the WBCSD Sustainable Mobility Project (SMP) forecast that the total global vehicle stock will increase from 683 million vehicles in 2000 to 2.0 billion vehicles in 2050. As shown in Figure 3-6, the proportion of vehicles in the developing world will increase from 21 percent (146.1 million vehicles) to 61 percent (1,216.9 million vehicles) during the period. This is an 833 percent increase in the developing nations outside the

34 *Driving Climate Change*

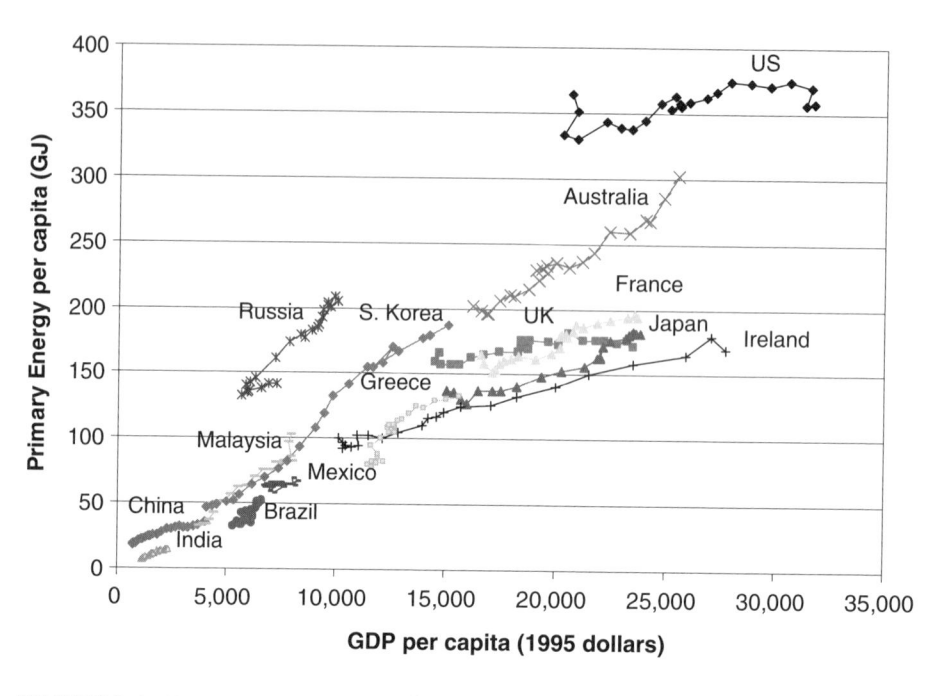

FIGURE 3-4. Energy use grows as a function of GDP.

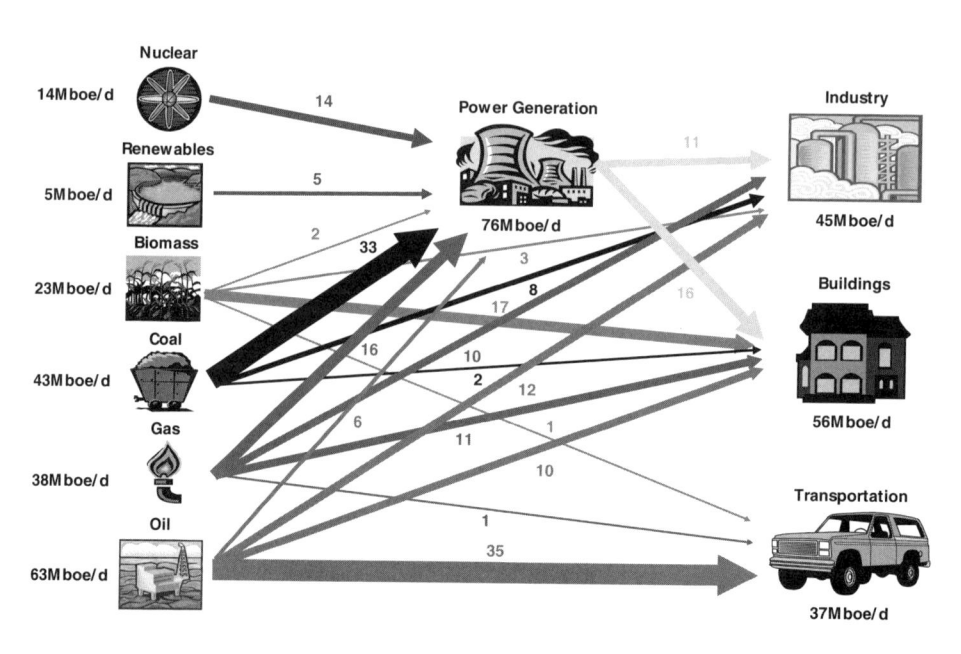

FIGURE 3-5. Global energy supply and demand (2002). *Source:* World Energy Outlook 2004.

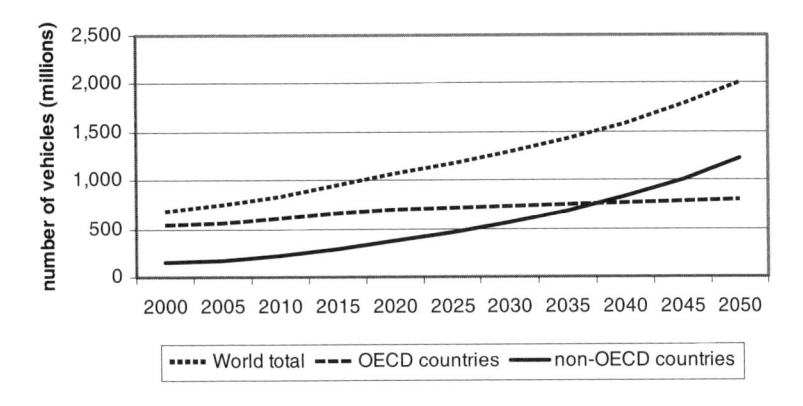

FIGURE 3-6. Forecast global vehicle stocks. *Source:* WBCSD SMP 2004.

Organization of Economic Cooperation and Development (OECD), compared with a 147 percent increase in OECD countries (WBCSD-SMP, 2004). This shift in the demand for vehicles toward developing countries raises some fairly significant questions regarding where vehicles will be made, the energy that will fuel them, and the policies and technologies that will be adopted in their manufacture.

Transportation Energy Policy Trends

Automotive exhausts, including emissions affecting local air quality (LAQ) and GHG emissions, have health impacts and costs. Most countries have already introduced lead-free gasoline, and many are currently in the process of greatly reducing the sulfur content in transportation fuels, especially diesel.

These actions have greatly improved LAQ in urban areas, but there is still much work to be done on the wider health impacts and climate change effects of transportation pollution. Gluskoter quotes Shafik in using Kuznet's curve to suggest that, on a historical basis, through the phases of economic development, air quality can be expected to worsen before it improves (Gluskoter, 1997). One might reasonably suppose that the same could apply to GHG emissions.

Europe is taking a lead in reducing GHG emissions as part of a program to address climate change issues; actions include the creation of the EU Emissions Trading Scheme. This is not an easy task. On a personal basis people have difficulty in making a link between their individual actions and the global climate change consequences. As a result, it will be difficult for the auto manufacturers, despite good intent, to keep to their voluntary GHG reduction targets. This is especially true where there is a public appetite for

36 Driving Climate Change

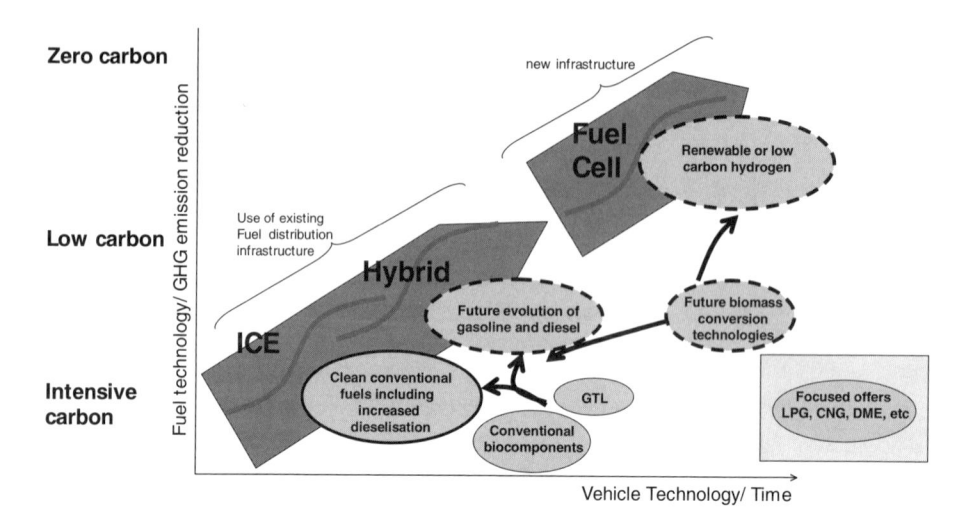

FIGURE 3-7. Future fuels pathway.

ever larger, more powerful, and more sophisticated vehicles without much policy incentive to restrict such indulgence by a relatively affluent population.

Public attitudes can be changed, however. An example has been the growth in dieselization in Europe. For many years in most European countries, especially France, diesel was cheaper than gasoline. In some countries, including the United Kingdom, legislation has been introduced that penalizes high CO_2 emitting cars, especially those that are provided as an element of an employee's remuneration package (DVLA, 2005; SMMT, 2005). Together, these measures have led to technology developments in both the mass and luxury car markets and also in fuels that make diesel cars an attractive option. Indeed, diesel cars now have approximately 50 percent of the light-duty vehicle market, and diesel is being imported to meet the shortfall in supply from European refineries.

BP has developed a future fuels pathway toward a lower CO_2 world, summarized in Figure 3-7. On the engine front, this shows the move from internal combustion engines to hybrid electric and fuel cell drivetrains. On the fuels side the path moves through dieselization to conventional biocomponents, to gas (or coal) to liquids, and on to advanced biomass conversion technologies and, perhaps in the long term, hydrogen.

In the short term, cleaner fuels that go some way to address LAQ problems and begin to make an impact on GHG emissions make those who have

TABLE 3-2. Energy Supply and Urban Planning

WBCSD Goals	Security of Supply	Urban Planning
GHG Emissions	Energy efficiency	Total energy approach
Local Air Quality	Alternative fuels	Total energy approach
Safety		Improved safety
Noise	Alternative fuels	General planning
Congestion	Modal shifts	Modal shifts
Social Equity	Modal shifts	Central to planning
Maintain Opportunity	Manage supply/demand balance	Opportunity to maintain/improve

the resources and political willpower to use them feel a little better. But it merely scratches the surface of the problems just outlined for the developing world.

For many countries, addressing climate change falls very low on their list of priorities. Innovative leadership is needed to create programs to reduce GHG emissions. By way of example, the issues that are of concern to China are security of energy supply and urban planning for their rapidly urbanizing population. In Table 3-2 these two policy drivers are mapped onto the WBCSD-SMP goals (2004). By addressing these drivers through strong policies, the wider issues such as global climate change and social equity are also met.

Technology Trends in Transportation

A variety of opportunities exist to address energy challenges in the transportation sector. They include use of clean burning, renewable fuels; implementation of vehicle fuel economy improvements; and consumer support of alternative transportation approaches such as walking and bicycling.

One example of an alternative fuel is ethanol produced from sugar cane. In Brazil the cost of cane-derived ethanol is approximately $35 per barrel of oil equivalent (boe). When oil prices exceed $40 per barrel, there is clearly a commercial market for Brazilian ethanol; recently this has been reflected in sharply increased prices for sugar on the international commodity exchanges. Other tropical countries may see the opportunity to establish cost effective domestic, renewable energy industries as well. Hopefully policy decisions that are taken in this regard will be taken for the right reasons, reflect truly sustainable values, and not jeopardize water supplies or essential food production.

Conventional biofuels have the disadvantage that they are not readily fungible in significant proportions with fossil fuels and thus may require costly, segregated facilities for storage and distribution. Moreover, at higher concentrations, they are unsuitable for much of the current vehicle fleet. Research is being conducted to address this problem through the use of advanced biofuels that are compatible for blending with conventional oil-derived fuels in any proportion and require no new infrastructure or engine technology. Indeed, the goal is to create biofuel substitutes with no discernable differences from fossil fuels.

Naturally there is a concern that in solving one problem we do not create another. In-depth analyses of potential biofuel production suggest that there is possibly a constrained land capacity to meet more than 20 percent of the demand for transportation fuel on a global basis. Such analysis makes allowance for land to maintain food production, water requirements, and land quality.

The fossil fuel that is often forgotten as a potential transportation fuel but is available in relative abundance, is coal. The rise in oil prices begins to make coal-to-liquids technologies attractive. In Shanxi province in China, there is extensive research on coal to methanol technology and the development of associated technologies that will be required to realize the benefits. This is an interesting example of bucking convention to make use of what is available locally.

Turning from the fuels to the vehicles that burn them, the advent of hybrid electric vehicles (HEVs) in the automotive market in recent years has generally improved the potential fuel economy of vehicles. The modular design of HEV powertrains enables several generations of development from the same platform. It is envisaged that HEVs will become progressively more electrified as new models are introduced over time. If forecast advances in battery technology are realized, the plug-in HEV with a 150-kilometer all-electric range will be available for city use in the relatively near future. This will have significant additional benefits in improving urban air quality and noise. With good planning and optimized electricity generation, it can also lead to more efficient use of energy and reduced overall GHG emissions.

The advent of HEV technology raises the possibility of all-electric vehicles for the mass market. What if a far more efficiently electrified society that optimizes the supply and demand balance in a way that includes supplying transportation needs could be developed? If electricity can be generated, stored, and transmitted cleanly and efficiently, there will be significant benefits in terms of GHG and LAQ emission reductions. There will also be other benefits—for example, a much quieter environment. The sources of electrical energy are many and can be chosen to suit the circumstances of the location. Coal, gas, liquid hydrocarbons, hydrogen, wind, solar photovoltaic, and thermal can all be used in the right circumstances and combinations. A challenge for the future is to take a holistic view that looks to a sustainable future, reflecting the various and diverse interests of the many relevant stakeholders.

Development of HEV technology by major automakers is one example of the successful effort by private companies, including oil companies, in promoting more efficient technology that can lead to reduced CO_2 emissions. This effort will be of little value, however, if the consumer does not also make a positive contribution in reducing demand. National, regional, and local governments have roles to play in promoting GHG reduction programs. They can formulate and implement policy that stimulates reduced consumer demand for energy. In the transportation field this can take many forms, including preferential taxation for low energy vehicles and more efficient fuels and lubricants, the provision of attractively routed and priced reliable public transport, the creation of private vehicle exclusion zones, and the provision of cycle paths and safe walkways for pedestrians. Encouraging people to walk and cycle will also have long-term benefits in terms of physical and mental health that can be translated into financial benefits and a general sense of public wellbeing. Provided they are implemented in a thoughtful and holistic way, actions to positively influence consumer behavior will contribute to all of the SMP goals: lower LAQ and GHG emissions, greater safety and social equity, reduced congestion and noise, and expanding mobility access.

Conclusion

There are three potential solutions to the stark energy challenges facing the world today:

- Increase the supply of energy by promoting technological developments, with an increasing emphasis on unconventional fossil fuel derivatives, renewable energy resources, and possibly nuclear power.
- Decrease demand through the widespread use of energy efficiency technologies coupled with changes in personal behavior, recognizing that the latter are perceived to be difficult to implement politically.
- A combination of the two, perhaps permitting the most effective progress toward the common goal of reducing GHG emissions. However, we must guard against the propensity for both sides of the equation to assume that the other is providing all the answers.

Analysis of the trends now underway suggests that the value that society puts on energy is going to change. Economists will argue that the marketplace will establish the price of energy, but this does not always account for uncertainty. Generally, uncertainty is not something that the public at large enjoys experiencing. In a world where energy security achieves greater prominence, it is quite conceivable that consumers will be prepared to pay more for assured "home-grown" access to energy.

For some the current trends are alarming; for others they open up all sorts of new opportunities. No longer will there be a "one size fits all" approach; what's right on the U.S. West Coast won't necessarily be right in the Midwest, let alone in China or India. Security of supply is becoming of increasing importance in energy discussions, and the result has been a diversity of response to this challenge.

For the optimists the changing value that will be associated with energy will unleash a raft of technologies that are waiting in the wings, and there will be others behind them. But to bring these forward, politicians and bureaucrats will need to set goal-based policies, as opposed to prescriptive route-based policies. Global and regional suppliers, whether of vehicles or the energy to power them, will need to produce products that are adaptable to the various markets that they aim to serve.

In meeting these changing and varied demands, Henry Ford's adage regarding the Model T car—"You can have any color you like as long as it's black"—just won't be good enough!

Author's Note

The opinions expressed in this paper are personal and do not necessarily represent BP's position, policy, or strategy.

References

BP (British Petroleum). *BP Statistical Review of World Energy.* London, UK: British Petroleum, June 2005.
Browne, John. "Beyond Kyoto." *Foreign Affairs*, Volume 83, Number 4, July/August 2004. pp. 20–32.
Driver and Vehicle Licensing Agency (DVLA). http://www.dvla.gov.uk/newved.htm. 2005.
Glenn, Jerome. *RSA Journal*, August 2005. p. 21.
Gluskoter, H. *Some Environmental Effects of Increased Energy Utilisation in the Twenty-first Century.* Reston, VA: U.S. Geological Survey, 1997.
International Energy Agency (IEA). *World Energy Outlook 2004.* Paris: IEA.
Pacala, S., and R. Socolow. "Stabilization Wedges: Solving the Climate Problem for the Next 50 Years." *Science*, August 13, 2004. pp. 968–972.
Population Reference Bureau. *World Population Data Sheet.* Washington, D.C.: Population Reference Bureau, 2004.
Prahalad, C. K. *The Fortune at the Bottom of the Pyramid.* Upper Saddle River, NJ: Wharton School Publishing, 2005. p. 12.
Society of Motor Manufacturers and Traders (SMMT). http://www.smmt.co.uk/co2/co2intro.cfm. 2005.
Times Concise Atlas of the World 2000. London: Times Books Group Ltd, 2000.
UN Department of Economic and Social Affairs Population Division. *World Population Prospects: The 2004 Revision.* New York: UN Department of Economic and Social Affairs Population Division, 2004.
World Business Council for Sustainable Development (WBCSD). *Facts and Trends to 2050, Energy and Climate Change.* Geneva, Switzerland: World Business Council for Sustainable Development, September 2004.
World Business Council for Sustainable Development, Sustainable Mobility Project (WBCSD-SMP). Mobility 2030. www.wbcsd.org/web/publications/mobility/exec-summary.pdf. 2004.

CHAPTER 4

Coordinated Policy Measures for Reducing the Fuel Use of the U.S. Light-Duty Vehicle Fleet

Anup P. Bandivadekar and John B. Heywood

The transportation sector is the biggest contributor to the emissions of carbon dioxide (CO_2) in the United States. Emissions of CO_2 from transport have grown by about 18 percent during the period from 1990 to 2002 (U.S. Department of Energy [DOE], 2004).

Increasing emissions of CO_2 from transportation present a big challenge from a climate change perspective. There exists no silver bullet for reducing petroleum fuel use of motor vehicles in the United States. There are, however, several policy measures available to affect the production and purchase of more fuel-efficient vehicles as well as reduce the amount of driving. Qualitative and quantitative analyses of individual policy options reveal the potential for a combination of such policies.

An integrated set of fiscal and regulatory strategies in the United States is essential to reduce petroleum consumption in transportation and transitions from the current greenhouse gas (GHG) emissions growth path to one that decreases risks from global climate change. This chapter elucidates policy options available to reduce the petroleum fuel use of the U.S. light-duty vehicle (LDV) fleet over the next three decades. More specifically, it identifies viable technology and policy options for making progress.

A policy package is proposed that combines market-based and regulatory measures to both pull and push advanced vehicle technologies into market, as well as reduce the carbon intensity of vehicle and fuel use. Such an approach aims at exploiting synergies between different measures, removing perverse incentives, and increasing political acceptability of the overall strategy by spreading the impact and responsibility. An integrated policy package that combines fuel economy standards, a fee and rebate scheme for new vehicles, fuel taxes, and increased renewable content in fuels is evaluated as an example. Such a coordinated set of policy actions might reduce the overall fuel use and GHG emissions of U.S. LDVs by 28 to 40 percent in 30 years from a no change, or status quo, scenario.

Light-Duty Vehicle Greenhouse Gas Emissions

Greenhouse gas emissions from motor vehicle use can be approximately estimated by Equation 4.1:

$$\text{GHG emissions} = \text{LPK} * \text{VKT} * \text{FI} \qquad (4.1)$$

Where:

GHG emissions = Greenhouse gas emissions (tons/year)
LPK = Liters of fuel consumed per kilometer per vehicle, generally reported as liters per 100 kilometers (L/100 km). One liter per 100 km is equivalent to 235 miles per gallon (mpg).
VKT = Fleet vehicle kilometers traveled per year (km/year)
FI = GHG intensity of fuel (GHG tons/liters of fuel)

Thus, GHG emissions from motor vehicles can be attributed to the amount of driving (VKT), fuel consumption (LPK), and the GHG intensity of the fuel (FI). The largest reductions in GHG emissions are achieved if all three of the factors are reduced. However, the three factors may interact with one another. For example, the carbon intensity of diesel as a fuel is slightly higher than gasoline, but diesel powered vehicles are typically 30 percent less fuel consuming than gasoline vehicles. As a result, diesel powered vehicles have significantly fewer GHG emissions relative to equivalent gasoline powered vehicles.

Vehicle fuel consumption of new vehicles, as measured in liters of fuel consumed per kilometer traveled, was reduced considerably in the 1970s and early 1980s due to the oil shocks of 1973 and 1979 and subsequent federal fuel economy standards. Since the early 1990s, however, fuel consumption has stagnated around 10 L/100 km (23.5 mpg) for new cars and 13.7 L/100 km (17.2 mpg) for new light trucks when adjusted for on-road performance (Heywood et al., 2004). The sales weighted fuel consumption of new vehicles has been lower during this period than in the 1980s as a result of the increasing number of light trucks in the new vehicle mix. Overall, the average on-road vehicle fuel consumption for the light-duty vehicle fleet has remained roughly steady at 11.5 L/100 km (20.5 mpg) during the last decade.

The lack of any significant reduction in vehicle fuel consumption during the last 20 years does not imply stagnation of technology. In fact, engine and vehicle technology has been improving steadily over this entire period. The technology is, however, "fungible" in that it can be used to enable other functions, such as increased amenities, vehicle power, size, and weight, rather than to improve fuel consumption performance (Plotkin, 2000). The Environmental Protection Agency (EPA) has completed an analysis of vehicle characteristics over the period from 1981 to 2003 that indicates the new 2003 LDV fleet could have achieved about 33 percent higher fuel economy if it had the same average performance and same distribution of weight as in 1981 (Hellman and Heavenrich, 2003).

The amount of vehicle kilometers traveled has more than doubled in the past 30 years (Davis and Diegel, 2003). This growth has been steady except for the years 1973, 1979, 1980, and 1990. The tremendous growth in VKT can be attributed to the following factors:

- *Increased number of vehicles:* The number of vehicles in the U.S. LDV fleet has increased from about 110 million vehicles in 1970 to about 230 million vehicles in 2003. Most of the growth has come in the light trucks segment, which now accounts for more than half of all sales as compared to about 15 percent of the sales in 1970. In general, light trucks consume more fuel relative to cars and, hence, have contributed significantly to the rising average fuel consumption of the LDV fleet.
- *Increased driving per year:* The average distance traveled per vehicle per year increased considerably from 1976 to 2001. This increased driving can be attributed to rising level of affluence, continuing urban sprawl, and the low costs of driving, among other factors.
- *Low cost of fuel:* The average fuel consumption of cars and trucks decreased from 1976 to 2001. When combined with flat costs of gasoline per km driven over this period (inflation adjusted), the net effect is a sharp drop in costs of travel per kilometer. The hypothesis that this has resulted in increased driving is known as the "takeback" or "rebound" effect. The rebound effect is estimated to be on the order of 20 percent (Greene et al., 1999; Greening et al., 2000).

The greenhouse gas intensity of fuel used in the LDV fleet has essentially not changed, since most vehicles run on gasoline. In the late 1970s, sales of new diesel cars increased rapidly to about 6 percent but fell in the early 1980s. The fraction of diesel vehicles in the new light truck sales has fluctuated around 3 to 6 percent in the past two decades. Apart from this, use of other fuels in LDVs is about 1 percent. Despite strong goals and incentives offered by Congress, alternative fuel vehicles have not succeeded (McNutt and Rodgers, 2004).

Considerable uncertainty exists about how technology will evolve over the next 20 to 30 years. While various studies differ in their estimates of the exact magnitude of fuel consumption reductions possible and the costs of doing so, the following conclusions can be drawn from these technology and cost assessments (Weiss et al., 2000; GM/ANL, 2001; An et al., 2001; NESCCAF, 2004):

- Mainstream gasoline internal combustion engines (ICEs) and vehicle technologies have significant potential to reduce energy consumption and GHG emissions. These technologies can improve at a rate of 1 to 2 percent annually over the next 20 years, which translates to up to 35 percent reduction in energy use at constant performance, size, and weight, at an additional cost of $500 to $1,500 per vehicle.
- Diesel vehicles are likely to be about 20 percent more efficient than gasoline vehicles in about two decades, but the difficulties in meeting stringent U.S. nitrogen oxide emissions standards, higher cost, and consumer perception are significant obstacles to their large scale adoption.
- Hybrid electric vehicles (HEVs), with batteries, electric motors, and ICEs, can provide an additional 30 percent benefit in energy reduction at an additional cost of $2,000 to $3,000 compared to the cost of conventional ICE vehicles.
- Fuel cell technology is currently very costly and probably a few decades away in terms of making a substantial contribution to GHG emissions reductions. However, in the longer term—roughly 30 to 50 years—fuel cells could make a difference if the hydrogen used in fuel cells is made from carbon neutral energy generation technologies such as renewable resources, nuclear power, or fossil fuels with carbon sequestration.

It is not clear if the current price of fuels in the U.S. market, even at levels above $2.00 per gallon of gasoline, can justify the development of new technologies for improving fuel consumption performance. It is possible that the current trend of sacrificing efficiency improvements for faster, more powerful, and bigger vehicles may continue.

Projections of LDV Fuel Use and GHG Emissions

The potential effects of new technologies on LDV fuel use can be evaluated based on a vehicle fleet simulation model developed by Bassene (Bassene, 2001; Heywood et al., 2004). The model examines car and light truck fleets based on vehicle sales, retirement, average fuel consumption, and miles driven per year. It allows the exploration of the sensitivity of fleet fuel use to growth in driving, vehicle ownership, and the share of light trucks in the fleet.

Different scenarios project the fuel use of LDVs under different market and policy conditions. Examining these scenarios allows us to understand the magnitude of technological and policy efforts that may be required to reduce fuel use of the LDV fleets to the levels achieved in 1990. Since most vehicle designs and production plans along with the Corporate Average Fuel Economy (CAFE) standard levels have been fixed until year 2007, the model scenarios begin in year 2008.

Description of Scenarios

There are four scenarios used in our analysis: no change, baseline, HEV, and composite. In the no change scenario, the new on-road car and light-duty truck fuel consumption remain at the levels projected for 2008 until 2035. These levels are 9.7 L/100 km for cars and 12.4 L/100 km for light trucks. This does not mean that vehicle technology will remain constant, but it is assumed that any improvement made in the fuel efficiency will be used to achieve better vehicle performance or compensate for additional vehicle weight resulting from new vehicle amenities and size. This has been the trend in the LDVs for at least the past 20 years. The no change scenario assumes that this trend will continue until 2035.

New vehicle sales are assumed to grow at a rate of 0.8 percent per year in the no change scenario, corresponding to the rate of growth of population. Average vehicle travel is assumed to grow at a rate of 0.5 percent per year, while the median age of all vehicles post year 2000 is assumed to be 15 years. In addition, the share of light trucks in new LDV sales is assumed to level off at 60 percent by year 2025.

In the baseline scenario, it is assumed that there is a modest, but steady increase in gasoline price, fuel economy standards, and competitive pressures that result in improved fuel economy. Fuel consumption of an average new gasoline ICE vehicle could decrease by about 35 percent in 20 years and 50 percent in 30 years, if vehicle performance characteristics are kept constant. This assumption is consistent with the results of recent MIT technology assessments (Weiss et al., 2000; Weiss et al., 2003). In the baseline scenario, it is assumed that only 50 percent of these efficiency improvements translate into reduction in fuel consumption. Thus, individual vehicle fuel consumption decreases by about 17.5 percent in 20 years and about 23.5 percent in 25 years.

The HEV scenario assumes ambitious fuel economy standards, coupled with economic incentives to push and pull advanced vehicle technologies—such as light-weighting, better aerodynamic designs, and hybridizing ICE vehicles assumed in baseline scenario—into the marketplace. Under this scenario, the simulation assumes that the fuel consumption of HEVs is 61.5 percent of the baseline gasoline ICE fuel consumption as shown in Figure 4-1. Two market penetration rates were examined for HEV sales as a fraction of all new vehicle sales to illustrate the impact of market penetration rates. These rates were assumed to be rising from about 1 percent in 2005 to 15 percent and 50 percent in 2035, under the low and high

46 *Driving Climate Change*

Relative On-Board Fuel Consumption

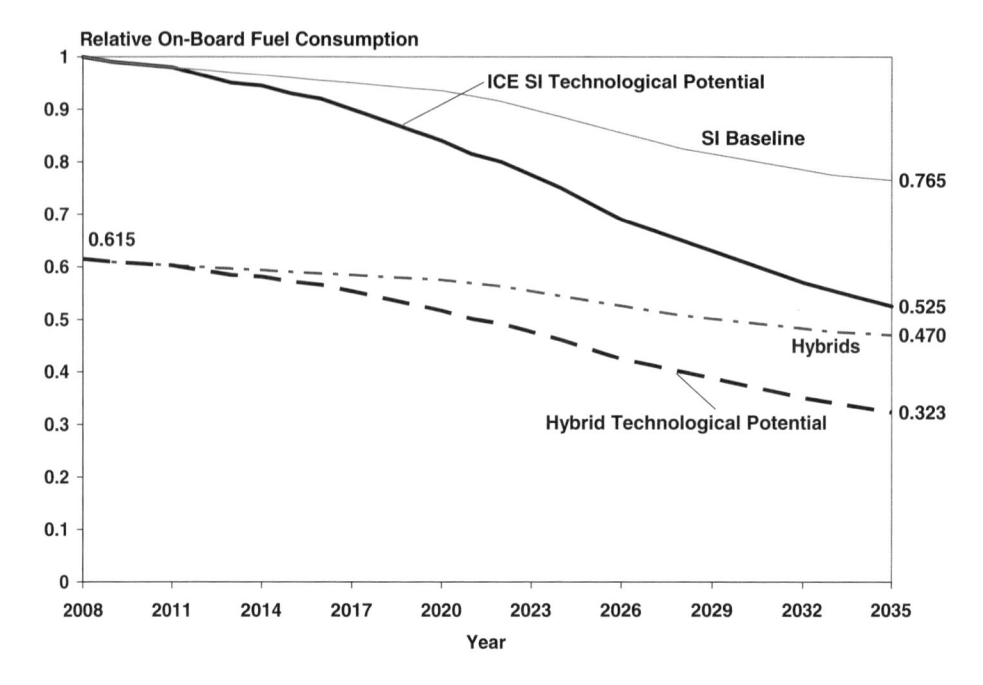

FIGURE 4-1. Relative improvements in fuel consumption for baseline scenarios.

assumptions, respectively. The high market penetration rate illustrates the upper bound in terms of reducing fuel use due to improvements in vehicle technology alone.

Under the composite scenario, it is assumed that in addition to all the factors present in the HEV scenario under the 50 percent penetration in the 2035 case, average per vehicle travel will stop growing beyond year 2008 and the rate of growth in sales of LDVs is halved to 0.4 percent.

Fuel Consumption under Different Scenarios

The projections of total fuel use under the no change, baseline, HEV, and composite scenarios are shown in Figure 4-2. The average fuel consumption of LDVs is shown in Figure 4-3. Under the no change scenario, the fuel consumption of the entire LDV fleet is projected to grow to 685 billion liters per year by 2020 and to 827 billion liters per year by 2035, or 11.8 and 14.2 million barrels per day, respectively.

Under the fuel consumption improvements assumed in the baseline scenario, growth in fuel use is considerably reduced. The total fuel use in year 2035 is projected to be 688 billion liters per year, or 11.8 million barrels per day. In the HEV scenario, the impact of 15 percent market penetration of HEVs is found to be small relative to the baseline improvements in ICE

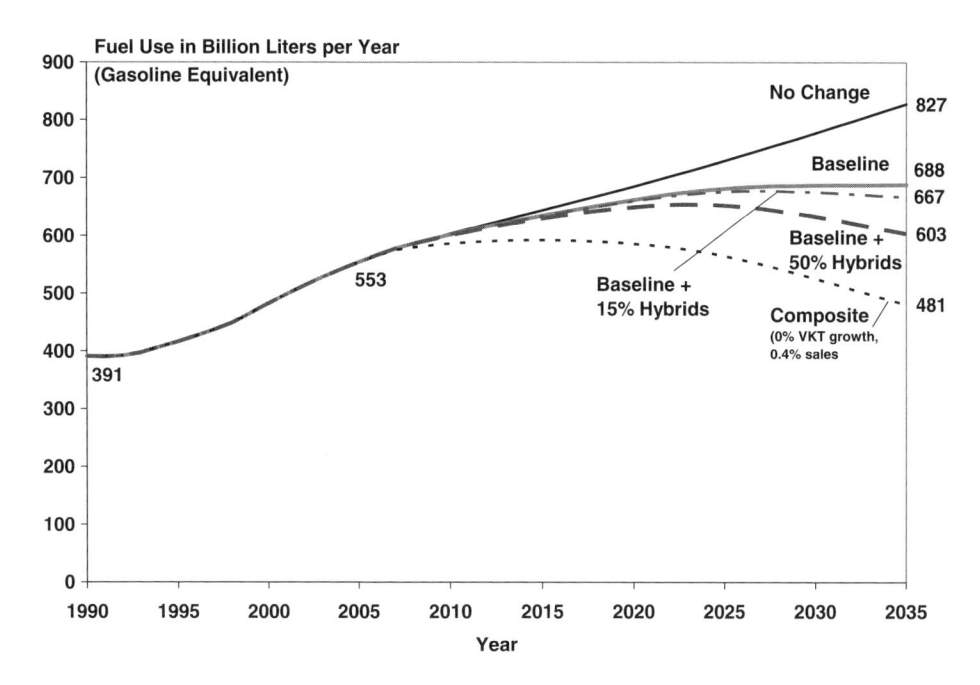

Reducing the Fuel Use of the U.S. Light-Duty Vehicle Fleet 47

FIGURE 4-2. LDV fuel use for various scenarios.

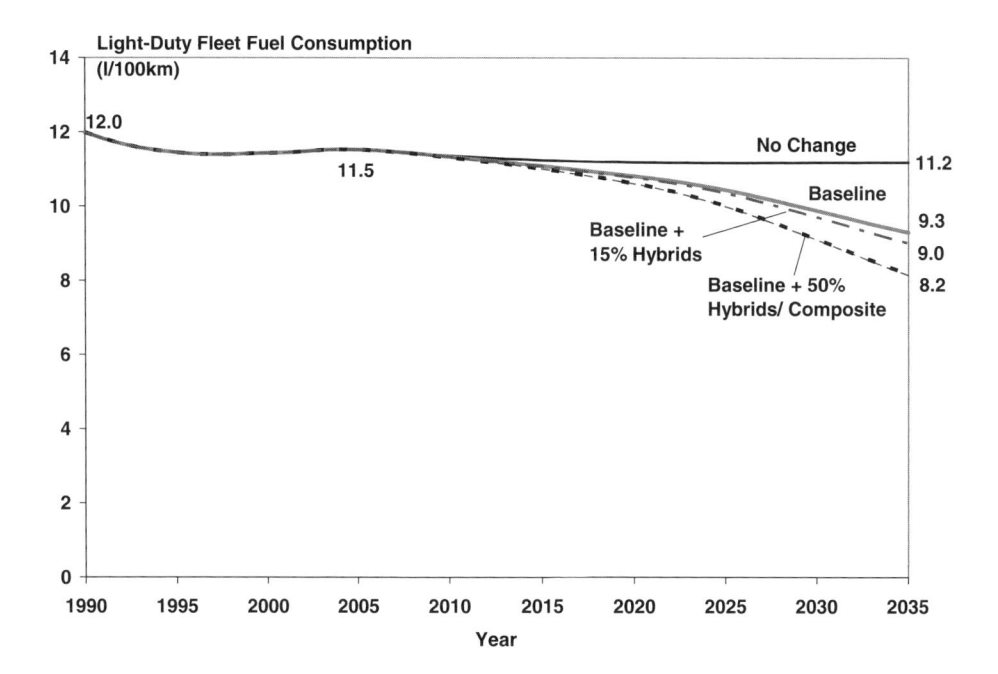

FIGURE 4-3. Light-duty vehicle fuel consumption for various scenarios.

48 *Driving Climate Change*

TABLE 4-1. LDV Fuel Use under Different Scenarios

Year	No Change		Baseline		Baseline + 15% Hybrids		Baseline + 50% Hybrids		Composite	
	Billion Liters per Year	MBD*	Billion Liters per Year	MBD	Billion Liters per Year	MBD	Billion Liters per Year	MBD	Billion Liters per Year	MBD
2005	554	9.5	554	9.5	554	9.5	554	9.5	554	9.5
2020	685	11.8	662	11.4	660	11.4	649	11.2	586	10.1
2035	827	14.2	688	11.8	667	11.5	603	10.4	481	8.3

* MBD: Million barrels per day.

technology alone. In the case of high penetration of HEVs into the LDV fleet, the total fuel use peaks at about 654 billion liters per year in year 2024 and then declines to about 603 billion liters per year by 2035. The average on-road fuel consumption of the fleet improves from 11.5 L/100 km, or 20.4 mpg, in 2005 to 8.2 L/100 km, or 28.8 mpg in 2035. This shows that rapid deployment of advanced vehicle technologies has significant potential to reduce fuel consumption in the next 30 years.

Finally, additional developments such as reduced rates of growth of vehicle sales and annual vehicle travel in the composite scenario show substantial benefits in terms of vehicle fuel use. This result is mainly due to the slowdown of growth in vehicle kilometers traveled from 7.4 trillion kilometers per year in the baseline scenario to about 5.9 trillion kilometers per year in the composite scenario by year 2035. The total fuel use in this scenario peaks at 592 billion liters per year in 2015 and decreases to 481 billion liters per year in 2035, which would be the same as the fuel use of LDVs in year 2000. Yet, this is still much higher fuel use than the 391 billion liters consumed in 1990.

These simulations show that improvements in vehicle technology and fuel consumption can play a key role in reducing the growth in LDV fuel use over the next 30 years. However, it will also take slowing down the growth in vehicle travel to achieve actual reductions in fuel use. Table 4-1 summarizes the results of these scenarios in more detail.

Effect of Delay

Delayed action scenarios were used to examine the consequences of postponing action by five or ten years on overall fleet fuel use and greenhouse gas emissions. These scenarios thus indicate the additional effort that would then be required to contain vehicle fuel use in the future as opposed to taking action immediately. These scenarios were evaluated for the HEV case as shown in Figure 4-4. As shown in the figure, the peak fuel use under the

Reducing the Fuel Use of the U.S. Light-Duty Vehicle Fleet   49

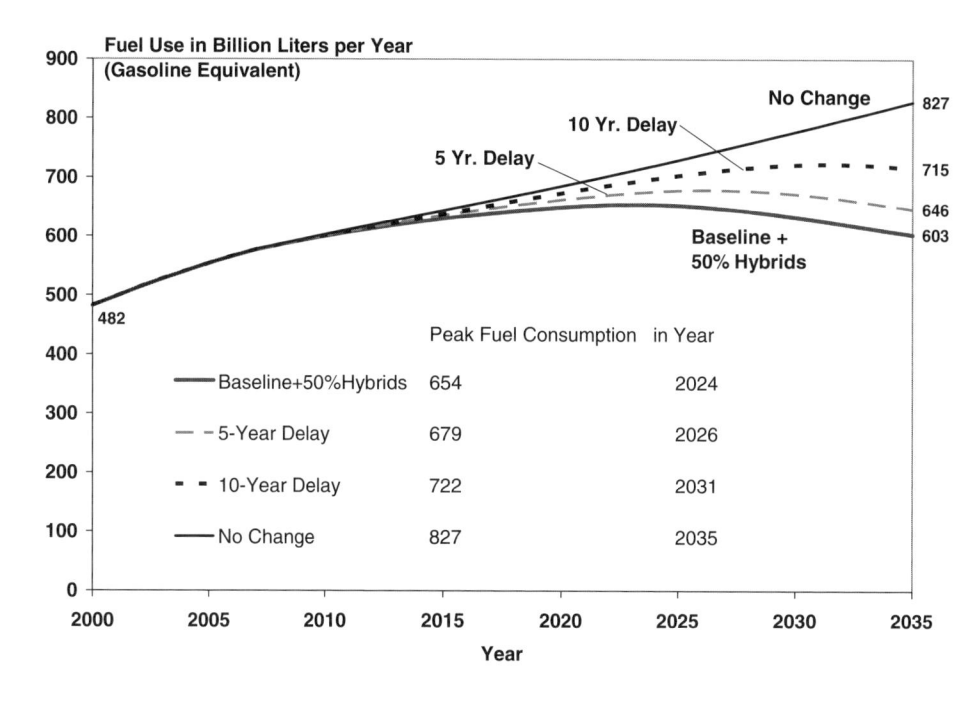

FIGURE 4-4. Effect of delay in action on light-duty vehicle fuel use (2000–2035).

50 percent HEV scenario is 654 billion liters in 2024. If action is delayed by five years, the peak in fuel use increases to 679 billion liters in 2026, whereas if action is delayed by ten years, the peak in fuel use increases to 722 billion liters in 2031.

It is clear that delayed action results not only in shifting the problem out in time, but also makes it more difficult to address. On the other hand, even small changes made sooner could result in larger benefits than more aggressive actions taken later. This also indicates that even if inherently low CO_2 emitting or nonpetroleum-based fuels were to become feasible in the future, the magnitude of the problem would be much more manageable if some action is taken now, as opposed to waiting for a cure-all.

Policy Measures to Reduce GHG Emissions

Increasing U.S. dependence on foreign oil and concerns about GHG emissions from motor vehicles are two important reasons for government intervention in the fuel use market. The DOE identifies different barriers to efficiency improvements in the U.S. transportation sector as underpriced fuel and services, imperfect information for consumers to make a rational choice about vehicle fuel economy, fungibility of technology, and risk averseness of the vehicle manufacturers (DOE, 2000). Different policy

TABLE 4-2. Policy Measures to Reduce Fuel Consumption of LDVs

Policy Measures	Type of Policy E R I	Anticipated Response/Action
CAFE Standards: As existing/Weight (E-CAFE)/Volume (VAFE)	•	Incorporate fuel efficient technologies, reduce average weight of vehicle fleet, reduce the spread between heavy and light vehicles
Tradable CAFE/Fuel Consumption Credits	• •	Increase flexibility for manufacturers and reduce cost of compliance with the CAFE standards
Feebates (A system of fees and rebates related to the fuel economy/ fuel consumption of the vehicles)	•	Establish fees for less fuel efficient vehicles and rebates for more fuel efficient vehicles to create incentive to produce and purchase more fuel efficient vehicles
Emissions/Carbon Tax (Economywide)	•	Provide incentive to purchase and use more fuel efficient vehicles by incorporating the externality costs
Fuel Tax	•	Increase the cost of operating the vehicle and reduce the vehicle miles traveled
Pay-at-the-Pump Schemes	•	Increase the cost of purchase and/or owning high fuel consumption vehicles or transfer it to the cost of motor vehicle use
Subsidies/Tax Incentives	• •	Provide incentive to purchase more fuel efficient vehicles
Government R&D Investment	•	Encourage more rapid development of fuel conserving technologies
Retiring Old Cars	• •	Provide incentive to purchase newer, more fuel efficient vehicles
Alternative Fuels (e.g., Cellulosic Ethanol/ Biodiesel)	• •	Displace (some) petroleum-based fuel used for transportation

An Economic Incentive (E), a Regulatory Requirement (R), a Public Investment (I).

measures have been proposed to overcome these barriers (OTA, 1994). The policy measures under consideration can be thought of as a means of providing an economic incentive (E), a regulatory requirement (R), a public investment (I), or some combination of these. They may be further classified as measures that provide incentives for more fuel efficient vehicles, measures that aim to change the cost structure of vehicle operation by increasing or converting some of the fixed or infrequently paid costs to usage costs, or measures aimed at shifting fuel use toward less-carbon-intensive fuels. Policy options selected for review are summarized in Table 4-2. A more detailed description of individual policies can be found in Bandivadekar and Heywood (2004).

Several other policy alternatives are available at the state or local level, such as increased investments in public transportation and transportation demand management (TDM) tools, including high occupancy vehicle (HOV) lanes, congestion charges, vehicle travel based fees, and telecommuting incentives. These options are not considered here, but they can be helpful in reducing the energy consumption of LDVs.

Qualitative Analysis of Individual Policy Options

The economic and societal impacts of government intervention in the market for fuel use assume multiple dimensions. The usefulness of individual policy measures cannot be judged on the basis of potential fuel use and greenhouse gas emission reductions alone. Apart from the fuel consumption of vehicles, other key issues under consideration include the following:

- *Vehicle performance:* It is expected that broadly popular vehicle performance measures such as acceleration, functional capacity, or the deployment of accessories and amenities will improve or at least remain constant in the pursuit of a more fuel-efficient fleet.
- *Safety implications:* It is generally accepted that if weight reductions occurred in the heaviest of LDVs, then overall safety should improve.
- *Mobility implications:* Implementation of certain strategies may change the purchasing, ownership, and usage patterns of LDVs. Consumers' essential mobility needs should be satisfied and the regressive effects of policy measures, if any, must be addressed. At the same time, the effect of different policies on other transportation issues, such as criteria emissions and congestion, must be considered.
- *Implementation issues:* The effectiveness of a policy measure will also depend on whether such a policy can be implemented successfully in practice. Generally, policy measures which give different stakeholders more flexibility for action should prove politically more acceptable.

Different policy options under consideration are evaluated across these different criteria in Tables 4-3 and 4-4. Quantitative estimates are provided wherever possible.

Careful observation of Tables 4-3 and 4-4 reveals that there are synergies between different policy measures. For example, while more fuel efficient vehicles may cause some increase in vehicle travel, this rebound effect could be offset by an appropriate increase in the fuel taxes. Also, an increase in fuel price at the pump makes it attractive for the automobile manufacturers to reduce fuel consumption in their vehicles, thus lowering the risks and costs associated with meeting CAFE standards. While the feebates and

52 Driving Climate Change

TABLE 4-3. Effectiveness of Policy Alternative to Reduce Fuel Consumption of the U.S. LDVs

Policy Measures	Cost/Cost Effectiveness (market or full societal benefits and costs?)	Scale of Applicability	Effectiveness in Addressing Energy Issues — Oil Use Reduction	Effectiveness in Addressing Energy Issues — Greenhouse Gas Reduction	Effectiveness in Addressing Other Transportation Issues — Emissions Reduction	Congestion Reduction	Effect on Vehicle Safety	Effect on Vehicle Miles Traveled (VMT)
CAFE	Costs of technological innovation and development necessary to meet the standards result in increased vehicle cost	Incremental gains from new vehicles. Currently affects cars and trucks separately.	Current savings of 2.8M barrels of oil/day (~14% of daily consumption). MPG gain from 20 to 30 saves more fuel than from 30 to 40.	In the short run, the greenhouse gas reduction is directly proportional to the reduction in oil use	Generally, better fuel economy means lesser emissions. Stricter emissions requirements may inhibit technologies like diesel, lean-burn.	Moderate worsening of congestion due to increased driving	Uncertain. It is likely that higher CAFE would reduce safety in vehicle-to-vehicle collision. This effect could be minimized by limiting weight & size reduction.	~1–2% increase in VMT for a 10% increase in fuel economy
Fuel Tax	Distributional effects: Regressive effects of fuel tax can be mitigated via other means such as explicit/earned income tax credits	Impact on entire fleet	Depends on the price elasticities of Demand. Short run estimates –0.1 to –0.4 (gasoline). Long-range estimates –0.2 to –1.0 (gasoline).	Same as above	Moderate improvement due to reduced driving	Moderate improvement in congestion due to reduced driving	Modest improvement in safety due to reduced driving	~1–3% reduction in VMT for a 10% increase in fuel prices
Feebates	Could be revenue neutral so that fees from more fuel consuming vehicles balance the rebates for more fuel efficient vehicles. Progressive?	Incremental gains from new vehicles	Savings due to improved overall fleet fuel economy	Proportional to the oil use reduction	Generally, better fuel economy vehicles cause lesser emissions	Moderate worsening of congestion due to increased driving	Small	~1–2% increase in VMT for a 10% increase in fuel economy

Alternative Fuels: Cellulosic Ethanol/ BioDiesel	Currently expensive as compared to Gasoline (~$2.70/gallon gasoline equivalent at the pump)	Potential to have a large scale fleet wide effect	Projections of 10+% of fuel displacement. Ambitious plans may displace a larger percentage.	40–70% reduction in full cycle CO2 emissions possible	Increase in lifecycle emissions likely, mainly due to use of fossil fuels in producing fertilizers and farm equipments. Biodiesel may have better characteristics than diesel.	No impact	No impact	No impact
PATP	PATP schemes involve transfer of insurance or registration fees to the pump (~$0.1 to 0.75 per gallon)	Impact on entire fleet	Depends upon the price elasticities of demand. Short run estimates –0.1 to –0.4, and Long range estimates –0.2 to –1.0 (gasoline)	~9 Million Metric Tons (MMT) per year of CO2 reduction for $0.10 per gallon of PATP charge; 32 MMT per year for $0.40 per gallon of PATP charge	Moderate improvement due to reduced driving	Moderate improvement in congestion due to reduced driving	Modest improvement in safety due to reduced driving, otherwise no direct effect	Some reduction in vehicle travel as a result of increased cost of driving
Retiring Old Cars	Financial incentives need to be created for replacement of older vehicles	Impact on replacement vehicles only	Depends on average fuel consumption of vehicle retired and average fuel consumption of vehicle replacing it, and the life remaining in the old vehicle	Proportional to oil use reduction	Newer vehicle purchase will result in emissions reductions	No change	As newer vehicles are more safe, overall safety may improve	Slight increase likely as newer vehicles tend to be driven more than older vehicles

TABLE 4-3. Effectiveness of Policy Alternative to Reduce Fuel Consumption of the U.S. LDVs—cont'd

Policy Measures	Cost/Cost Effectiveness (market or full societal benefits and costs)	Scale of Applicability	Effectiveness in Addressing Energy Issues		Effectiveness in Addressing Other Transportation Issues			
			Oil Use Reduction	Greenhouse Gas Reduction	Emissions Reduction	Congestion Reduction	Effect on Vehicle Safety	Effect on Vehicle Miles Traveled (VMT)
RD&D	Public investment of several hundred millions of dollars per year	Across the next generation of vehicles	Significant long-term impacts possible	Significant long-term impacts possible	Significant long-term impacts possible	No change	Significant long term impacts possible	No direct effect. More fuel efficient technology will encourage more driving.
Manufacturer Tax Incentives	Costs will be of the tune of $2 billion over ten years	Incremental gains from new vehicles	Savings due to improvement in new fleet fuel economy	Proportional to oil use reduction	Positive effect through manufacture of vehicles with better emissions performance	No effect	No effect	No effect

Reducing the Fuel Use of the U.S. Light-Duty Vehicle Fleet 55

TABLE 4-4. Considerations for Implementation of Individual Policy Options to Reduce GHGs

Policy Measures	Rate of Implementation	Scale of Implementation	Flexibility	Political Acceptability	Level of Co-operation Needed Between Agencies	Technological Change	Degree of Lifestyle Change Required	Other Factors
CAFE	Standards must give manufacturers sufficient time to respond. Widespread penetration of new technologies requires 10–15 years.	Standards need to be set at a level where the marginal cost of additional fuel savings equals marginal benefits from savings to the consumer	Details of standards are important and complicated. Many possible approaches. Current standards distinguish between light trucks and passenger cars.	CAFE standards are by and large the politically most acceptable means. Automobile manufacturers oppose increase in CAFE.	While EPA does the testing of car fuel economy, NHTSA is actually responsible for CAFE and highway safety issues	Consumer demand for fuel-efficient vehicles is low at low fuel prices. CAFE drives improvement in technological efficiency. However, also encourages vehicle sales mix change.	Small or uncertain. MAY require shifting towards lighter/"less bigger" vehicles.	—
Fuel Tax	Immediate impact on implementation, but implementation needs to be gradual	Tax equal to amount of externality generated by the fuel use ($0.02 to $0.50 per gallon). Actually, not as straightforward.	Fuel Tax is a means of decreasing the incentive to drive more	Political acceptability is currently poor	Minimal. A fuel tax collection system is already in place	Diminished pressure for technological advances. Encourages change in behavior.	Change driving habits. Evaluate other modes of transport.	—

56 Driving Climate Change

TABLE 4-4. Considerations for Implementation of Individual Policy Options to Reduce GHGs–cont'd

Policy Measures	Rate of Implementation	Scale of Implementation	Flexibility	Political Acceptability	Level of Co-operation Needed Between Agencies	Technological Change	Degree of Lifestyle Change Required	Other Factors
Feebates	Feebate levels must give manufacturers sufficient time to respond.	~5–10% of vehicle price; level of fees and rebates need to be adjusted frequently to maintain the program revenue neutral, may require a pool of money for rebates	Feebates aim at sale of new personal vehicles, based on fuel efficiency, fuel economy or emissions of carbon dioxide	Revenue neutral nature has political appeal	Feebate monitoring mechanism has to be established	Increased incentive to bring fuel efficient vehicles to market	Little or no impact	Consumer and Manufacturer response to feebates is unknown. However, consumer response estimated to be smaller as compared to the manufacturers' response.
Alternative Fuels: Cellulosic Ethanol/ BioDiesel	Introduction of alternative fuels must be gradual	Limited by the production capability, and the amount of subsidy needed	Large amount of land needed for biomass production on a regular basis	Large-scale use of alternative fuels could mean increased oil security. "Biodiversity" may be an issue.	Fuel safety and supply network must be established	Development of Alternative Fuel Vehicle (AFV) technology	Little or no impact	Consumer response to large scale changes in type of fuel unknown.
PATP	Immediate impact upon implementation, but implementation needs to be gradual	Implementation will vary from state to state	PATP charges could be based on insurance, inspection/ maintenance, registration fees individually or as a combination	Revenue neutral nature has political appeal, PATP often associated with insurance reforms, a very sticky issue. Opposition from trial lawyer groups	Insurance, inspection/ maintenance, registration fees are mostly state affairs, however nationwide changes required for significant impact	Encourages more fuel efficient vehicles	Insurance premiums will be correlated to fuel use	It will be possible to insure everybody who purchases fuel.

Retiring Old Cars	Rate of implementation will vary from state to state	Small, and restricted to older vehicles	Limited	Potential regressive effects need to be considered	Coordination between revenue, transportation and environmental departments required	Not technology forcing, but will increase rate of technology penetration	No change	A small increase in new vehicle sales
RD&D	Long-term precompetitive projects (up to 10 years of development times)	A public-private partnership to conduct joint research program with an intention to innovate	Possible to set ambitious long-term goals	Strong political support for projects such as Partnership for the New Generation of Vehicles (PNGV) and FreedomCar	DOE, DOT, and government labs need to coordinate with industry	Technological breakthroughs possible in long run	None	Benefits of RD&D are not always seen in short term
Manufacturer Tax Incentives	Will only affect a part of new vehicle market gradually over 10-15 years.	Credits would cover "substantial percentage" of capital investment (~0.5–1.0 billion for a new factory ?)	Incentives should be performance based and technology neutral	Support from domestic automakers and unions	Revenue services, DOE and DOT will need to coordinate, some monitoring mechanism needs to be established	Encourages deployment of more fuel efficient technologies in the new vehicles	None	Automobile manufacturers have been vocal about consumer tax incentives, but not all that much about manufacturer tax credits?

CAFE standards apply only to new vehicles, fuel taxes and alternative fuel use requirements have fleetwide impact. While introduction of more fuel efficient technology might cost more initially, the rebates given to the more fuel efficient vehicles can reduce the economic burden on consumers. At the same time, the fees on vehicles with high fuel consumption will not only discourage consumers from buying those vehicles, but also provide incentives to the vehicle manufacturers to produce more fuel efficient vehicles. While the cost of renewable alternative liquid fuels may currently be higher than gasoline, regulations requiring increased renewable fuel content along with government purchasing of the alternative fuel vehicles can provide economies of scale and the learning needed to reduce the cost associated with alternative fuels.

Rationales for Combinations of Policy Measures

Clearly, there is no single agreed upon policy measure that would significantly reduce the fuel use of LDVs, and differences over policy measures are likely to persist (McNutt et al., 1998). Our assessment is that the vehicle fuel use problem can best be addressed by a carefully selected combination of policy measures that shares the responsibility among all stakeholders. There is a twofold argument for combining policy measures to reduce fuel consumption of LDVs. The first is that increasing vehicle fuel consumption is a market failure that necessarily requires regulatory and fiscal responses. The second is that without such an integrated approach, a policy proposal may not have the necessary broadbased support to move forward. Both of these arguments are explored in the next two sections.

Market Failure or a Failed Market?

Greene (1998) claims that the market for fuel economy is inherently sluggish for two primary reasons. To start with, consumers have imperfect information of the net present value of fuel savings achieved from higher-fuel-economy vehicles and no reasonable way of comparing it to the additional cost it imposes at the time of vehicle purchase. Moreover, fuel consumption is only one of many characteristics that consumers care about when buying a vehicle.

In addition, according to Greene, unless there are clear signals that consumers demand better fuel consumption performance, manufacturers are likely to be reluctant to invest in major technological changes aimed at reducing fuel consumption. In other words, the risk of providing better fuel consumption at an additional cost may be too high for the automobile manufacturers.

More than two decades ago, the National Research Council's Committee on Nuclear and Alternative Energy Systems (CONAES) noted the following (NRC, 1980):

The willingness to invest in capital substitutions for energy and to practice energy conservation clearly rises or falls with changes in the anticipated price of energy. Conservation of energy represents a middle- to long-range investment; if the investment is to be made, the signals the economy reads from prices for energy must be unambiguous, and the trends reasonably predictable over the lifetimes of normal investments.

However, because even accurate, widely noted market signals are sometimes insufficient to guide market decisions in the direction of energy conservation—as, for example, when the total cost of owning and operating a particular facility, appliance, or process is relatively insensitive to energy efficiency—prices alone cannot carry the burden of effective conservation policy.

The cost of fuel use is small, although not negligible, when compared to the total operating costs of a vehicle, and relatively large improvements in fuel economy, which involve additional upfront costs, are needed to reduce these costs further. Thus, the amount of fuel savings may be an insufficiently attractive proposition for the consumers to demand less-fuel-consuming vehicles. The 1980 NRC CONAES study said further, "Where energy prices are insufficient to induce the appropriate, economically rational responses from consumers—as they are, for example, in the case of the automobile—they could be supplemented by nonprice measures."

In other words, while price signals are necessary, they may not be sufficient to induce the technological changes required to substantially reduce the fuel consumption. On the other hand, if regulatory standards are set without providing the market incentives, then the manufacturers have to bear the risks of producing vehicles with characteristics that consumers may be unwilling to accept. A National Research Council study on the effectiveness and impact of CAFE standards commented on this issue in its findings (NRC, 2002):

There is a marked inconsistency between pressing automotive manufacturers for improved fuel economy from new vehicles on one hand and insisting on low real gasoline prices on the other. Higher real prices for gasoline—for instance, through increased gasoline taxes— would create both a demand for fuel-efficient new vehicles and an incentive for owners of existing vehicles to drive them less.

Thus, while increasing fuel economy standards alone would be a more effective policy than not acting at all, a combination of an increase in gasoline tax and increased fuel economy standards would be a significantly more effective approach (Gerard and Lave, 2003).

Political Appeal of an Integrated Policy Approach

It is extremely difficult to measure the value of all the different externalities caused by fuel use. The escalating fuel use of LDVs presents a classic common problem. If the aim of policy were economic efficiency alone, then

TABLE 4-5. Types of Policies Based on Costs and Benefits of Policy

Types of Regulatory Activity		Costs	
		Widely Distributed	Concentrated
Benefits	Widely distributed	Majoritarian	Entrepreneurial
	Concentrated	Client	Interest-group

getting the prices right would help but might not completely solve the problem as explained in the previous section. In practice, the policy process has aims beyond economic efficiency, such as equity and access with respect to mobility. Different policy approaches are criticized for different reasons. For example, one argument against the CAFE standards is that they constrain the automobile manufacturers too much. Gasoline taxes are criticized as having regressive economic effects, and so on.

Among other factors, the success of a proposed policy depends upon the real and perceived distribution of costs and benefits resulting from the policy (Wilson, 1980). Such costs and benefits are not always monetary and perceptions of the fairness of a policy often affect whether the stakeholders find the policy legitimate and persuasive. According to Wilson, public policies can be classified into different categories depending on the distribution of costs and benefits resulting from the implementation of the policy as shown in Table 4-5.

CAFE standards, for example, can be described as entrepreneurial because the costs of meeting the regulations fall largely on the automobile manufacturers. Although the monetary costs may ultimately be passed on to consumers, the risks involved in the process are borne solely by automobile manufacturers. The benefits, on the other hand, are seen by society in the form of reduced fuel use and GHG emissions. It should come as no surprise, therefore, that automobile manufacturers oppose the CAFE standards as the only means to reducing fuel use.

The stakeholders in this problem include vehicle purchasers and users, the automobile and petroleum industries, and governments at different levels. A policy package that attempts to spread the costs and benefits among different stakeholders is likely to have a broader political appeal and could be perceived as a more fair approach to fuel use regulation. Such a multidimensional policy approach seeks to generate positive commitment from all the stakeholders, without exposing any one set of stakeholder groups to a large risk.

Development of a Sample Policy Package

A conclusion endorsed by most participants at the 1997 Asilomar conference on Policies for Fostering Sustainable Transportation Technologies was that the overall strategy for meeting environmental quality and energy system goals must include a creative and flexible blend of regulation, pricing reform, incentives, and consumer education (Lipman et al., 1998). The aim of such a policy must be to reduce individual vehicle fuel consumption, slow the growth in vehicle travel, and reduce the carbon intensity of fuel used (BEST, 2001).

The previous section showed qualitatively that useful synergies exist between different policy measures. Agras and Chapman (1999) claim that using a combination of increased gasoline taxes and CAFE standards is more effective than using either policy individually. DeCicco and Lynd (1997) discuss scenarios that combine vehicle fuel economy improvements along with increased use of cellulosic ethanol. The combined impacts of policies are not necessarily additive, although some previous analyses assume that they are additive (NRTEE, 1998). The extent of cross-elasticity or cross-coupling of different measures is highly uncertain. The effect of policy measures affecting the same aspect of emissions could be considered multiplicative to avoid double counting (Greene and Schafer, 2003). DeCicco and Gordon (1995) affirm that the effect of a small increase in gasoline tax when coupled with an increased fuel economy standard will be limited to a reduction in vehicle travel, and the fuel economy standards will override the effects of improved fuel economy resulting from increased gasoline tax.

As an example of an integrated policy approach, a proposal that combines several different policy measures is presented here, and its potential impact on vehicle fleet fuel use is described. While this represents one possible example of a policy package, various other synergistic combinations with different policy options could be used. The policy package example described below combines measures to reduce vehicle fuel consumption, slow the growth in vehicle travel, and increase the renewable content of the fuel.

Sample Policy Package

Vehicle manufacturers could be required to meet CAFE standards in the current or modified form as part of a policy package that aims at both pushing and pulling advanced vehicle technology and renewable fuels into the market. Key considerations would be the extent of changes in the form of the CAFE program, as well as the aggressiveness of the proposed standard. A possible increase in CAFE standards could be based on baseline or optimistic HEV scenarios as discussed previously. The fuel economy levels corresponding to these scenarios are shown in Table 4-6. These fuel economy levels assume that about half of the realizable improvement in advanced engine and vehicle technology would be utilized in improving the vehicle fuel consumption, as shown earlier in Figure 4-1.

TABLE 4-6. CAFE Standard Levels under Baseline and Optimistic Scenarios

| Year | Baseline Scenario |||| Optimistic Scenario ||||
| | Cars || Light Trucks || Cars || Light Trucks ||
	Miles per Gallon	Liters per 100 km	Miles per Gallon	Liters per 100 km	Miles per Gallon	Liters per 100 km	Miles per Gallon	Liters per 100 km
2020	30.5	9.0	24.1	11.4	31.9	8.6	25.2	10.9
2030	35.4	7.8	28.0	9.8	41.5	6.6	32.8	8.4

A revenue neutral feebate program can encourage the manufacture and purchase of more fuel efficient vehicles. Such a program consisting of fees for gas guzzlers and rebates for gas sippers at the time of vehicle purchase could complement the CAFE program. A moderate fee or rebate rate of $25,000 per gallon per mile driven is proposed. This is roughly equivalent to a fee/rebate range of +$400 to –$1,500 per vehicle.

Gasoline taxes could be increased by about $0.10 per gallon every year, or roughly $0.03 per liter per year. Equivalent tax credits could be granted to consumers to achieve revenue neutrality and minimize regressive impacts. Such a form of tax shifting may encourage reduction in vehicle kilometers traveled without causing a financial burden to the vehicle users. Without such compensation, generating political support for this measure might prove difficult.

The renewable content in fuels could be increased by mandating an increasing amount of biomass-based liquid fuels blended in gasoline. This mandate may require a biomass-based liquid fuel blend in gasoline of 4.5 percent by 2025 and 7 percent by 2035 on a volumetric basis. In a more aggressive action, these requirements could be doubled to 9 percent by 2025 and 14 percent by 2035. These levels correspond to a 0.25 to 0.5 percent increase per year in the volume of biomass-based fuel blended in gasoline. Fuels with high renewable content could also receive preferential tax treatments with respect to gasoline or diesel, which should encourage fuel suppliers to make a shift toward renewable fuels. For calculations shown in this chapter, it is assumed that this requirement is fulfilled through cellulosic ethanol.

This analysis assumes that the policy package will be implemented starting in 2008 and continuing through 2035. The policies can be combined in different proportions, and the sensitivity of different combinations is evaluated through the four policy package scenarios shown in Table 4-7. Policy scenarios 1 and 3 are based on fuel consumption improvements as per the baseline, whereas policy scenarios 2 and 4 are based on fuel consumption improvements as per the optimistic HEV scenario.

TABLE 4-7. Policy Combinations Examined

Policy Measures	Policy Scenarios			
	1	2	3	4
CAFE Standards	Baseline	Optimistic	Baseline	Optimistic
Gasoline Tax Increase per Year	10¢ per gallon	10¢ per gallon	10¢ per gallon	10¢ per gallon
Cellulosic Ethanol Content Increase per Year	0.25%	0.25%	0.5%	0.5%

Calculation of Expected Impact

The anticipated impact of such an integrated policy package is estimated by the following multistep process. A vehicle fleet model is used to evaluate the effect of improved vehicle fuel consumption on fuel use (Heywood et al., 2004). No changes in vehicle sales or vehicle travel growth rates are assumed. The fuel use and vehicle travel of cars and light trucks are calculated separately. The price of gasoline is assumed to remain at $2.50 per gallon until 2007, when a $0.10 per gallon per year increment in gasoline taxes is applied. The effect of fuel prices on the driving distances can be calculated as shown in Equation 4.2 (Hayashi et al., 2001).

$$D_{t+1} = \left[1 - E_{vmt\_fuelp}\left(1 - \frac{P_{t+1}}{P_t}\right)\right]D_t \quad (4.2)$$

Where:

D_t is the driving distance in year t
P_t is the gasoline price
E_{vmst_fuelp} is the elasticity of vehicle travel with respect to fuel price

Past estimates of elasticity of vehicle travel with respect to fuel price have varied widely in both the short term from –0.09 to –0.2, and in the long term from –0.2 to –0.5 (Goodwin, 1992; Haughton and Sarkar, 1996; Greene and DeCicco, 2000; Nivola and Crandall, 1995). The elasticity of vehicle travel with respect to fuel price is assumed here to be –0.2, which is a low-end estimate for the long-term effect and a high-end estimate of the short-term effect. Thus, a 10 percent increase in gasoline price decreases vehicle travel by 2 percent over a one-year period.

The effect of decreased fuel consumption on vehicle travel is estimated using a takeback factor of –0.2. This takeback is assumed to affect all vehicles and thus may overestimate the impact of the rebound effect. In quantitative terms, a 10 percent decrease in fuel consumption is assumed to cause an increase in vehicle travel of 2 percent over a period of one year.

Note that the gasoline tax increase and rebound effect estimates tend to offset one another.

The effect of a vehicle feebate on vehicle fuel consumption is not modeled explicitly. Instead, it is assumed that the feebate neutral point will be established at the level of the CAFE standards. The feebates will then provide the necessary incentive for the consumers to purchase more fuel efficient vehicles and reduce the risk to the vehicle manufacturers of meeting fuel economy standards. In practice, the feebates are likely to provide an additional incentive to the vehicle manufacturers to produce more fuel efficient vehicles. Thus, the impact on vehicle fuel consumption is underestimated in this analysis.

Increasing the proportion of cellulosic ethanol blended into gasoline is assumed to displace gasoline use. The amount of gasoline displaced as a result of blending ethanol can be calculated as shown in Equation 4-3.

$$V_{g\_d} = \left(\frac{E_{ethanol} * p}{E_{gasoline} * (1-p) + E_{ethanol} * p} \right) \qquad (4.3)$$

Where:

V_{g_d} is the fraction of gasoline displaced
$E_{ethanol}$ is the energy content of cellulosic ethanol in MJ/liter
$E_{gasoline}$ is the energy content of conventional gasoline in MJ/liter
p is the percentage of cellulosic ethanol blended in gasoline by volume

Note that the energy content of ethanol is about two-thirds that of conventional gasoline. Therefore, a 10 percent by volume blend of cellulosic ethanol reduces the consumption of gasoline by about 6.8 percent.

Results

Figures 4-5 and 4-6 show the effect of policy combinations on LDV fleet fuel consumption and travel, respectively. The reduction in average fuel consumption of new cars and trucks in scenarios 1 and 3 is about 23.5 percent, and that in the overall fleet fuel consumption is over 16 percent. As a result of increased gasoline tax, the total car travel in 2035 is only slightly higher than the current level. The total vehicle travel by light trucks continues to increase but at a slower rate. The reduction in overall vehicle travel from the no change scenario is about 10 percent.

Fuel use under different policy scenarios is shown in Figures 4-7 and 4-8. Under policy combination 1, the total fuel use of LDVs peaks at 606 billion liters per year (10.4 million barrels per day) in 2023 and gradually reduces to 590 billion liters per year (10.1 million barrels per day) in 2035. This is still slightly higher than the current LDV fuel use of about 554 billion liters in 2005. Table 4-8 summarizes the results of the different

Reducing the Fuel Use of the U.S. Light-Duty Vehicle Fleet   65

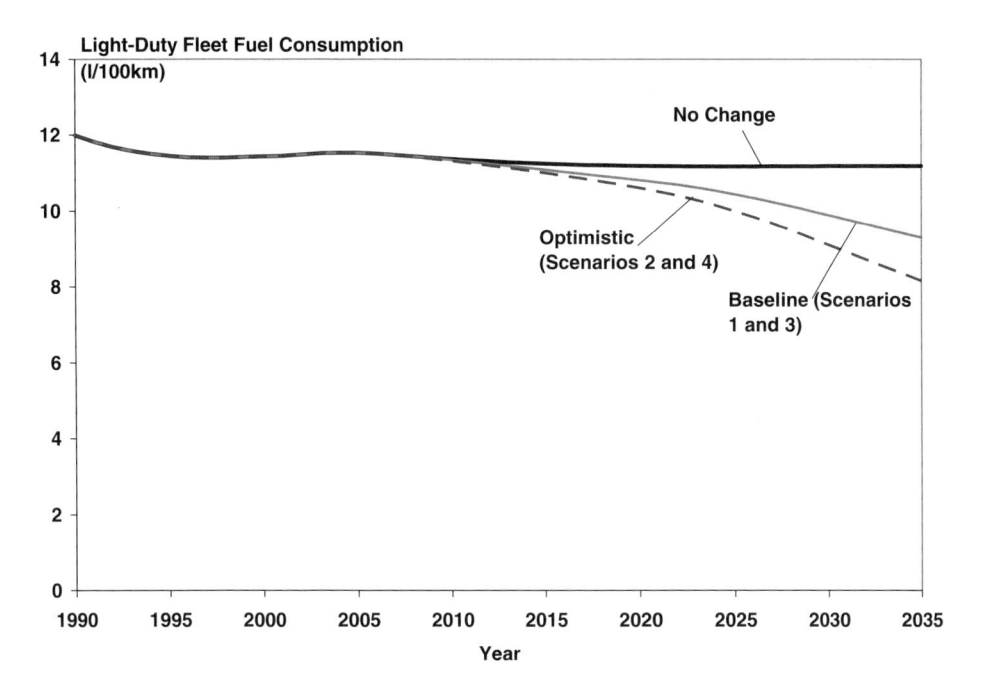

FIGURE 4-5. Average fuel consumption of light-duty vehicle fleet (1990–2035).

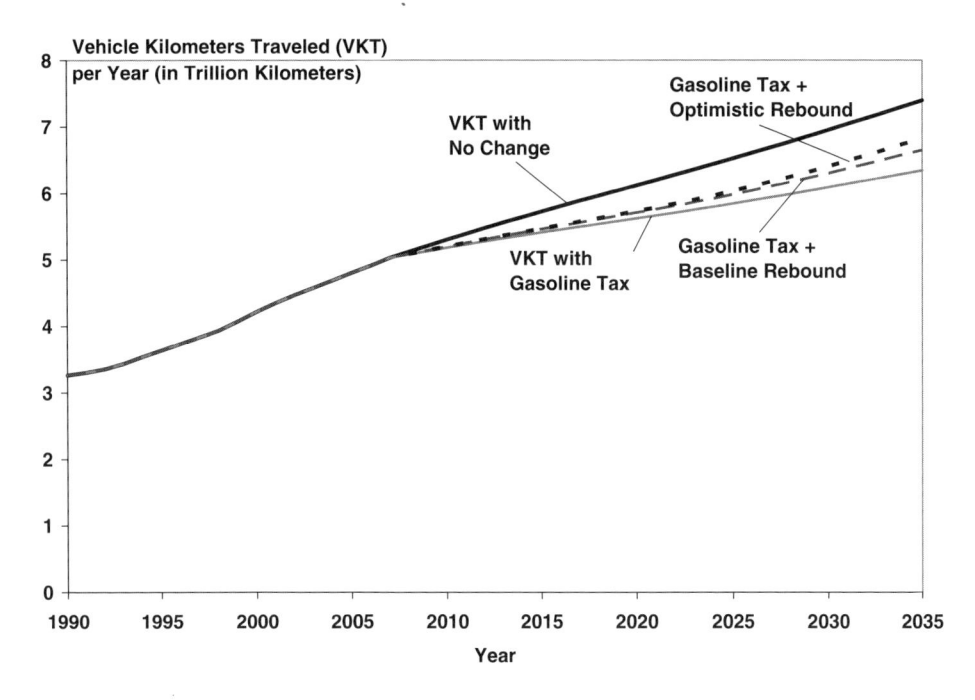

FIGURE 4-6. Total light-duty fleet travel (1990–2035).

66 *Driving Climate Change*

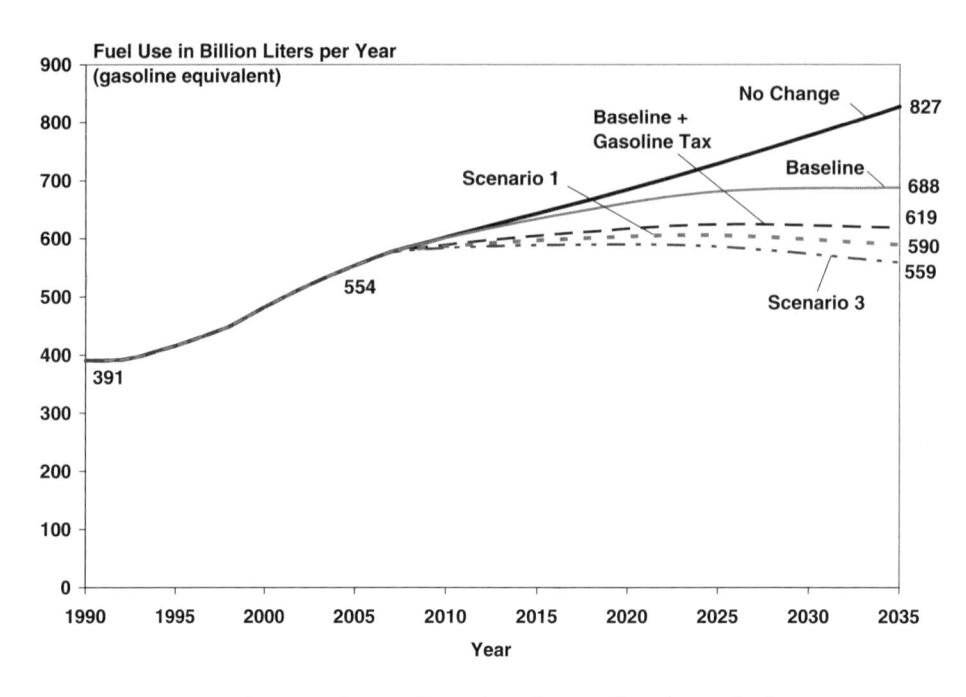

FIGURE 4-7. Fuel use for policy based on baseline fuel consumption improvements.

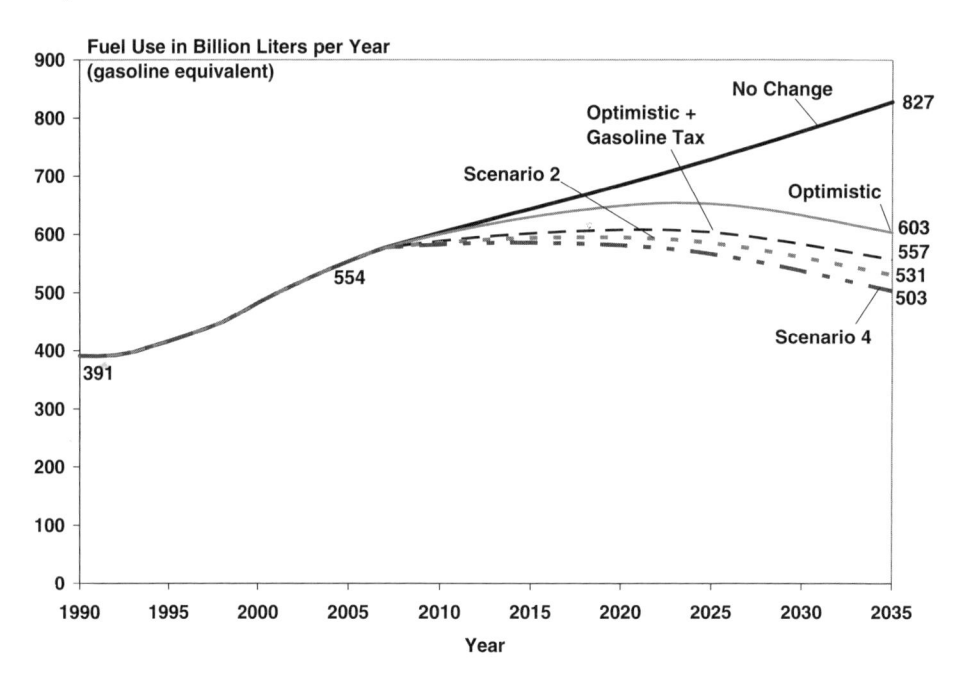

FIGURE 4-8. Fuel use for policy based on optimistic fuel consumption improvements.

TABLE 4-8. LDV Fleet Fuel Use under Different Policy Scenarios

Fuel Use (in Billion Liters per Year)*	No Change Scenario	Policy Scenarios			
		1	2	3	4
2010	603	587 (2.7)	586 (2.9)	584 (3.1)	583 (3.3)
2020	685	604 (11.8)	595 (13.1)	591 (13.8)	581 (15.1)
2035	827	590 (28.7)	531 (35.9)	560 (32.4)	503 (39.2)

* Numbers in brackets indicate percentage reduction in fuel use from no change.

scenarios. Since fuel taxes on vehicle travel and the increased ethanol content in gasoline affect the entire fleet, changes in fuel use can be seen almost immediately.

Table 4-8 illustrates that the potential exists to reduce the fuel use of LDVs by 12 to 15 percent by 2020 and by as much as 28 to 40 percent by 2035 relative to the no change scenario. However, an integrated set of fiscal and regulatory measures designed to affect vehicle fuel consumption, vehicle travel, and the nonpetroleum content in fuels would need to be implemented soon in order to achieve these results.

The analysis shows that raising fuel prices in the short term may well achieve significant reductions in fuel use. In the long run, however, it will be necessary that the improvements in vehicle technology which reduce the fuel consumption of new vehicles penetrate into the entire vehicle fleet. Over a 15- to 25-year period, this improvement in technology can deliver significant benefits. It should be noted that this is neither a surprising nor a new conclusion (Wildhorn et al., 1976). It does, however, reinforce the notion that both market-based and regulatory instruments aimed at pulling and pushing more fuel efficient technology into the market are needed.

Sensitivity Analysis

The effect of variations in elasticity of vehicle travel with respect to gasoline prices and amount of rebound effect were tested for scenario 1 as shown in Table 4-9 and Figure 4-9.

As seen from the range of results, the rebound effect has a small impact on total fuel use for the improvements in fuel economy considered here. However, the impact of the gasoline tax increase is quite sensitive to the elasticity of vehicle travel. The difference between fuel use for elasticity of travel equal to –0.2 and –0.3 is of the order of 7 percent.

Challenges in Implementing a Coordinated Policy Package

As noted by Fulton (2001), a comprehensive policy package may be able to combine the best elements of policies aimed at different drivers of GHG emissions from motor vehicles. At the same time, it may be difficult to

68 *Driving Climate Change*

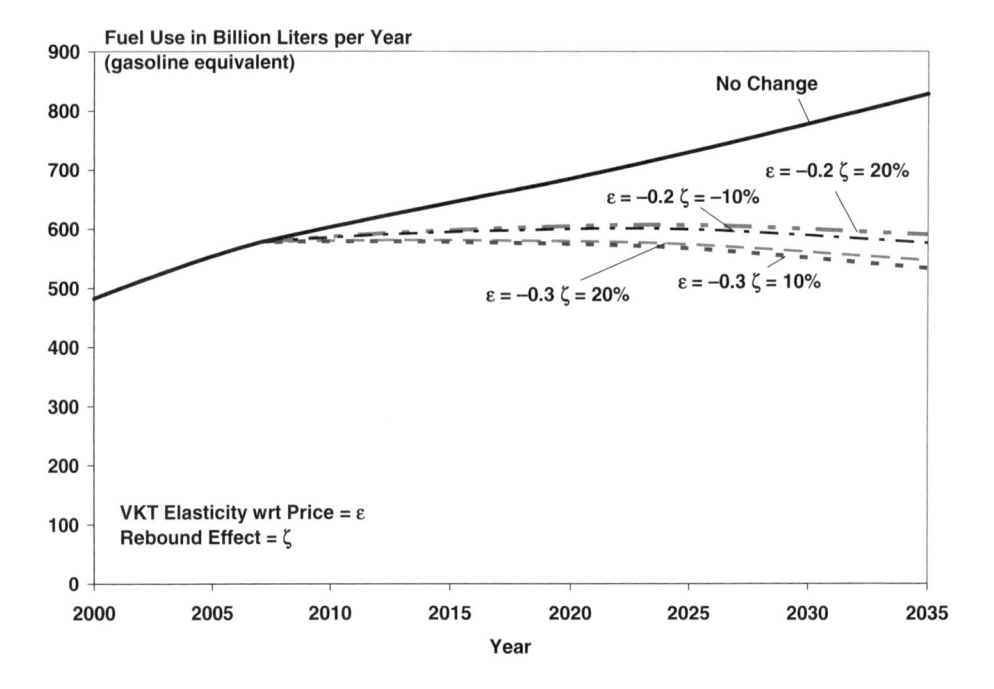

TABLE 4-9. Sensitivity to Vehicle Travel Elasticity and Rebound Effect for Scenario 1

Year	Fuel Use (in billion liters per year) (Numbers in brackets indicate million barrels per day)			
	VKT Elasticity = −0.2 Rebound Effect = 20%	VKT Elasticity = −0.3 Rebound Effect = 20%	VKT Elasticity = −0.2 Rebound Effect = 10%	VKT Elasticity = −0.3 Rebound Effect = 10%
2010	587 (10.1)	580 (9.9)	586 (10.1)	579 (9.9)
2020	604 (10.4)	579 (9.9)	599 (10.3)	574 (9.8)
2035	590 (10.1)	546 (9.4)	576 (9.9)	533 (9.1)

FIGURE 4-9. Sensitivity to vehicle travel elasticity and rebound effect for scenario 1.

implement such a policy package. This is true because progress on transport related policy is usually made one step at a time, and it may not be possible to consider legislatively all the different aspects of a policy package together. Further, the authority to deal with different aspects of fiscal and regulatory policies related to transport lies with different institutions, and overcoming institutional obstacles could be a more difficult task than formulating the policy package. Nevertheless, if a combination of different policy approaches is not considered, then it may be even more difficult to generate commitments from different stakeholders.

Attempts to develop a comprehensive policy of the type outlined here could turn into an ambitious effort to influence many aspects of motor transport. For example, in brainstorming activities at the Organization of Economic Cooperation and Development (OECD) on policies to achieve environmentally sustainable transportation, anywhere from 14 to 88 different policy instruments were suggested by different country groups (OECD, 2002). Therefore, attention should be focused on a small number of policy options, which nevertheless affect all the different aspects of vehicle fuel use. Also, many different small or large coalitions may come together to oppose a comprehensive policy package. It is necessary, therefore, to develop transparent policy measures. Thus, the role of public education and feedback in bringing about the necessary participation must not be neglected.

It is also possible that different policy options may affect different automotive manufacturers differently, and there may be some wealth transfer between different vehicle manufacturers. Fuel economy standards or feebates can be designed to minimize such competitive impacts (McNutt and Patterson, 1986; Davis et al., 1995).

Conclusion

This analysis indicates that fuel use and GHG emission reductions from U.S. LDVs cannot be achieved in practice by regulations alone. Neither will the current market forces bring about the necessary technological change needed to reduce fuel use significantly.

To reduce LDV fuel use, transportation policies will have to integrate fiscal and regulatory measures. A carefully designed policy package can both pull and push more fuel efficient transportation technology into the market, as well as moderate growth in vehicle travel. A set of policies that combine a steady increase in CAFE standards, a moderate but steady rise in gasoline taxes, economic incentives for purchasing more fuel efficient vehicles, and increased renewable content in fuels would be required to achieve these goals.

The technological change needed to bring about GHG emissions reductions can come through incremental improvements in mainstream internal combustion engines, transmissions, and key vehicle technologies coupled with the development and deployment of battery energy storage systems, electric motors, and ICEs integrated into advanced hybrid electric drivetrains. Biofuels, such as efficiently produced ethanol, also have the potential to displace 5 to 10 percent of transportation fuel use, but may require some cost support. If implemented appropriately, this could result in a 3 to 7 percent reduction in GHG emissions from LDVs.

Postponing action on reducing LDV fuel use not only shifts the problem forward in time, but it also results in a higher level of fuel use than if actions are taken immediately. Since the time delays involved in vehicle fleet turnover are of the order of 15 years, urgent action is needed to address the challenge posed by steadily increasing fuel use and GHG emissions from LDVs in the United States.

References

Agras, J., and D. Chapman. "The Kyoto Protocol, CAFE Standards, and Gasoline Taxes." *Contemporary Economic Policy*, Vol. 17, No. 3, July 1999. pp. 296–308.

An, F., J. DeCicco, and M. Ross. "Assessing the Fuel Economy Potential of Light-Duty Vehicles." SAE Technical Paper No. 2001-01-2482, 2001.

Bandivadekar, A., and J. Heywood. "Coordinated Policy Measures for Reducing the Fuel Use of the U.S. Light-Duty Vehicle Fleet." MIT LFEE Report 2004-001-RP, 2004. Available online at http://lfee.mit.edu/public/LFEE_2004-001_RP.pdf.

Bassene, S. A. "Potential for Reducing Fuel Consumption and Greenhouse Gas Emissions from the U.S. Light-Duty Vehicle Fleet." MIT S.M. in Technology and Policy Thesis, September 2001.

Better Environmentally Sound Transportation (BEST). Response to BC Government Discussion Paper: "Options to Reduce Light Duty Vehicle Emissions." Vancouver, Canada, January 2001.

Davis, S. C., and S. W. Diegel. *Transportation Energy Data Book: Edition 23*. Oak Ridge, TN: Oak Ridge National Laboratory, October 2003.

Davis, W. B., M. D. Levine, and K. Train. *Effects of Feebates on Vehicle Fuel Economy, Carbon Dioxide Emissions and Consumer Surplus*. Washington, D.C.: U.S. Department of Energy, Office of Policy, DOE/PO-0031, February 1995.

DeCicco, J. M., and D. Gordon. "Steering with Prices: Fuel and Vehicle Taxation as Market Incentives for Higher Fuel Economy." In Sperling and Shaheen, eds., *Transportation and Energy: Strategies for a Sustainable Transportation System*. Washington, D.C.: ACEEE, 1995. pp.177–216.

DeCicco, J. M., and L. Lynd. "Combining Vehicle Efficiency and Renewable Biofuels to Reduce Light-Vehicle Oil Use and CO_2 Emissions." In DeCicco and Delucchi, eds., *Transportation, Energy and Environment: How Far Can Technology Take Us?* Washington, D.C.: ACEEE, 1997. pp. 75–108.

DOE. *Annual Energy Outlook 2004 with Projections to 2025*. Washington, D.C.: Energy Information Administration, U.S. Department of Energy, January 2004.

DOE Interlaboratory Working Group. *Scenarios for a Clean Energy Future*. ORNL/CON-476, LBNL-44029, and NREL/TP-620-29379, November 2000.

Fulton, L. *Saving Oil and Reducing CO_2 Emissions in Transport: Options and Strategies*. Paris: OECD/IEA, 2001.

Gerard, D., and L. Lave. "The Economics of CAFE Reconsidered: A Response to CAFE Critics and a Case for Fuel Economy Standards." AEI-Brookings Joint Center for Regulatory Studies, Working Paper 03-10, September 2003.

General Motors, Argonne National Lab (GM/ANL). *Well-to-Wheel Energy Use and Greenhouse Gas Emissions of Advanced Fuel/Vehicle Systems—North American Analysis*, Vol. 2. http://www.transportation.anl.gov/software/GREET/publications.html, 2001.

Goodwin, P. B. "A Review of New Demand Elasticities with Special Reference to Short and Long Run Effects of Price Changes." *Journal of Transportation Economics and Policy*, Vol. 26, No. 2, May 1992. pp. 155–169.

Greene, D. L. "Why CAFE Worked." *Energy Policy*, Vol. 26, No. 8, 1998. pp. 595–613.

Greene, D. L., and A. Schafer. "Reducing Greenhouse Gas Emissions from U.S. Transportation." Prepared for the Pew Center on Global Climate Change, May 2003.
Greene, D. L., and J. M. DeCicco. "Engineering-Economic Analyses of Automotive Fuel Economy Potential in the United States." *Annual Review of Energy and the Environment*, Vol. 25, 2000. pp. 477–536.
Greene, D. L., J. R. Kahn, and R. C. Gibson. "Fuel Economy Rebound Effect for U.S. Household Vehicles." *Energy Journal*, Vol. 20, No. 3, 1999. pp. 1–31.
Greening, L. A., D. L. Greene, and C. Difiglio. "Energy Efficiency and Consumption—The Rebound Effect—A Survey." *Energy Policy*, Vol. 28, 2000. pp. 389–401.
Haughton, J., and S. Sarkar. "Gasoline Tax as a Corrective Tax: Estimates for the United States, 1970–1991." *The Energy Journal*, Vol. 17, No. 2, 1996. pp. 103–126.
Hayashi, Y., H. Kato, R. Val, and R. Teodoro. "A Model System for the Assessment of the Effects of Car and Fuel Green Taxes on CO_2 Emissions." *Transportation Research Part D*, Vol. 6, 2001. pp. 123–139.
Hellman, K. H., and R. M. Heavenrich. *Light-Duty Automotive Technology and Fuel Economy Trends: 1975 Through 2003*. Washington, D.C.: Advanced Technology Division, Office of Transportation and Air Quality, U.S. Environmental Protection Agency, EPA420-R-03-006, April 2003.
Heywood, J. B., M. A. Weiss, A. Schafer, S. A. Bassene, and V. K. Natarajan. "The Performance of Future ICE and Fuel Cell Powered Vehicles and Their Potential Fleet Impact." SAE Technical Paper # 04-P254, 2004.
Lipman, T., D. Dantini, and D. Sperling, eds. "Policies for Fostering Sustainable Transportation Technologies: Conference Summary." University of California, Davis, UCD-ITS-RR-98-8, May 1998.
McNutt, B., and D. Rodgers. "Lessons Learned from 15 Years of Alternative Fuels Experience: 1988–2003." In Sperling and Cannon, eds., *The Hydrogen Energy Transition: Moving Toward the Post Petroleum Age in Transportation*. Burlington, Massachusetts: Elsevier Academic Press, 2004. pp. 181–190.
McNutt, B., and P. Patterson. "CAFE Standards—Is a Change in Form Needed?" SAE Technical Paper # 861424, 1986.
McNutt, B., D. Greene, T. Cackette, L. Lave, and S. Peake. "Policies: Regulation." In Lipman, Dantini, and Sperling, eds., *Policies for Fostering Sustainable Transportation Technologies: Conference Summary*. University of California, Davis, UCD-ITS-RR-98-8, May 1998.
Nivola, P. S., and R. W. Crandall. *The Extra Mile: Rethinking Energy Policy for Automotive Transportation*. Washington, D.C.: Brookings Institution, 1995.
Northeast States Center for a Clean Air Future (NESCCAF). *Reducing Greenhouse Gas Emissions from Light-Duty Motor Vehicles*. Boston, MA: September 2004.
National Research Council (NRC), Board on Energy and Environmental Systems. *Effectiveness and Impact of Corporate Average Fuel Economy (CAFE) Standards*. Washington, D.C.: National Academy Press, 2002.
National Research Council (NRC), Committee on Nuclear and Alternative Energy Systems. *Energy in Transition 1985–2010*. San Francisco: W. H. Freeman and Co., 1980.
National Roundtable on the Environment and the Economy (NRTEE). *Greenhouse Gas Emissions from Urban Transportation*. Backgrounder, Ottawa, Canada, 1998. Available online at: http://www.nrtee-trnee.ca/Publications/PDF/BK_Urban-Transportation_E.pdf. Last accessed May 2004.
OECD. *Policy Instruments for Achieving Environmentally Sustainable Transport*. Paris: Organization for Economic Co-Operation and Development, 2002.
Office of Technology Assessment (OTA). *Saving Energy in U.S. Transportation*. Paper OTA-ETI-589. Washington, D.C.: OTA, July 1994.
Plotkin, S. *Technologies and Policies for Controlling Greenhouse Gas Emissions from the U.S. Automobile and Light Truck Fleet*. Argonne, Illinois: Center for Transportation Research, Argonne National Laboratory, January 2000.

Weiss, M. A., J. B. Heywood, A. Schafer, and V. K. Natarajan. *Comparative Assessment of Fuel Cell Cars.* MIT Laboratory For Energy and the Environment Report, MIT LFEE 2003-001 RP, February 2003.

Weiss, M. A., J. B. Heywood, E. M. Drake, A. Schafer, and F. AuYeung. *On the Road in 2020: A Life-Cycle Analysis of New Automobile Technologies.* MIT Laboratory for Energy and the Environment Report, MIT EL 00-003, October 2000.

Wildhorn, S., B. Burright, J. Enns, and T. Kirkwood. *How to Save Gasoline: Public Policy Alternatives for the Automobile.* Cambridge, MA: Ballinger Publishing Company, 1976.

Wilson, J. Q. "The Politics of Regulation." In Wilson, J. Q. ed., *The Politics of Regulation.* New York: Basic Books Inc., 1980. pp. 357–394.

CHAPTER 5

Carbon Burdens from New Car Sales in the United States

John DeCicco, Freda Fung, and Feng An

In mature industrialized economies, cars typically account for the largest portion of transport-related greenhouse gas (GHG) emissions, and nowhere is this more true than it is in the United States. Cars—referring to all light-duty vehicles (LDVs), including passenger cars, light-duty trucks, and sport utility vehicles (SUVs)—emit just over 60 percent of the carbon dioxide (CO_2) from the U.S. transportation sector, which itself accounts for one-third of U.S. energy-related CO_2 emissions overall (Davis and Diegel, 2004). Because the United States is the world's largest emitter of GHGs, its cars account for a significant portion of the world's heat-trapping emissions— about 5 percent of the global energy-related CO_2 emissions. For perspective, U.S. cars alone emit more than the total CO_2 emissions of all but four countries—China, Russia, Japan, and India—exceeding the nationwide emissions of many large countries such as Germany and Brazil.

In tabulations by agencies such as the U.S. Environmental Protection Agency (EPA) and the International Energy Agency (IEA), the transportation portion of a GHG inventory includes only emissions from a vehicle during its operation, mainly its tailpipe CO_2 emissions from fuel combustion, plus trace gas emissions (EPA, 2005). A full accounting would incorporate the entire vehicle lifecycle, called the full fuel cycle, including emissions from supplying the fuel and manufacturing the cars and their components. Manufacturing, for example, accounts for about 11 percent of total automobile lifecycle emissions. The remaining 89 percent is proportional to the amount of fuel consumed, although roughly 30 percent of these emissions occur upstream in the fuel supply chain.

Automotive carbon burdens are the results of decisions made by public officials who finance and shape much of the transportation system, oil companies that supply motor fuel, automakers that build cars, and consumers who drive them. Because car design is such a key determinant of CO_2 emissions rates, automakers' product strategies have a profound impact on the sector's inventory. The carbon burden concept isolates this impact by focusing on the new vehicle market and controlling for other factors, such as amount of driving or carbon content of fuel, that influence automobile CO_2 emissions largely through decisions made by parties other than automakers.

The carbon burden concept underscores how the ultimate effectiveness of actions taken to cut CO_2 emissions must be evaluated according to tons of carbon reduced, rather than the type of technology or fuel used (DeCicco et al., 2005). Similarly, the effectiveness of measures to reduce petroleum dependence is ultimately measured in terms of barrels of oil consumption avoided. Focusing on these metrics of barrels of oil and tons of carbon provides a robust framework for assessing strategies to reduce auto sector oil demand and GHG emissions.

This chapter reviews the historical U.S. light vehicle carbon emission trends from all LDVs, new and used, over the period from 1970 through 2003, and highlights new vehicle CO_2 emission rates and related factors for each major automaker for the more recent period from 1990 through 2003. Primary data sources for our analysis include the National Highway Traffic Safety Administration (2005) and Hellman et al. (2004). It examines only tailpipe CO_2 emissions so that the findings can be directly compared to the findings of studies by other agencies monitoring GHGs. Thus, carbon burden is expressed as the expected annual direct CO_2 emissions averaged over a vehicle lifetime. The calculations assume that all vehicles are driven 12,000 miles per year, emit 19.4 pounds of CO_2 per gallon of fuel burned, and have an average 15 percent shortfall in fuel economy relative to the laboratory test values used for Corporate Average Fuel Economy (CAFE) compliance.

Trends in U.S. Automotive CO_2 Emissions

The total CO_2 emissions, as well as the average emission rates of all vehicles in each automaker's fleet, have continued to rise despite notable changes in factors thought to influence emissions. In particular, the past five years saw much higher gasoline prices than the period from 1990 to 1998 as well as notable developments in technology, such as the introduction of hybrid electric vehicles (HEVs). Examining aggregate emissions trends shows that annual sales of even a million HEVs—which some analysts foresee as early as 2010—would not suffice to offset even half the increase in CO_2 emissions and oil consumption observed in the auto market between 1990 and 2003.

The new fleet average CO_2 emission rate per vehicle is shown as the thicker line in Figure 5-1. It had been rising prior to the 1973 oil embargo,

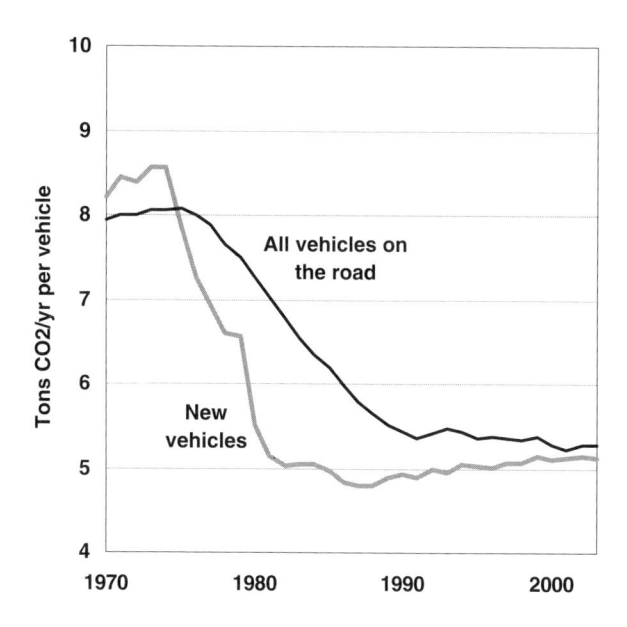

FIGURE 5-1. Average CO_2 emissions rates of U.S. light-duty vehicles, new fleet and on-road stock. *Source:* DeCicco et al. (2005), as derived from U.S. government statistics. *Notes:* CO_2 emissions measured by metric tons. One metric ton equals 1.102 short tons.

reaching a peak of 8.6 metric tons of CO_2 per year (TCO_2/yr) in 1973 and 1974. It then plummeted as fuel economy rose in response to gas lines, high fuel prices, general fears of fuel shortages, and the imposition of CAFE standards. The subsequent fuel economy decline due to the shift from cars to trucks pushed the new fleet average CO_2 emissions rate slowly upward from its historical low of 4.8 TCO_2/yr in 1987 and 1988. The thinner line in Figure 5-1 shows the average emissions rate of the total LDV stock, which lags that of new vehicles due to stock turnover. The stock average CO_2 emissions rate continued to decline into the 1990s, but subsequently has stagnated at about 5.3 TCO_2/yr.

Figure 5-2 shows the growth of U.S. automotive CO_2 emissions along with a key factor behind rising emissions—namely, growth in light-duty vehicle miles traveled (VMT). As shown by the thicker line, total LDV emissions reached 317 million metric tons of carbon per year (MMTc) in 2003. This CO_2 emissions level is equivalent to 8.6 million barrels of gasoline consumption per day, or 132 billion gallons per year. The 2003 level represents a net growth of 64 percent since 1970 and a 25 percent increase since 1990, a common base year for climate policy. Nevertheless, as Figure 5-2 shows, this growth in carbon emissions is much less than the 160 percent jump in VMT from 1970 to 2003. These trends illustrate how a decrease in

76 *Driving Climate Change*

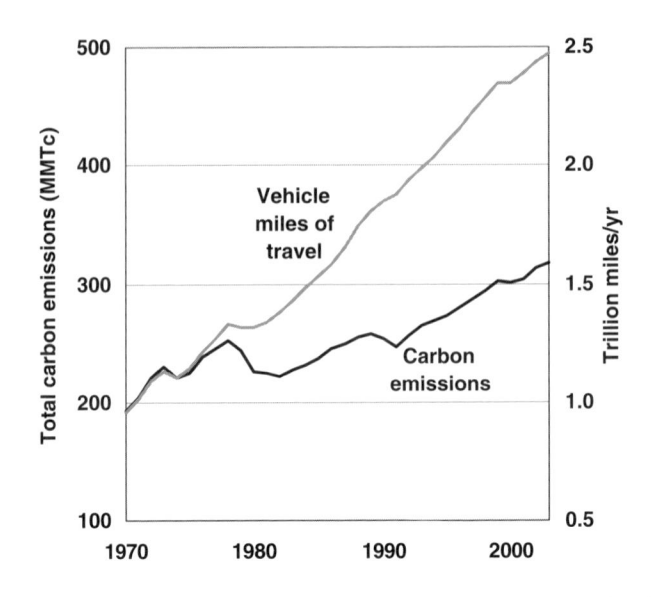

FIGURE 5-2. Vehicle miles of travel and total CO_2 emissions of light-duty vehicles in the United States. *Source:* DeCicco et al. (2005), as derived from U.S. government statistics.

CO_2 emissions rates, itself driven by the increase in fuel economy following the 1970s oil shocks, can at least partly offset the effects of increased driving.

Carbon Burdens of Major Automakers

The Big Six, including the six largest automakers in the U.S. market—General Motors (GM), Ford, DaimlerChrysler (DCX), Toyota, Honda, and Nissan—had an 87 percent market share and accounted for 88 percent of the new fleet carbon burden in 2003. The next six firms, measured by total U.S. sales—Volkswagen, Hyundai, Mitsubishi, BMW, Kia, and Subaru—had a combined market share of 12 percent in 2003 and accounted for nearly all of the remaining new fleet carbon burden. The ranking of new fleet carbon burdens by firm follows market share, with GM accounting for the largest total carbon burden.

Focusing on the Big Six, Figure 5-3 breaks down each firm's carbon burden growth from 1990 to 2003 into two components: sales increase and fuel economy decrease. The average fuel economy for all Big Six automakers decreased from 1990 to 2003, largely due to each firm's rising proportion of trucks in its total sales, known as the truck fraction, resulting in increased fleet-average CO_2 emissions rates. All automakers significantly expanded their light truck offerings, with the overall light truck fraction

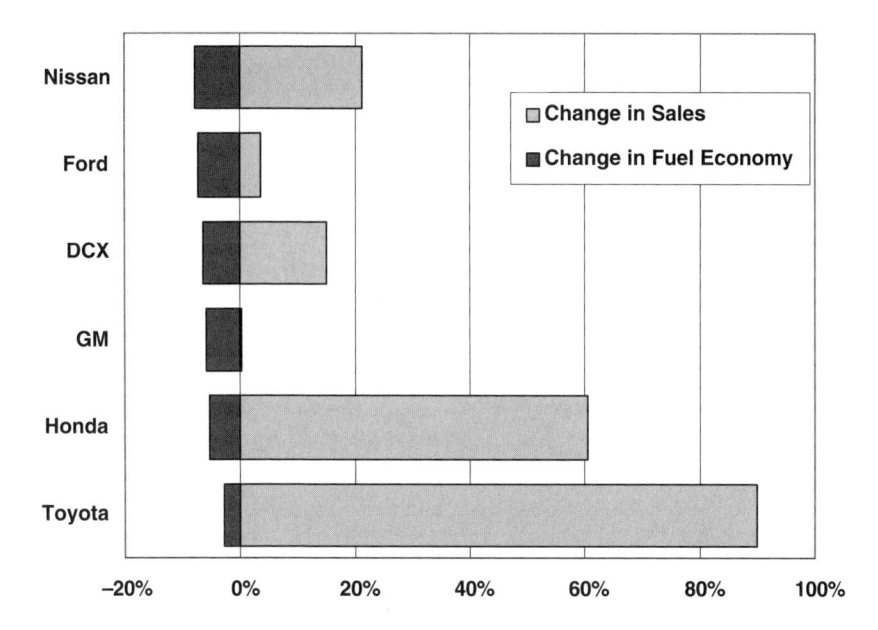

FIGURE 5-3. Breakdown of growth in Big Six carbon burdens, 1990–2003. *Source:* DeCicco et al. (2005) as derived from National Highway Traffic Safety Administration (2005).

growing 21 percentage points over this 14-year period. Nissan had the largest increase in its average CO_2 emissions rate due to the combined effect of rising truck fraction and declining truck fuel economy. Toyota's 95 percent increase in new vehicle carbon burdens was the greatest among the Big Six, but its fuel economy declined the least. Thus, Toyota's carbon burden increase was due predominately to sales success.

Two other trends have been serving to increase CO_2 emissions by U.S. cars and light trucks. One is the growing reliance on flexible-fuel vehicle (FFV) credits by the Big Three (GM, Ford, and DaimlerChrysler). Federal law gives automakers extra CAFE credits for selling FFVs, regardless of whether alternative fuel is actually used, enabling the companies to sell less fuel-efficient vehicles overall. The other adverse trend is an apparent increase in sales of heavier light trucks between 8,500 and 10,000 pounds gross vehicle weight. These vehicles are mainly three-quarter and one-ton pickup trucks, but include a growing number of the largest SUVs, such as the Hummer H2 and some models of the Chevy Suburban and Tahoe and their GMC brand variants. Because such vehicles have been exempt from CAFE regulation, data are not available to quantify the additional carbon burdens associated with their sales. In any case, the actual carbon burdens of the Big Three are even larger than reported here based on data for only the CAFE-regulated under-8,500 pound fleet.

The following sections provide summaries of the key findings on new fleet CO_2 emissions trends for each of the Big Six, plus shorter summaries for other automakers, during the period from 1990 through 2003.

General Motors

As the largest firm in the U.S. market, GM's model year 2003 fleet accounted for 29 percent of the total carbon burden from new LDV sales. This share is higher than GM's 28 percent market share because the average CO_2 emissions rate of its products is somewhat higher than the market average. Over each of the last four years before 2003, GM improved the fuel economy of its new car fleet, which was 5 percent higher in 2003 than it was in 1990.

Light truck sales at GM rose from 28 percent in 1990 to 56 percent in 2003 while showing no significant fuel economy improvement. Mainly as a result of this shift, GM's new fleet-average CO_2 emissions rate was 6.3 percent higher in 2003 than it was in 1990, reaching 5.4 TCO_2/yr per vehicle, about 5 percent higher than the market average in 2003. The company's market share dropped 6.8 percentage points during the period from 1990 through 2003, but GM still had the largest carbon burden overall, standing at 6.4 MMTc as of 2003.

Ford

Ford's market share dropped over 4 percentage points from 1990 through 2003, standing at 21 percent as of 2003. The company's 2003 total fleet carbon burden of 5.0 MMTc accounted for 23 percent of the market total. Light trucks rose from 35 percent of Ford's sales in 1990 to 59 percent as of 2003, while the average fuel economy of Ford's light trucks dropped by 2 percent compared to 1990.

Ford's SUV fuel economy dropped in 2003 after having risen for two years following the company's now-abandoned July 2000 pledge to improve fuel economy. Ford has also relied heavily on FFV credits. The combined effect of these factors pushed Ford's new fleet-average CO_2 emissions rate to a level 7.7 percent higher in 2003 than it was in 1990. That made Ford the company with the worst fleet average CO_2 emissions rate, 5.6 TCO_2/yr per vehicle, as of 2003, nearly 9 percent above the market average and 3.5 percent higher than GM.

DaimlerChrysler

DCX has become the automaker with the greatest dependence on light trucks. The truck fraction of its total LDV sales increased 24 percentage points since 1990 to reach 74 percent in 2003. While its average new car fuel economy revealed no obvious trend, DCX's light truck fuel economy rose 6 percent from 2001 through 2003, although it was still down a net 2 percent in 2003 from its 1990 level.

The combined share of Chrysler and Mercedes-Benz products in the U.S. market was 13.5 percent in 1990. Chrysler sales did very well in the mid-1990s, and the combined share was around 17 percent at the time of the merger with Mercedes-Benz to form DCX. Market share dropped after the merger, however, and was down to 12.4 percent as of 2003. Mainly as a result of the extensive shift to light trucks, but also due to the adverse effect of FFV credits, DCX's new fleet average CO_2 emissions rate ended up 6.8 percent higher in 2003 than the average level of the premerger firms in 1990. That level of 5.5 TCO_2/yr per vehicle was second highest in the market, just behind Ford. DCX's total fleet carbon burden of 3.0 MMTc in 2003 accounted for 14 percent of the market total.

Toyota

Toyota's market share gained 4 percentage points between 1990 and 2003, closing the period at just under 12 percent of the market. The company's new fleet average CO_2 emissions rate was up 2.9 percent over the period, the smallest increase among the Big Six. It stood at 4.6 TCO_2/yr in 2003.

Driven mainly by increased sales, Toyota's new fleet carbon burden saw net 95 percent growth from 1990, reaching 2.3 MMTc in 2003, 11 percent of the market total. The company's higher CO_2 emissions rate was caused by the 15-percentage-point increase in the truck fraction of Toyota's new vehicle sales. Toyota's average light truck fuel economy in 2003 was the same as it was in 1990, despite an extensive expansion of the company's lineup into SUVs and larger and more powerful trucks generally. Toyota's average new car fuel economy improved 4.9 percent but not nearly enough to compensate for the shift to trucks. Toyota's HEV sales were still too small as of 2003 to have a perceptible effect on its fleet-average CO_2 emissions rate and carbon burden.

Honda

Honda remained the fuel economy leader among the Big Six. As of 2003, only Volkswagen had a higher new fleet fuel economy overall among the top 12 automakers. Honda gained two percentage points of market share from 1990 to 2003, but its rapidly growing truck fraction resulted in a 5.7 percent rise in the company's average CO_2 emissions rate. That put its new fleet average CO_2 emissions rate at 4.3 TCO_2/yr in 2003. Honda's average truck fleet fuel economy dropped by 8 percent from 1997 through 2003, while its average new car fuel economy rose by 7 percent from 1990 through 2003.

Since entering the light truck market in 1997, Honda's truck share grew at an average 5.6 percentage points per year, reaching 39 percent in 2003. Driven by both market share gain and a rising CO_2 emissions rate due to higher truck fraction, Honda's 2003 carbon burden reached 1.7 MMTc, contributing 8 percent of the overall new LDV market carbon burden. Honda was first to introduce an HEV to the U.S. market, but as of 2003, like Toyota, its HEV sales were too small to significantly impact the company's fleet average CO_2 emissions rate.

Nissan

Nissan saw great fluctuations in its sales over the period from 1990 through 2003, although by 2003 its market share was about the same as in 1990 at 5 percent. During this time, the truck fraction of Nissan's sales grew from 25 to 36 percent, while its average new light truck fuel economy dropped 13 percent.

The net effect was that Nissan's new fleet average CO_2 emissions rate rose 8.4 percent between 1990 and 2003, the largest increase among the Big Six. That put its CO_2 emissions rate at 4.9 TCO_2/yr per vehicle as of 2003, still about 5 percent below the market average of 5.1 TCO_2/yr for the Big Six manufacturers. The 1.0 MMTc carbon burden of Nissan's 2003 new fleet was responsible for 4.8 percent of the total new LDV fleet carbon burden in 2003.

Other Firms

The collective sales of the next six largest automakers nearly tripled between 1990 and 2003, when their combined market share reached 12 percent and they accounted for 11 percent of the new LDV fleet carbon burden. Highlights of the carbon-burden related performance of the individual companies are as follows:

- Volkswagen more than doubled its market share from 1990 through 2003 while improving fuel economy and cutting its fleet-average CO_2 emissions rate by 3.3 percent. Corresponding to its high average fuel economy, Volkswagen's 2003 new fleet-average CO_2 emissions rate was the lowest among the 12 automakers examined here.
- Hyundai nearly tripled its market share from 1990 through 2003, but it had the worst increase, 16 percent, in new fleet-average CO_2 emissions rate among major automakers, shifting it from having the lowest new fleet CO_2 emissions rate in 1990 to the third lowest in 2003.
- Mitsubishi saw its market share generally decline from 1990 through 1998, but it had rebounded by 2003. Following the truck trend, the fuel economy of its sales mix declined to the point that the company's new fleet average CO_2 emissions rate increased 6 percent from 1990 through 2003.
- BMW improved its average fuel economy from 1990 through 2003 by more than any other firm, reducing its new fleet CO_2 emissions rate by

12.7 percent over the period in which it achieved a nearly fivefold increase in U.S. sales.
- Kia has steadily gained sales since entering the U.S. market in late 1993, but its new fleet CO_2 emissions rate rose 27 percent as it increased its truck sales and converged toward the market average.
- Subaru posted little net change in market share from 1990 through 2003 and its new fleet-average CO_2 emissions rate increased 3 percent over the period.

Notable Trends Influencing Carbon Burdens

The specific trends for each automaker discussed above reflect a number of broader trends in the U.S. auto market. A notable trend has been the ongoing erosion of the market share held by the Big Three. This occurred even as those same firms led the market into the light truck segments, largely through the popularity of SUVs. The resulting general shift to trucks became the main factor behind falling fuel economy and rising average CO_2 emissions rates. The trend toward use of FFVs has also contributed to the worsening CO_2 emissions rates of the Big Three.

While the latter years of the analysis period saw the introduction of HEVs, their sales remained small as of 2003. The fuel economy values of the HEVs introduced to date provide a basis for estimating the fuel saving characteristics of this technology as actually deployed. Such estimates provide a perspective on the likely CO_2 emissions impact of the nascent trend toward HEVs, which can be compared to the ongoing trend of shifting production from vehicles classified as cars to those classified as trucks.

The Steady Rise of Light Trucks

The rise in light truck sales started in the 1980s and has progressed in several waves. First was the introduction of the minivan in 1984, followed by the modern SUV beginning around 1989 and increasing rapidly throughout the 1990s. A growing popularity of various "crossover" vehicles, with designs that blend the traits of traditional body styles, has been the most recent trend. Examples of crossovers include car-based SUVs—such as sport wagons, which once might have been called station wagons—as well as minivan/SUV combinations and pickup/SUV blends.

Automakers are classifying nearly all of these new designs as trucks in order to ease their compliance obligations with CAFE standards. Because light trucks are held to a lower standard—20.7 miles per gallon (mpg) as of model year 2003, compared to 27.5 mpg for cars—this strategy helps in several ways. Simply moving a vehicle from a car fleet to a truck fleet subjects it to a lower standard. Then, because vehicles that were once cars or

82 Driving Climate Change

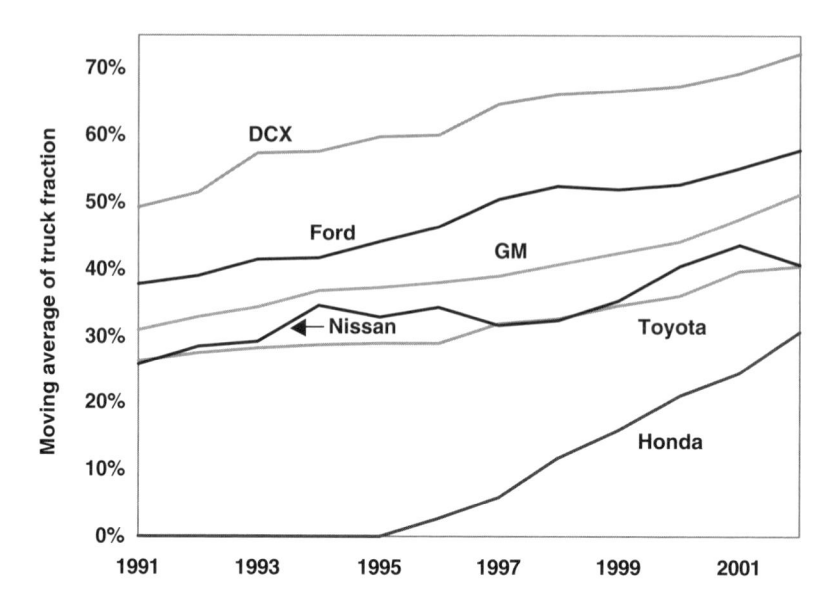

FIGURE 5-4. Light truck fractions of Big Six sales, 1990–2003. *Source:* DeCicco et al. (2005) as derived from National Highway Traffic Safety Administration (2005).

derived from cars are generally more fuel efficient than trucks, the averaging approach on which CAFE is based enables an automaker to sell more of the large trucks on which profit margins have been so high. Finally, because vehicles shifted into the truck category are typically less efficient than the average car, the company's remaining cars can more easily meet the car standard.

Since 1988, when new fleet fuel economy peaked, the market share of vehicles classified as light trucks has climbed from 30 to 51 percent. It is this shift that largely accounts for the 7 percent increase in new fleet-average CO_2 emissions rate over this period. Truck fractions of new vehicle sales have been climbing for all major automakers and there is no sign that the car-to-truck shift had saturated as of 2003. All of the Big Six have rising truck fractions, as shown in Figure 5-4. DCX has the highest fraction of trucks in its fleet and Honda's truck fraction has been growing most rapidly. The overall truck share has been growing in an essentially linear fashion since 1980, when it was only 17 percent. Extrapolating the average gain of 1.5 percentage points per year would put the new light truck share at 60 percent by 2010.

The fuel economy of trucks relative to cars declined, even as truck sales share rose. The main reason for this disparity was the fact that CAFE standards for light trucks were raised much less than those for cars. The result has been a rising excess of light truck CO_2 emissions rates over that of cars, as shown in Figure 5-5. In 1975, when fuel economy recordkeeping

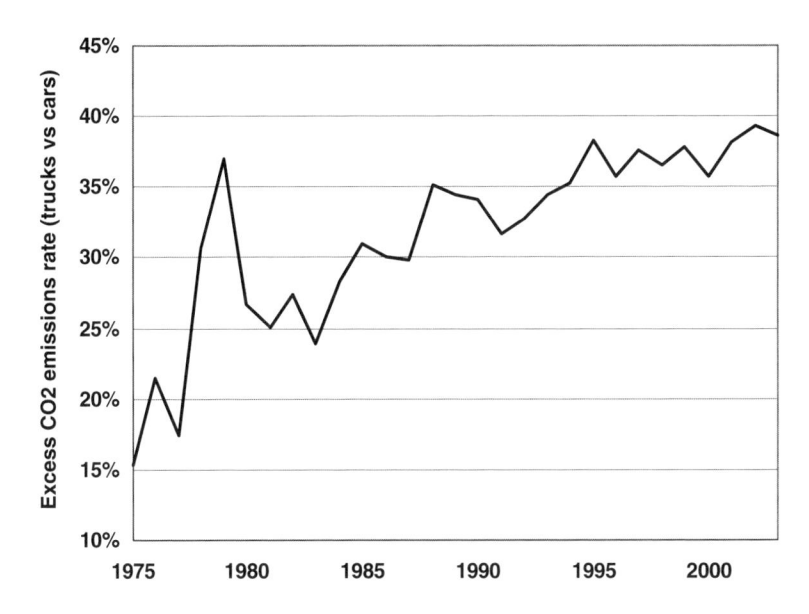

FIGURE 5-5. Excess CO_2 emissions rate of light trucks over cars, 1990–2003. *Source:* DeCicco et al. (2005) as derived from Hellman and Heavenrich (2004). *Notes:* Excess CO_2 emissions rate is defined as the percentage by which the average new light truck CO_2 emissions rate exceeds that of new cars.

started and trucks were in fact much more utilitarian in design and use, the average light truck emitted just over 15 percent more CO_2 per mile than the average car. In 2003, light trucks on average emitted 39 percent more CO_2 per mile than cars. Therefore, light trucks accounted for 59 percent of the new fleet carbon burden in 2003, disproportionately more than their 51 percent sales share. If the trucks that have substituted for cars during the period of market shift had met the passenger car CAFE standard, U.S. automobiles would have consumed 536,000 fewer barrels per day of gasoline and emitted 20 MMTc per year less carbon as of 2003.

Flexible Fuel Vehicles

The Alternative Motor Fuel Act of 1988 (AMFA) established incentives for automakers to sell dual-fuel vehicles, including FFVs, by earning credits applicable to meeting CAFE standards. Most FFVs have been designed to burn a blend of 85 percent ethanol and 15 percent gasoline (E85). For the purpose of CAFE calculation, such an FFV's E85 fuel economy is defined as the value measured on a test run with E85 divided by 0.15—that is, it represents only the number of miles per gallon of the petroleum-derived portion of the fuel. The CAFE calculation further assumes that the FFV operates on E85 50 percent of the time, burning gasoline the other 50 percent of the time. The net result is that an FFV's fuel economy is counted as being a factor of roughly 1.65 times its fuel economy on gasoline. For example, an 18 mpg Chevy Tahoe FFV has its fuel economy tallied as roughly 30 mpg for purposes of calculating GM's light truck CAFE compliance. Because FFVs rarely if ever burn E85, this crediting inflates an automaker's compliance fuel economy and enables the firm to sell less fuel-efficient vehicles while still meeting the standard. The result is higher gasoline consumption and higher CO_2 emissions than if FFV credits were not used.

Each of the Big Three has been making increased use of FFV credits. In GM's case, the credits pushed the company's combined car and light truck CO_2 emissions rate 2 percent higher than it would otherwise have been in model year 2003. Ford makes the most extensive use of FFV credits to date, inflating its combined CAFE by an estimated 1.1 mpg in 2003 and making its new fleet-average CO_2 emissions rate 3 percent higher than if it had met CAFE standards without credits. More extensive use of FFV credits inflated DCX's 2003 combined CAFE by 1 mpg, pushing its average CO_2 emissions rate 4.4 percent higher than it otherwise might have been. None of the Asian or other automakers have exploited this dysfunctional aspect of federal policy to date.

A 2002 report to Congress foresaw the increases in petroleum consumption and CO_2 emissions resulting from the FFV credit policy (U.S. Department of Transportation, 2002). Nevertheless, the FFV credits were subsequently extended by Congress under the Energy Policy Act of 2005, which also provided mandates and other incentives for greatly expanding the availability of ethanol fuel. Future analysis will be needed to assess whether enough ethanol fuel with certifiably low CO_2 emissions will be sold to offset the emissions increases that have already been caused by the use of FFV credits to date.

HEVs in Context

For many observers, the introduction of HEVs has been among the most exciting and hopeful trends to emerge in the auto market from an environmental perspective. Toyota pioneered the technology with the Prius, introduced in Japan in late 1997 and first sold in the United States in 2001. The company has subsequently introduced hybrid electric versions of its Lexus RX series and Highlander SUVs, and has announced ambitious plans for widespread availability of HEVs throughout its product lineup. Honda was first to market HEVs in the United States with its Insight in 2000. The Civic Hybrid was introduced in the spring of 2002 as an early model year 2003 vehicle. Subsequently, Honda introduced a hybrid electric version of its Accord sedan. Ford was first to market with a hybrid SUV, the Escape Hybrid, in model year 2005. GM has produced and sold limited numbers of hybrid electric versions of its Silverado and Sierra pickup trucks. All major automakers have now announced plans for expanded HEV models over the next few years.

HEV sales were still very small in 2003. However, the initial models offered provide a basis of comparison regarding the fuel savings and CO_2 emissions reductions that might be seen through greater use of the technology. Since all of the Big Six automakers have seen their fleet average CO_2 emissions rates increase due to rising light truck sales fractions, it is instructive to compare the likely impact of HEVs in lowering CO_2 emissions rates to the increase in emissions already caused by the car-to-truck shift.

For example, to compensate for its 2.9 percent increase in fleet-average CO_2 emissions rate over the period from 1990 through 2003, Toyota would have to sell 150,000 HEVs that achieve the same average fuel savings as the Prius and Lexus RX400h, or 8 percent of its total sales at 2003 volumes. This is quite likely given Toyota's announced plans, and it may become the first company to offset its truck-driven carbon burden increase by using HEVs. This achievement, however, would only trim its fleet average CO_2 emissions rate to what it was in 1990.

To take another example, Honda had a 5.7 percent increase in its fleet-average CO_2 emissions rate from 1990 through 2003. To compensate for that impact, Honda would have to sell over 300,000 HEVs, or 22 percent of its 2003 sales, with the same average fuel savings as the Civic and Accord Hybrids. Similarly, Ford would have to sell over 650,000 HEVs, or 20 percent of its 2003 sales, with the same average fuel savings as the Escape Hybrid in order to compensate for the 7.7 percent increase in its fleet-average CO_2 emissions rate, due mainly to its shift to trucks compounded by its use of FFV credits.

Reducing Automotive Carbon Burdens

Many actors are involved in the decisions that determine what kind of cars are built and sold, how much they are driven, and how they are fueled. Thus, cutting the carbon burdens of cars is a shared responsibility, though the auto industry is a dominant player.

The past several years have seen shifts in automakers' public positions on global warming. Not long ago, many automakers, particularly the Big Three, denied that a problem existed and carried out campaigns to undermine U.S. support for climate protection actions. Now, all firms profess a desire to help solve a very real problem. In 1998, major automakers made voluntary agreements with the European Union to cut their fleet-average CO_2 emissions rates. A recent report, *Mobility 2030*, endorsed by the Big Six companies in the U.S. market plus Renault and Volkswagen, recommended a goal of limiting GHG emissions to sustainable levels (World Business Council for Sustainable Development, 2004). Automakers have started reporting emissions from their fleets and factory operations and they now regularly publicize new technologies and other activities promising emissions reductions.

Nevertheless, automakers have yet to make significant progress in cutting CO_2 emissions in the United States, the world's largest auto market. With few exceptions, their strategies have made emissions worse. Major product trends, such as the shift to light trucks, have been driving CO_2 emissions rates higher. Some policies rationalized under the guise of reductions, such as the FFV credits, are actually aggravating the adverse CO_2 emissions trends as well as increasing U.S. oil consumption. Although HEVs offer a ray of hope, assessing their influence using the metric of fleetwide carbon burden indicates that any one technology, even an advanced and promising one, can do little to offset broad market trends that continue to push emissions upward.

The missing part of the auto industry's role in cutting carbon burdens is a constructive stance on public policy. Government intervention is essential for resolving the inherent tension between market forces and nonmarket concerns such as global warming and energy security. As this analysis has shown, technology strategies alone are unlikely to address the auto sector's CO_2 emissions problem. Automakers need to embrace balanced but meaningful regulation in order to be true to their promises to address these public concerns. There is no other way to break out of the competitive box that binds product strategies and design priorities to offering consumers almost every variation imaginable, but doing very little to address the huge, nonmarket problems of global warming and oil dependence.

Automakers rightly point out that lack of customer interest is a barrier to higher fuel economy, in contrast to when CAFE standards were established during the oil crisis in the 1970s. Indeed, an extensive public education effort to make fuel efficiency matter more to consumers is needed as part of a broader public strategy to realign market signals and establish U.S. leadership in addressing oil consumption and global warming. The auto industry's cooperation and expertise could help guide such endeavors, but effective steps seem unlikely until automakers take a more positive approach in helping establish a binding U.S. greenhouse gas reduction policy. A good faith effort on the industry's part would open the door to developing more comprehensive solutions for the cars versus climate challenge.

References

Davis, S. C., and S. W. Diegel. *Transportation Energy Data Book: Edition 24.* Report ORNL-6973. Oak Ridge, TN: Oak Ridge National Laboratory, December 2004.

DeCicco, J., F. Fung, and F. An. *Automakers' Corporate Carbon Burdens: Update for 1990-2003.* New York: Environmental Defense, August 2005.

Hellman, K., and R. M. Heavenrich. *Light-Duty Automotive Technology and Fuel Economy Trends: 1975 Through 2004.* Report No. EPA 420-R-04-001. Ann Arbor, MI: U.S. Environmental Protection Agency, Office of Transportation and Air Quality, Advanced Technology Division, April 2004.

National Highway Traffic Safety Administration. *Summary of Fuel Economy Performance.* Washington, D.C.: National Highway Transportation and Safety Administration, March 2005.

U.S. Department of Transportation. *Report to Congress: Effects of the Alternative Motor Fuels Act CAFE Incentives Policy.* Washington, D.C.: U.S. Department of Transportation, March 2002.

U.S. Environmental Protection Agency. *Inventory of U.S. Greenhouse Gas Emissions and Sinks: 1990–2003.* Report No. EPA 430-R-05-003. Washington, D.C.: U.S. Environmental Protection Agency, April 2005.

World Business Council for Sustainable Development. *Mobility 2030: Meeting the Challenges to Sustainability.* Geneva: World Business Council for Sustainable Development (www.wbcsd.org), 2004.

CHAPTER 6

Reducing Vehicle Emissions Through Cap-and-Trade Schemes

John German

Global warming is a worldwide problem that is growing in importance. Carbon released during fossil fuel burning is the primary contributor to greenhouse gas (GHG) emissions, and cap-and-trade programs are actively being developed worldwide to provide a sound economic framework for reducing carbon emissions. Cap-and-trade programs have been used successfully in several emissions-related undertakings, such as the U.S. effort to control acid rain by limiting sulfur dioxide emissions and the transition from leaded to unleaded gasoline. Today, carbon trading programs are already being implemented for selected sectors, such as electric utilities.

Encouraged by the past success of emissions trading programs, many energy and environmental specialists are looking to implement carbon trading across all carbon sectors. Most of these plans implicitly assume that motor vehicles can easily be incorporated into global cap-and-trade programs that already exist for other energy sectors. However, there has been remarkably little analysis of the mechanisms needed to incorporate vehicles into cross-sector trading programs. This chapter examines whether or not this assumption is realistic and suggests an alternative method for vehicles that will likely work better. The focus here is on light-duty vehicles (LDVs), while recognizing that transportation includes other vehicles and systems that are likely to require their own focused analyses.

Previous Studies

Two recent studies have already examined the creation of GHG cap-and-trade programs to reduce GHG emissions from motor vehicles. Robert R.

Nordhaus and Kyle W. Danish published the most recent study in May 2003 for the Pew Center (Nordhaus and Danish, 2003). Called "Designing a Mandatory Greenhouse Gas Reduction Program for the U.S," this was a comprehensive study covering all carbon sectors and included a detailed assessment of different ways to incorporate transportation into a trading program. The Pew study examined four cap-and-trade structures:

- An upstream program applying, for example, to fuel producers
- A fully downstream program applying, for example, to vehicle owners or manufacturers
- A hybrid sectoral program with tradable standards. This program combines a downstream cap-and-trade program for large sources in the electricity and industrial sectors with enhanced product efficiency standards for energy end users, such as carbon dioxide (CO_2) per mile standards for vehicles. Vehicle manufacturers could trade between their own product lines, with each other, and with firms subject to the downstream cap-and-trade program.
- A hybrid sectoral program with capped tradable standards. This is similar to the hybrid sectoral program with tradable standards, except that a cap would be set on the total emissions from vehicles. Thus, manufacturers would have to account for the total projected emissions associated with each product they sold, not just for vehicle efficiency.

Each of these structures was evaluated for environmental effectiveness, cost-effectiveness, administrative feasibility, distributional equity, and political acceptability. Table 6-1 presents a summary of the findings in the Pew study.

The authors of the Pew study concluded the following:

> The analysis would argue against an economy-wide downstream cap-and-trade program (as difficult to administer), a stand-alone large source cap-and-trade program (as incomplete), and a GHG tax that is not part of a larger tax reform initiative (as unviable politically). The analysis does suggest that the comprehensive, upstream cap-and-trade approach and the sectoral hybrid approach are the most viable alternatives for a domestic GHG reduction program.

The study goes on to state that if the sectoral hybrid approach is taken, "then careful attention will have to be given to minimizing economic costs and administrative complexity, and assuring that the program can be effectively enforced." The authors simply assumed these concerns could be dealt with, however. No attempt was made to address specific design issues.

The second, earlier, report by Steve Winkelman, Tim Hargrave, and Christine Vanderlan, of the Center for Clean Air Policy (CCAP), in April 2000 focused on ways to incorporate transportation into GHG trading (Winkelman et al., 2000). The policies examined were similar to the later

Reducing Vehicle Emissions Through Cap-and-Trade Schemes 91

TABLE 6-1. Summary of Pew 2003 Study Results with Respect to Vehicles

	Environmental Effectiveness—Coverage, Certainty, Enforceable	Cost-Effectiveness—Flexibility, Predictability, Long-Term Incentives	Administrative Feasibility—Administrative Cost, Adaptability	Distribution Equity	Political Acceptability
Cap & Trade—upstream	Good	Hypothetically least cost if includes flexibility measures	Good	Depends on allocation and auctioning provisions	Works by limiting fuel availability and raising fuel cost
Cap & Trade—downstream			Prohibitive administrative cost		
Sectoral Hybrids—tradable standards	Must expand coverage (vessels, locomotives, HD, aircraft, buses) Emission reductions uncertain	No incentive to reduce end-use Vehicle fuel sales must be exempted from upstream cap & trade Likely considerably more costly than upstream cap & trade	Must translate mpg into annual CO2 (annual VMT assumptions, timing) Requires continuously promulgating adjustments Double-counting risks Evasion if upstream trading allowed	Possible concern	Avoids gasoline price increases
Sectoral Hybrids—capped tradable standards	Emission reductions more certain, although still rely heavily on estimates		Raises issues with allowance allocation, shutdowns, new market entrants, mfr. market share shifts, overall sales		

92 *Driving Climate Change*

TABLE 6-2. Direct and Indirect Influences on Transportation GHG Emissions (from the CCAP Study)

Entity/Factor	Vehicle Miles Traveled	Vehicle Efficiency	Fuel Carbon Content
Consumers	Travel Decisions	Consumer Preferences Vehicle Maintenance	Consumer Preferences
Vehicle Manufacturers	(Indirect influence: vehicle efficiency impact on driving costs)	Vehicle Technology	Vehicle Technology
Fuel Producers	Fuel Price	NA	Product Mix
Land Use & Transportation Infrastructure	Available Travel Options	NA	NA

Pew 2003 study; upstream, downstream, and hybrid approaches. The CCAP analysis also investigated the influences of fuel producers, vehicle manufacturers, customers, and land-use policies on the three aspects of carbon emissions: vehicle miles traveled (VMT), vehicle efficiency, and fuel carbon content.

Table 6-2 summarizes the direct and indirect influences of these different factors, illustrating how downstream, upstream, and hybrid programs affect carbon emissions. For example, vehicle manufacturers are the primary influences on vehicle technology, while fuel producers have more leverage on VMT.

The authors found that "the hybrid approach is aimed at combining the benefits of the upstream and downstream systems in a synergistic way. It appears to fall short of this goal, however, because additional complexities are introduced without any clear environmental benefit." The study recommended combining land use policies and an upstream trading system with carbon efficiency standards similar to the current Corporate Average Fuel Economy (CAFE) standard, although inclusion of carbon efficiency standards was based primarily upon arguments that there were other reasons for improving vehicle efficiency than just carbon emissions.

While both of these studies explore the relative merits of the various GHG cap-and-trade systems for transportation, they fall short in addressing the major implementation challenges associated with each alternative. The following sections explore in detail the major policy options in order to highlight obstacles that must be addressed in cap-and-trade systems. Based on an understanding of these issues, an alternative regulatory framework is suggested that has the potential to achieve the same environmental goals while avoiding much of the administrative complexity of current cap-and-trade systems.

Upstream Trading

Upstream trading refers to trading between producers of carbon-based fuels or products. For the transportation sector this would be fuel refiners and importers of refined petroleum fuel. In a pure upstream emissions trading system, fuel producers would be required to hold GHG emission allowances for each ton of CO_2 equivalent emissions produced each year. In addition, fuel producers would be required to hold allowances for each ton of emissions produced due to consumption of the fuels they sell.

Such a system reduces emissions via two mechanisms. First, emissions can be reduced as a direct result of the allocation of emission allowances. Each fuel producer can reduce emissions by reducing fuel output, manufacturing process emissions, or average fuel GHG intensity measured by fuel carbon content, or by purchasing additional allowances from the allowance market. In all of these cases, the total quantity of GHGs emitted to the atmosphere is reduced, although reductions may come from a nontransportation sector in the last case. Second, direct activities to reduce emissions on the part of fuel producers can reduce fuel demand by increasing the cost of fuel production. Demand will decrease as a result of the price elasticity of fuel, although the effect may be small.

The advantage of upstream trading systems is that administration is simplified due to the relatively small number of regulated firms in the upstream industry. There are approximately 175 petroleum refiners, 200 oil importers, and 900 gasoline pipelines in the United States. Data about their operations are readily available. This makes upstream trading schemes reasonably simple to administer at low cost, while providing comprehensive coverage. However, while fuel producers can affect the GHG emissions from their own operations, they have no direct control over fuel consumption in an upstream system. The only consumption control fuel producers have is on price. An upstream trading program has the effect of raising fuel prices. This increased fuel price signal creates incentives to produce and use low carbon fuels, reduce vehicle use and maintain vehicles in good condition, and purchase carbon-efficient vehicles.

The increased price will be limited by the cost of reducing carbon in other sectors. For example, if it costs $50 to reduce a ton of carbon in the electric sector, this translates to just $0.13 per gallon of gasoline. Thus, if the fuel price needed to meet the carbon cap on fuel producers exceeds $0.13 per gallon, fuel producers will simply buy credits from the electricity sector.

Even at $2.50 per gallon, gasoline costs are relatively low compared to fuel prices outside the United States and to historical trends in the United States. Gasoline costs are still a relatively minor economic factor in vehicle ownership. A $0.13 per gallon increase will only have a small impact on VMT and vehicle purchase decisions, and virtually none on low carbon fuels or vehicle technology.

The problem with vehicle technology is that the fuel savings are largely offset by the cost of the technology. The net benefit over a wide range of vehicle technology is less than $200. In addition, the average customer only values the fuel savings for his or her ownership period, which is roughly 50,000 miles, so the net benefit valued by the customer is frequently less than zero. When compared with the multitude of important tradeoffs facing customers in their purchase decision and the emotional factors involved, most customers treat fuel economy technology as a very low priority.

The dynamics of an upstream cap-and-trade system for transportation can be illustrated with a simple example. If we assume that regulated fuel producers will not reduce fuel production, each firm is then faced with three compliance strategies: reduce emissions associated with fuel production, produce and sell cleaner fuels, or purchase emissions allowances from other sectors. If the marginal cost of reducing fuel-related emissions exceeds $0.13 per gallon, a fuel producer minimizing marginal abatement cost will simply buy credits from the electricity sector. Even if one assumes that the entire increase in marginal cost is passed on to consumers, a $0.13 per gallon increase will be unlikely to change driver behavior. Because fuel cost represents only a relatively minor economic factor in vehicle ownership, it is unlikely to have a large impact on VMT and vehicle purchase decisions, and virtually none on low carbon fuels or vehicle technology.

Overall, upstream trading schemes have low administrative costs, but they promise little direct reduction of motor vehicle fuel consumption. Emissions reductions, especially in the near term, are likely to come from other sectors where the marginal cost of GHG abatement is lower. Of course, from a global climate change perspective, this is fine because the atmosphere doesn't care if CO_2 reductions come from vehicles or somewhere else. Due to the oil intensive nature of transportation, however, this may not be the best solution for energy security and trade deficits. Moreover, upstream trading programs are likely to be hampered by political barriers to increasing fuel price. While upstream trading has important benefits and should not be discarded, a supplemental program targeting motor vehicle fuel consumption is likely to be desirable.

Downstream Trading

Downstream trading schemes shift the responsibility for carbon emissions from fuel producers to vehicle owners or operators. The narrowest downstream trading scheme would be similar to the rationing coupons issued in World War II, except that the coupons could be freely traded. The problems with this approach are obvious. There are over 200 million vehicles on U.S. roads, with allocations and trading provisions needed for all. This is a huge administrative burden, and there is no low-cost technology for monitoring vehicle emissions. There are also privacy concerns with mandatory monitoring of individual vehicles. This approach is simply not feasible politically or administratively.

The more practical alternative for downstream trading schemes is to bring vehicle manufacturers into the programs. Vehicle technology is one of the major factors in reducing carbon emissions, and fuel price is a relatively weak lever to bring technology to market. Incentives and mandates affecting vehicle technology could have a major effect on vehicle GHG emissions. There are numerous advantages to this approach:

- In theory the mechanism is simple, requiring only that vehicle manufacturers turn in allowances for imputed lifetime emissions.
- It avoids fuel price increases, which would be politically sensitive.
- There are few automotive manufacturers, so the administrative costs are low.
- Vehicle manufacturers have a great deal of control over the installation of fuel efficiency technology.
- Vehicle manufacturers can influence fuel type.
- Vehicle manufacturers can influence purchase decisions with vehicle pricing and marketing, although manufacturers are limited in what they can do in isolation from changes in customer preferences and the broader carbon-related decision-making context.

It clearly makes sense to try to include vehicle manufacturers in trading programs. Some people stop here and just assume this is the best approach. Even those who acknowledge the problems with economic costs, administrative complexity, and double counting, such as the Pew report, often assume that the program can still be effectively enforced with careful attention to the structure. There are problems with downstream trading schemes focused on vehicle manufacturers, however, which may not be easily remedied.

Problem 1—Double Counting

Perhaps the largest problem with downstream trading schemes is the timing of the allocations and credits to vehicle manufacturers. Fuel producers and other upstream allocations are done for the current year. This is also true for downstream allocations and credits for electric utility companies. However, allocations for vehicle manufacturers are based on projected carbon for the vehicle lifetime. Technology added by vehicle manufacturers now will accrue actual carbon reductions in the future over the vehicle lifetime.

One consequence is double counting of credits. For example, assume that manufacturers improve fuel efficiency or sell alternative fueled vehicles. The carbon from these vehicles will be lower in the future. Manufacturers are given credits for the future reduction in carbon, which they can trade to someone else. In the future, fuel producers will receive credits when vehicles use less fuel, but these are the same credits that were already taken by vehicle manufacturers. The same credits, therefore, are traded twice. Note that increasing efficiency in the other parts of transportation would create similar problems. For example, if the fuel efficiency of freight trucks or the system/logistical efficiency improved, the carbon reductions would also be realized through long time horizons, often longer than light duty. Carbon reduction credits earned by the freight industry should not be double-counted by fuel suppliers in the future, either.

There are three possible ways to eliminate double counting. The first is to switch to upstream trading with fuel producers. This is not really a solution. It simply avoids downstream trading by reverting to upstream trading, with the problems discussed above.

Second, upstream trading with fuel producers could be eliminated and the trading system restricted solely to downstream trading with motor vehicle manufacturers. This would focus on efficiency technology for LDVs, which is likely to be a stronger lever than trying to reduce VMT or change product mix through higher fuel prices. Unfortunately, LDVs consume less than half of the petroleum fuel produced. It would not be desirable to eliminate all other transportation sectors from upstream trading just to focus on vehicle efficiency in LDVs. One possible solution would be to exempt vehicle fuel sales from the upstream cap-and-trade system for other fuel users. This requires forecasting future VMT, scrappage rates, and in-use efficiency. If these forecasts are low, then overall carbon emissions will exceed the cap. When combined with the elimination of incentives to reduce vehicle end-use, this is not likely to be a desirable option.

Finally, the downstream trading scheme could be modified to include a hybrid program, whereby allocations are split between vehicle manufacturers and fuel producers. This could provide some incentives for manufacturers to improve efficiency and for fuel producers to reduce use. While this approach is attractive in theory there are a number of problems, which are evaluated later in this chapter.

Problem 2—Manufacturer Allocations

All the normal problems with allocating versus auctioning carbon caps apply to trading with vehicle manufacturers. However, there are two additional considerations that apply to vehicles. The first is how to handle existing vehicles on the road in the carbon allocations. Who is responsible for the existing fleet of vehicles? The second issue is whether the manufacturers should be held responsible for the total lifetime carbon from their vehicles, or just the carbon intensity measured by CO_2 emissions per mile resulting from the operation of their vehicles.

The total carbon approach would allocate a set amount of carbon that could be emitted from each manufacturer's fleet over its lifetime. John DeCicco suggested this approach in his paper, "An Oil Consumption Cap-and-Trading Scheme for Light Duty Vehicles" (DeCicco, 1993). The advantage of this system is that it holds manufacturers responsible for customer purchase decisions and use, creating incentives for manufacturers to reduce carbon using all of the possible levers, including technology, vehicle characteristics, lower-carbon fuel, and reduced VMT.

The problem with the total carbon approach is that it holds manufacturers responsible for customer purchase decisions and use, as well as sales and market shifts. The compliance burden on manufacturers increases if their sales increase, if the market shifts to larger vehicles, or if people drive more. There are also very anticompetitive consequences. Increasing sales makes it more difficult to meet the requirements, while decreasing sales yields windfall credits without any improvement in efficiency. The net effect is to tend to freeze manufacturers into their existing market share. A total carbon system would be even worse than the Uniform Percentage Increase (UPI) method for CAFE standards. The problems with UPI were discussed at length in a National Academy of Science report on CAFE (National Research Council, 2002).

Downstream trading based on the carbon intensity of vehicle driving would hold manufacturers responsible for the average CO_2 per mile of their vehicles, not the total carbon per fleet. This would be similar to combining CO_2 emissions standards with broader trading. This is equitable and provides a good lever for efficiency technology. It has little influence on VMT or the type of vehicle purchased by consumers, however, and it does not control total carbon emissions. Manufacturers can earn credits even if total carbon increases due to higher sales or increasing in-use driving. While it would be desirable to combine a carbon intensity system for vehicle manufacturers with an upstream system for fuel producers, there are several problems with this approach as well, which are discussed later.

Problem 3—Accounting

To avoid double counting, vehicle efficiency improvements must be subtracted from future carbon allocations for other sectors. As carbon intensity allocation is likely to be the only workable system for vehicles, future changes in vehicle carbon emissions must be estimated annually by predicting yearly scrappage rates, annual VMT by vehicle age, in-use fuel economy shortfall resulting from a gap between certification test results and average in-use fuel economy, and the carbon content of the fuel. If vehicle carbon emissions are estimated incorrectly, then extra burden may be put on other sectors or the desired carbon reduction may not be achieved.

It is difficult to forecast the future and the assumptions may prove to be very inaccurate. For example, lifetime VMT per vehicle may change. This will cause vehicles to use more or less carbon compared to the original accounting. Also, in-use efficiency may change due to factors such as more congestion, higher speeds, and more sprawl, with similar impacts on total carbon emissions. To further compound the problem, most of these factors are not well known. For example, estimates of the in-use fuel economy shortfall are now based on 25-year-old data and scrappage rate estimates are based on very limited and imprecise surveys. In addition, vehicle use characteristics in terms of lifetime and annual VMT are likely to vary by vehicle type and manufacturer.

There are further issues with alternative fueled vehicles. How is double counting handled with respect to alternative fuel producers? How are credits determined for flexible fuel vehicles that can run on more than one fuel? The future use of alternative fuels on flexible fuel vehicles and the actual GHG emissions impact must both be accounted for with reasonable accuracy.

Another important issue is that vehicle accounting is not compatible with the rest of the trading system, due to the mismatch in the timing of the carbon reductions. For example, if carbon prices are low, manufacturers are encouraged to buy credits instead of using advanced technologies. However, the efficiency of the current in-use fleet doesn't change, and the credits are used to increase vehicle emissions in future years as the fleet turns over. Thus, the carbon ceiling for the current year goes down. The future vehicle fleet will use more fuel and less carbon will be allowed, which will cause oil producers to exceed their allocations in the future and force them to buy credits. This will drive up the price of carbon in the future.

On the other hand, if carbon prices are high, manufacturers are encouraged to exceed requirements and sell credits. Again, the efficiency of the current fleet doesn't change, so the carbon ceiling for the current year goes up. The future vehicle fleet will use less fuel and more carbon will be allowed, which will drive down prices in the future.

The mismatch in the timing of the carbon reductions results in an artificial cycling of both carbon availability and pricing, based on vehicle turnover. This will make it difficult for other sectors to manage their allocations, especially the oil producers.

Additional Concerns and Questions

There are a number of other potential problems that a downstream vehicle trading program must address, including the following:

- How should equipment other than the LDV fleet be handled, including pickups and sport utility vehicles over 8,500 pounds gross vehicle weight, heavy trucks, farm equipment, buses, lawnmowers, motorcycles, and construction equipment?

- If a vehicle efficiency standard is used alone, how should the lack of environmental certainty be addressed?
- There is a need to consider how to monetize non-GHG considerations, such as energy security and the trade deficit, into a vehicle trading system designed to reduce GHG emissions.

Upstream/Downstream Hybrid

The analysis in the last section suggests that a compromise approach between upstream and downstream vehicle trading programs might be to split the carbon allocation between vehicle manufacturers and fuel producers. In theory, this could provide some incentives both for manufacturers to improve efficiency and for fuel producers to reduce carbon use.

To help visualize how such a hybrid would operate, let us assume a 2015 target of 200 million tons (mmT) CO_2 reductions from vehicles and that the responsibility for achieving these reductions is split equally between vehicle manufacturers and fuel producers. Table 6-3 provides a summary of the assumptions and issues in this example to make it easier to follow the discussion.

The first step is to calculate baseline whole lifetime CO_2 emissions for the total number of vehicles sold annually. For cars and light-duty trucks, baseline CO_2 emissions over the vehicle lifetime are approximately 1,074 mmT per model year. This result is obtained by multiplying 17 million new LDV sales per year by 150,000 lifetime miles traveled per

TABLE 6-3. Hybrid Program Summary

	Vehicle Manufacturers	*Fuel Producers*
Baseline	Vehicle sales * Lifetime VMT / in-use MPG * carbon content	Gallons sold * carbon content
2005—million metric tons CO_2	17 * 150 / 21.0 * 19.5 / 2.205 = 1074 mmT	1982 mmT (inc. rail, bus, freight, ship, boat, air, 75-05 LD)
2020—each reduce 100 mmT	17 * 150 / 23.2 * 19.5 / 2.205 = 972 mmT	How is LD handled versus other transportation sectors?
	What if: sales change, lifetime VMT increases, in-use FE shortfall changes	How are vehicle efficiency improvements handled in the future?
	• Actions by one will reduce emissions of the other without any action, although offset in time by fleet turnover	
	• If want to influence both manufacturers and oil producers, must hold both accountable for total reductions	
	• Actions by each still influences requirements for the other	

vehicle, divided by an average in-use fuel economy of 21 miles per gallon (mpg), multiplied by 19.5 pounds of CO_2 per gallon of gasoline, divided by 2,205 pounds per metric ton. For fuel producers, baseline CO_2 emissions are simply the carbon content of the fuel sold. Per the Annual Energy Outlook 2005 with Projections to 2025 (EIA, 2005), total CO_2 emissions for all transportation sources are 1,982 mmT for 2005.

For 2020, additional assumptions are that new vehicle sales remain constant, in-use VMT doesn't change, and the average carbon content of in-use fuel doesn't change. With these assumptions it is possible to calculate the level of in-use mpg needed to reduce 100 mmT from vehicles. Next, manufacturers can back-calculate the new vehicle efficiency needed using estimates of scrappage rates and VMT/year by vehicle age, assuming that the relationship between fuel economy tests and in-use mpg doesn't change. Changes in the carbon content of in-use fuel can also be included in the model for new vehicle efficiency, although this raises the issue of whether vehicle manufacturers or fuel producers should receive credit. Of course, if any of these six assumptions are wrong it means that the projected savings will not equal the actual savings.

For fuel producers, there are some additional considerations. For the 2005 baseline year, LDVs emitted 1,074 mmT CO_2 and all transportation sources emitted 1,982 mmT, which means that over 900 mmT were generated by other sources, such as rail, buses, freight, shipping, boats, airplanes, construction, and lawnmowers. Should the fuel for LDVs be separated from other uses and, if so, how? Should the 100 mmT reduction for vehicles be included in a larger, overall reduction for all transportation? What is the baseline for the fuel producers? Should the 100 mmT reduction in CO_2 be compared to 2005 CO_2 emissions or to a "business as usual" base case for 2020? This last is a critical issue, as VMT has been steadily increasing and will continue to do so barring some catastrophic event.

Assuming that all the accounting issues can be managed, there is still a major problem with interactions between actions taken by vehicle manufacturers and fuel producers. For example, if vehicle manufacturers take steps to improve the efficiency of their vehicles from 2005 through 2020 such that the in-use vehicle fleet achieves a 100 mmT reduction in CO_2 emissions in 2020, then fuel producers don't have to do anything to reach their 100 mmT reduction goal. The vehicle manufacturers would have already accomplished the entire reduction.

This can be corrected by doubling the reduction required from fuel producers so that they will be held to a 100 mmT reduction in addition to the 100 mmT required from the vehicle manufacturers. This would force fuel producers to take steps to reduce carbon content in the fuel, carbon from refining, and transporting fuel, or raise the price of fuel by limiting quantities or buying credits from other sectors. Reducing the carbon content of fuel or raising fuel prices would reduce vehicle CO_2 emissions, both directly and indirectly by encouraging the purchase of more efficient vehicles and reducing VMT. Now the vehicle manufacturers can wait for the steps taken by the fuel producers to reduce vehicle CO_2 emissions by 100 mmT, without significant action on the manufacturers' part.

Actions taken by vehicle manufacturers and fuel producers will always reduce the emissions from the other without any action being taken. This interaction is offset in time by fleet turnover, making it virtually impossible to determine what the effects will be. This interaction between vehicle manufacturers and fuel producers makes it virtually impossible to determine separate allocations. If the goal is to involve both vehicle manufacturers and fuel producers to achieve a 200 mmT CO_2 reduction, both must be held accountable for the full amount of the 200 mmT. This would not be an enforceable system, as it would not be possible to allocate shortfalls between the vehicle manufacturers and fuel producers.

Incorporating Vehicles into a Carbon Trading Program

A single example is discussed in this section to help illustrate the difficulty in incorporating vehicles into an overall carbon trading program. The example is drawn from the U.S. Climate Stewardship Act of 2003 (U.S. Senate, 2003). Jonathan Hughes, who is conducting research on vehicle trading schemes for the UC Davis Institute of Transportation Studies, suggested the fuel economy credit conversion methodology presented in the sidebar (Hughes, 2005).

Sidebar: Hybrid Upstream Emissions Trading System

An upstream trading system for transportation was proposed in the McCain-Lieberman Climate Stewardship Act of 2003 and is currently under discussion by the California Climate Action Team in the state of California. These systems have the benefit of administrative simplicity due to the relatively small number of firms that would be regulated and a high potential for environmental effectiveness due to broad coverage, certainty, and enforceability. However, incentives to reduce fuel consumption via the indirect mechanism of price signals are less than those for systems specifically targeting VMT reduction or fuel economy improvements. In order to promote improvements in vehicle fuel economy, a pure upstream system could be modified to incorporate vehicle manufacturers. As an example, the Climate Stewardship Act of 2003 would allow vehicle producers that more than comply with the CAFE standards to sell excess credits to a central GHG allowance market. However, the provision would require a complex accounting methodology to convert improvements in fuel economy to GHG emission allowances. In order to avoid double counting and estimation issues, allocations of allowances to vehicle manufacturers for improved vehicles

> would need to occur annually over the vehicle lifetime. In addition, such a system must reduce the annual allocation of emission allowances to the allowance market and to fuel producers by an amount equivalent to the annual amount awarded to automakers in order to avoid double counting of emissions reductions.

The advantage of this system is better accounting and control of the emissions reductions. Instead of allocating the entire credit to the manufacturer when the vehicle is produced, it would allocate credits annually over the vehicle lifetime as the carbon savings occur. The annual allocation to fuel producers would be reduced by the annual amount awarded to vehicle manufacturers. This would ensure that emissions reductions would not be double counted.

This is the best approach proposed to date for a hybrid upstream/downstream vehicle program. Nonetheless, there is still an issue with allocation between vehicle manufacturers and oil producers, although there is no longer a possibility of double-counting credits. The allocation is done annually, which means scrappage rates, VMT by vehicle age, in-use fuel economy shortfall, and fuel carbon content must be calculated for each model year. If the estimates are not accurate, it will benefit one of the parties and make it harder for the other. Also, changes in these variables will affect the allocations to vehicle manufacturers and fuel producers, changing the cost of complying with trading requirements.

Allocating credits annually substantially reduces the incentive for vehicle manufacturers to participate. Manufacturers will have to utilize engineering resources and spend capital up front to implement efficiency improvements, but the credits will be allocated over the 25- to 30-year vehicle life. While this is also true for other sectors, especially the electric sector, vehicle manufacturers are unique in that they do not capture the savings from the future reduction in fuel use and they would not be required to participate in carbon trading. Further, the amount of the future credits would be uncertain, as they depend on assumptions about future lifetime VMT and fuel carbon content, which are likely to be inaccurate. Thus, there would likely be little motivation for vehicle manufacturers to significantly improve vehicle efficiency.

Another problem is that offering an alternative fueled or flexible fueled vehicle does not do any good if the fuel is not available. On the other hand, offering alternative fuels does not do any good if vehicles are not available. Both are needed to move the market toward lower carbon fuels. The system does not address this problem.

Finally, the system does not reduce overall carbon emissions. Oil producer allocations are reduced, but this is offset by allocations to the vehicle manufacturers. In sum, the total number of allocations does not change. This is also true if the vehicle credits are given to vehicle manufacturers when the vehicles are sold instead of when the fuel is used. Allowable CO_2

emissions will increase in the baseline year when the vehicle manufacturers are allowed to sell CAFE credits into the system, as the reductions in vehicle CO_2 only occurs in the future. Then, in the future, overall CO_2 allowances are reduced corresponding to the reduced CO_2 allocation to fuel producers. In sum, over the vehicle lifetime, the initial increase in credits and the future reductions in allocations will exactly offset each other, assuming all the factors were estimated correctly. There is no net decrease in CO_2 emissions.

One argument in support of a hybrid vehicle trading system is that, even if it doesn't reduce overall carbon emissions, it could help to reduce the overall cost by encouraging fuel efficiency technology. However, this system has no explicit mechanism to minimize GHG reduction costs in transportation by selecting between fuel and vehicle technologies that offer lower marginal costs. It just requires that any improvements made by vehicle manufacturers be subtracted from future fuel producer allocations. The cost control is entirely on the side of the vehicle manufacturers.

Another argument in support of a hybrid vehicle trading system is that there are other benefits to reducing oil consumption, such as energy security, trade deficits, and the effect of oil price shocks on the economy. However, creating a very complex trading system, with no mandatory participation by vehicle manufacturers, is unlikely to be the optimum solution.

Conclusion

Previous studies have identified most of the problems with trying to incorporate vehicles into carbon trading programs, but none are comprehensive. The 2003 study by the Pew Center, for example, simply presented the advantages and disadvantages of all the different options. The CCAP study in 2000 was based primarily on arguments that there were other reasons for improving vehicle efficiency than just carbon emissions. Neither study tried to solve the problems from integrating vehicle manufacturers into overall carbon trading programs, which are overwhelming. Some of the key problems are outlined below.

Double counting must be avoided. This is not a problem if only fuel producers or vehicle manufacturers are included in a trading program. Vehicle manufacturers have little impact on VMT and fuel producers have little impact on vehicle technology, so it is desirable to include both. Systems that provide allocations to vehicle manufacturers must subtract this amount from fuel producer allocations.

Currently, only vehicles with gross vehicle weight ratings less than 8,500 pounds are subject to the fuel economy testing necessary for proper emissions accounting. This requires that fuel producer allocations be divided between LDVs and all other transportation uses. It also raises the question as to how the other transportation uses should be handled in the trading system.

Proper allocation requires accurate estimates of vehicle scrappage rates, VMT by vehicle age, average carbon content of in-use fuel, and in-use mpg compared to test results. Except for the average carbon content in fuel, none of these factors is well understood. None of the factors can be forecasted with any accuracy.

Actions taken by vehicle manufacturers to improve efficiency do not affect current year carbon emissions but only future emissions. Other sectors are dealing with current year emissions. This time offset creates multiple problems in accounting and operation of incentives.

If allocations are given to manufacturers for the lifetime estimated emissions when the vehicle is built, it results in an artificial cycling of both carbon availability and pricing, based on vehicle turnover. This will make it difficult for other sectors to manage their allocations, especially the oil producers. This can be fixed by allocating manufacturer credits annually over the vehicle lifetimes as the carbon savings occur. However, allocating credits annually instead of when the vehicle costs are incurred substantially reduces the incentive for manufacturers to participate.

Actions taken by vehicle manufacturers affect the allocation for fuel producers and vice versa. This makes it impossible to set separate allocations for manufacturers and fuel producers. Both must be held accountable for the entire reduction in carbon emissions, but currently there is no known way to administer such a program.

Handling of alternative fueled vehicles is problematic. Vehicle manufacturers are needed to produce the vehicle and fuel producers must make the fuel available, but there is no way to split the carbon allocation between manufacturers and producers. Flexible fueled vehicles create an additional problem, which is accounting for the amount of the alternative fuel that will actually be used.

Maybe there is a way to solve all of the problems and make a workable hybrid vehicle trading system, but ten years of effort by many different organizations has yet to yield a good system. Different systems solve some of the problems, but the overall complexity is overwhelming.

Even if the problems could be solved, vehicles would still not reduce overall carbon emissions. To avoid double counting, the vehicle manufacturer allocations must be subtracted from the fuel producer allocations. Thus, the primary justifications for creating a hybrid vehicle trading system are to reduce the overall costs of reducing carbon emissions and to capture other benefits for reducing fuel use beyond just GHGs, such as energy security, trade deficit, and oil price shocks. Incorporation of vehicles into an overall carbon trading system is a very complex and likely unworkable way to try to capture these benefits. The sidebar offers a proposal that could avoid several of the problems identified in this chapter.

Sidebar: A Better Approach

The primary arguments for creating a hybrid vehicle trading system are that it could reduce the overall costs of reducing carbon emissions and that there are other reasons for reducing fuel use beyond just greenhouse gases.

These same advantages could be obtained with a lot less complexity by creating a stand-alone incentive program for vehicle efficiency. This would still be based on vehicle carbon-intensity incentives based on CO_2 emissions per mile. If desired, the incentives could be class-based to address customer choice, safety, and intermanufacturer equity concerns. Due to the limited number of manufacturers who control the large majority of the market, trading between manufacturers is not likely to be very successful. Thus, the system should allow manufacturers to buy and sell efficiency credits to and from the government at a fixed rate. This rate could be set based on the going carbon trading rate plus monetization of benefits to the nation for conserving energy and reducing oil consumption.

Such a system would provide certainty on the monetary value for improving efficiency and would allocate the full value immediately, increasing the incentive for manufacturers to bring technology to the market. It would also be far simpler to administer and would keep the credits out of the overall sector carbon trading system, avoiding most of the problems with incorporating vehicles into an overall trading system.

One unavoidable problem is that the improvements in vehicle efficiency would still need to be subtracted from the fuel producers' allocation. Otherwise, oil producers would have windfall benefits from vehicle manufacturer actions. The government would need to monitor actual efficiency improvements and in-use VMT and use this data to adjust carbon caps for the fuel producers.

References

DeCicco, John M. "An Oil Consumption Cap and Trading Scheme for Light Duty Vehicles." Policy brief prepared for the American Council for an Energy-Efficient Economy. Washington D.C., April 1993.

Hughes, Jonathan. Personal communication. December 2005.

National Research Council. Effectiveness and Impact of Corporate Average Fuel Economy (CAFE) *Standards*. Washington, D.C.: National Academy Press, 2002.

Nordhaus, Robert R., and Kyle W. Danish. "Designing a Mandatory Greenhouse Gas Reduction Program for the U.S." Prepared for the Pew Center on Global Climate Change. May 2003.

U.S. Energy Information Administration (EIA). *Annual Energy Outlook 2005 with Projections to 2025*. Report # DOE/EIA-0383. Washington, D.C.: EIA, January 2005.

United States Senate. U.S. Climate Stewardship Act of 2003, S. 139. Washington D.C., 2003.

Winkelman, S., T. Hargrave, and C. Vanderlan. *Transportation and Domestic Greenhouse Gas Emissions Trading*. New York: Center for Clean Air Policy, April 2000.

CHAPTER 7

North American Feebate Analysis Model

Alexandre Dumas, David L. Greene, and André Bourbeau

Canadian automobile manufacturers and the Canadian government recently signed an agreement to reduce greenhouse gas (GHG) emissions from passenger cars and light trucks by 5.3 megatonnes (MT) in 2010. The agreement is a key component of Canada's plan to meet its commitments under the international Kyoto Protocol. In 2003, transportation accounted for 25.7 percent of GHG emissions in Canada (Environment Canada, 2004), and this is expected to grow to 27.0 percent by 2020 (Natural Resources Canada, 1999). Seventy percent of transportation emissions are allocated to passenger transportation, a sector where emissions continue to grow.

If the current voluntary approach to mitigating passenger transportation emissions is not effective, one of the options to mitigate emissions would be to use economic instruments to provide an incentive for the purchase of fuel-efficient vehicles as a potential alternative or complement to regulation. Feebates are one possible economic instrument that could be established as a cost effective-way to curb GHG emissions from new light-duty vehicles (LDVs). Feebates are a market-based system in which every vehicle is either subject to a fee or rebate when purchased, depending on its fuel economy.

This chapter presents the results of recent analyses of financial incentives for fuel-efficient vehicles using a model of the North American vehicle market. It focuses mostly on the impacts of feebate policies implemented solely in Canada on the North American vehicle market. Successful implementation of a Canadian feebate system might induce some states in the United States to adopt similar systems and could possibly lead to a harmonized North American feebate policy. The potential gains in efficiency and the effectiveness of greater harmonization are explored.

Over the longer term, widespread use of advanced technologies offers perhaps the best potential for significant reductions in transport-related emissions, including GHG emissions. However, the progress of technological change is often slow and entails many stages. It is possible to mandate changes, but regulatory standards need to be frequently tightened to create a sustained incentive for fuel economy improvement. Market-based instruments, such as increased vehicle or fuel prices, cannot only be a powerful driver of change, but they can also provide a sustained incentive for the market to adopt advances in energy efficient technology.

In a recent analysis of feebates for the United States, Greene et al. (2005) concluded that feebates could effectively correct an imperfect market for fuel economy in which consumers appear to undervalue the discounted fuel savings over the full lifetime of a vehicle. That study found that a well-designed feebate policy could induce an economically efficient level of fuel economy with relatively little cost.

Like other studies of feebates for the U.S. market, Greene et al. found that the overwhelming majority of fuel economy improvements brought about by feebate policies, 90 percent or more, would be a result of the adoption of energy efficient technologies and efficient vehicle designs. Only 10 percent or less would be due to a shift in sales toward higher fuel economy makes and models. However, the Canadian market is smaller than the U.S. market and a feebate policy in Canada alone would have much less leverage for inducing manufacturers to adopt fuel economy technologies. Implementing a full-scale feebate program solely in Canada raises new questions about the costs to producers and consumers, as well as the effectiveness in increasing fuel economy.

The analysis conducted so far in Canada indicates that feebates could be designed to be an environmentally effective and economically efficient way to reduce GHG emissions from the transportation sector. However, these conclusions are only valid if the assumptions underlying the analysis are correct. In particular, great uncertainty remains due to poor knowledge of Canadian new vehicle price elasticities and perceived value of fuel savings. There are also knowledge gaps related to the impact of feebates on the used vehicle market. For instance, feebates might induce owners of vehicles subject to a fee to keep their vehicles on the road marginally longer or induce prospective buyers to try to avoid the fee by importing nearly new vehicles from markets where no feebate is imposed. Finally, this study did not consider how a feebate program would be implemented. Thus, the current analysis could not support the immediate implementation of a feebate program without first addressing these issues.

Analyzing Feebates in the North American Market

Although there have been previous analyses of feebate systems for the United States (e.g., Davis et al., 1995, and Greene et al., 2005) and Canada (e.g., HLB, 1999), only HLB Decision Economics, Inc., a consulting company with offices throughout North America, analyzed the potential effectiveness of a feebate system that applied solely to the Canadian vehicle market. This analysis was conducted by HLB in 1999 for the Canadian government in support of the National Climate Change Table process. The effect of a feebate policy in Canada is most appropriately represented as a change in the North American market demand for fuel economy. The task is to represent how manufacturers will respond when demand for fuel economy in the Canadian market increases, while demand in the U.S. market remains unchanged. This calls for two modifications of the single market model of Greene et al. First, Canadian and U.S. motor vehicle demand must be represented separately so that a feebate system can be imposed in one market without affecting demand in the other. Second, decision criteria must be specified for manufacturers to use in deciding whether or not to redesign a vehicle in response to the change in Canadian demand for fuel economy.

The separation of U.S. and Canadian markets was accomplished by first developing separate vehicle sales and attributes databases for the United States and Canada. Country-specific nested multinominal logit (NMNL) choice models were calibrated to the two databases (Train, 1986). Manufacturers were assumed to redesign a vehicle specifically for the Canadian market, if the sales of that vehicle in Canada exceeded a minimum level necessary to achieve scale economies in production.

When a vehicle sold in both markets did not have sufficient Canadian sales to justify redesign only for sale in Canada, a manufacturer was assumed to design a single vehicle for both markets, taking into consideration the increased demand for fuel economy in Canada. Since a feebate system will induce some increase in demand for fuel economy relative to the no-policy case, manufacturers will choose some increase in fuel economy but not nearly as much as if the feebate system applied in both markets. Domestic models lacking adequate sales in the combined North American market are not redesigned. Imports even with low-volume sales are redesigned to simulate a foreign manufacturer's option to substitute a different European or Asian design with higher fuel economy.

Structure of the North American Feebate Analysis Model

Like the U.S. feebate model of Greene et al., the North American Feebate Analysis Model (NAFAM), used to produce the analysis in this chapter, assumes that manufacturers will implement fuel economy technology on vehicles in a way that maximizes consumer satisfaction. Customer satisfaction is represented by consumer's surplus, the economist's monetary measure of well being. The consumer's surplus can be expressed as the difference between the total satisfaction a consumer obtains from the attributes of a new vehicle (hence, the maximum price he would be willing to pay) and the total price the consumer has to pay. Hence, the total utility of a new vehicle is the price paid plus the consumer's surplus.

Decision variables facing manufacturers are the fractional changes in fuel economy for every LDV sold in North America. Choosing a change in fuel economy affects consumer satisfaction in three ways: it provides fuel savings, it increases vehicle purchase price, and it reduces the fee or increases the rebate applicable to the vehicle in question.

Although there are many possible forms of feebates (Davis et al., 1995), the simplest and perhaps most interesting is a constant dollar rate per liter per 100 kilometers (L/100 km) of fuel consumption. This formulation values each liter of fuel saved at the same amount, regardless of which vehicle consumes the fuel. It assumes only that all vehicles are driven the same number of kilometers (km) per year. A single pivot point can be specified for all LDVs, or different pivot points can be assigned to different vehicle classes. This form of feebate is illustrated in the equation below for a vehicle model i in class j by a rate, R, and a pivot point, C', that determines the fuel consumption number above which a fee must be paid and below which a rebate is received. This formulation implies that fees will be negative and rebates positive.

$$F = R(C'_j - C_{ij})$$

The utility a consumer derives from a particular vehicle is assumed to be a function of its attributes. Among these attributes are purchase price and fuel economy. For purposes of the feebate analysis, it is assumed that all attributes except price, fee or rebate, and fuel economy are constant at the base year levels. Consumers are assumed to value fuel economy increases according to their perception of the value of fuel that will be saved as they use the vehicle. Perfectly rational consumers would measure this by the expected discounted present value of fuel saved over the full life of the vehicle. There is evidence that consumers do not actually make such assessments (Turrentine and Kurani, 2005). The NRC (2002) fuel economy study considered two alternatives for valuing fuel savings: full lifetime discounted present value fuel savings and a three-year simple payback. The same two conventions are used here.

The ability of manufacturers to supply fuel economy is represented by fuel economy technology cost curves. Curves describing the total cost of fractional improvements in fuel economy from a base level can be constructed from data on specific technologies, their costs and impacts on fuel economy (Greene and DeCicco, 2000). Cost is measured in terms of retail price equivalent, an estimate of the incremental price the purchaser of a car would pay based on fully burdened manufacturing costs plus manufacturer's profit and retailing cost and profit. Manufacturers are assumed to choose the fractional increases in fuel economy that maximize the total change in consumers' surplus for each vehicle, subject to whatever feebate policy may be in place.

Two alternative sources of fuel economy cost information were used in this study. The National Research Council (NRC, 2002) study produced three fuel economy cost curves—optimistic, average, and pessimistic—for four classes of passenger cars and six types of light trucks. The NRC average curves are used in this study. The NRC fuel economy cost curves do not distinguish between cars produced in North America and imported vehicles. A Transport Canada study conducted by the U.S. consulting firm EEA, Inc., provided data from which cost curves can be estimated for imported and North American–manufactured small cars, large cars, compact trucks and standard trucks (EEA, Inc., 2005). The EEA fuel economy cost curves have the added advantage of being calibrated to the year 2003, the same year as the sales and base fuel economy data used. Therefore, they are used for the majority of the policy cases presented in this study.

Canadian and U.S. Light-Duty Vehicle Markets

Some background on the distinctions between the Canadian and U.S. LDV markets is needed in order to understand the effect of feebates in the North American market and to identify implementation strategies. In North America, approximately 17.2 million LDVs with a gross vehicle weight rating of less than 3,856 kilograms (kg) were sold in 2003. Most of these vehicles, roughly 15.7 million, were sold in the United States. Table 7-1

TABLE 7-1. Light-Duty Vehicle Sales Market Shares—2003

	Canada	U.S.
Passenger Cars	56.5%	50.6%
Subcompact	7.7%	6.9%
Compact	31.4%	20.6%
Midsize	13.5%	16.9%
Large	3.9%	6.2%
Light Trucks	43.5%	49.4%
Small SUV	2.7%	2.8%
Medium SUV	13.9%	18.7%
Large SUV	1.0%	6.0%
Minivan	14.8%	6.4%
Large Van	1.2%	0.7%
Small Pickup	3.4%	5.1%
Large Pickup	6.4%	9.7%

Source: K.G. Duleep, Energy and Environmental Analysis Inc. Based on U.S. sales data coming from the National Highway Traffic Safety Administration's fuel economy database and on Canadian sales data coming from Transport Canada's Vehicle Fuel Economy Information System.

112 *Driving Climate Change*

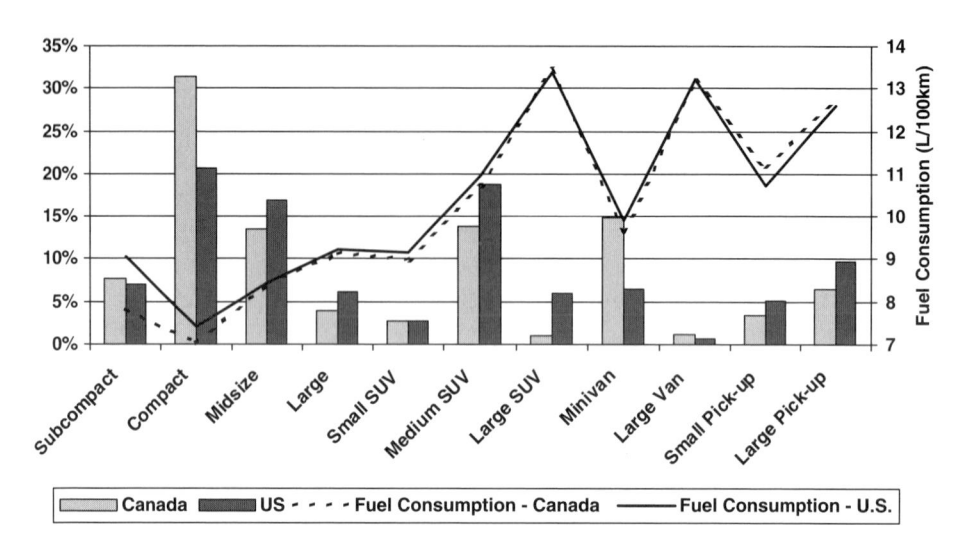

FIGURE 7-1. Average fuel consumption and vehicle sales by class—2003. *Source:* K. G. Duleep, Energy and Environmental Analysis Inc. Based on U.S. sales data coming from the National Highway Traffic Safety Administration's fuel economy database and on Canadian sales data coming from Transport Canada's Vehicle Fuel Economy Information System.

shows the distribution of sales in Canada and the United States among LDV categories. Overall, about 51 percent of North American LDV sales were cars, 27 percent sport utility vehicles (SUVs), 14 percent pickup trucks, and 8 percent vans/minivans.

There are major similarities between the U.S. and Canadian markets, particularly in the range of LDV models for sale. However, there are also some differences in choices of models and classes. Figure 7-1 shows Canadian and U.S. sales further disaggregated into vehicle size classes, using a market distinction based on the interior volume for cars and gross vehicle weight rating for light trucks. These distinctions are similar to U.S. Environmental Protection Agency size classes but customized for the development of the NRC fuel economy cost curves. Compact cars, minivans, and midsize SUVs—with 31 percent, 15 percent, and 14 percent of total sales, respectively—are especially attractive to Canadian vehicle purchasers, while compact cars, midsize cars, and SUVs are the most popular vehicles in the United States, with 21 percent, 17 percent, and 19 percent of total sales, respectively. Compact vans and minivans are relatively more popular in Canada, while in the United States, larger vehicles such as mid- and large-size cars, medium and large SUVs, and pickup trucks are more popular.

The average fuel consumption for all 2003 model year LDVs sold was 9.0 L/100 km in Canada and 9.8 L/100 km in the United States, but consumption among the 11 classes differed substantially in both countries. The full range in Canada and the United States varied from 3.2 to 22.0 L/100 km. Figure 7-1 further illustrates the variability of average fuel consumption, both between countries and vehicle classes.

Despite substantial adoption of technologies capable of increasing energy efficiency, fuel consumption for LDVs in both countries remained relatively stable between 1990 and 2003 as consumers increasingly favored greater performance and weight over saving gasoline. Between 1990 and 2003, the fuel consumption of the average car improved by about 6 percent in Canada, from 8.2 to 7.6 L/100 km. However, during the same period the average fuel consumption for LDVs sold in the United States increased by 4 percent, from 9.3 to 9.7 L/100 km.

Canadian and U.S. fuel economy standards have not significantly changed in recent years. The Canadian Corporate Average Fuel Consumption (CAFC) standards for LDVs have tracked the U.S. Corporate Average Fuel Economy (CAFE) standards for many years. As in the United States, Canadian standards differ for cars and light trucks, defined as all classes of SUV, vans, and pickup trucks. In 2003, the CAFC standards were 8.6 L/100 km for cars and 11.4 L/100 km for light trucks. Fleet average fuel consumption in 2003 was substantially below these values, at 7.6 L/100 km for cars and 10.7 L/100 km for light trucks.

The 5.3 MT reduction in GHG emissions called for by the Canadian industry/government voluntary agreement is roughly equivalent to improving fuel consumption to 7.4 L/100 km for all light-duty vehicles. This goal will be useful in comparing the environmental effectiveness of various feebate policy options.

Manufacturer and Consumer Decision Making

A key variable in the NAFAM analysis is the degree to which North American manufacturers and importers to the North American market will add fuel economy technologies to vehicles in response to a Canadian feebate policy. Although the NAFAM represents country-specific vehicle demand, vehicle manufacturers typically design one configuration of each make to sell in both countries in order to achieve economies of scale. The NAFAM simulates economies of scale by specifying sales thresholds over which a vehicle will be redesigned in response to a feebate and below which it will not. These thresholds were set at different levels for domestic and import vehicles. In the case of vehicles built in North America, sales must be above 20,000 units in Canada in order for manufacturers to consider a redesign in response to a Canadian feebate program.

For imported vehicles, Canadian sales must be above 2,000 units in order for manufacturers to consider a redesign in response to a Canadian feebate program. Although one of the key assumptions of the NAFAM is that no LDVs are introduced or removed from the market, the lower threshold for import vehicles reflects the fact that import manufacturers might

TABLE 7-2. NAFAM—Country-Specific Assumptions

	Canada	U.S.
Fuel Price (CAN¢/L)	80.0	47.6
Fuel Price (US$/gal)	2.51	1.50
Vehicle Lifetime (years)		
Cars	15	16.9
Light Trucks	15	15.5
New Vehicle Distance Traveled (km/yr)	23,500	25,106
Annual Rate of Decline	4%	4%
Discount Rate	10%	6%

choose to respond to the feebate by substituting a more fuel efficient design already being sold in another market, thereby avoiding the problem of scale economies.

Diesel-powered vehicles and hybrid electric vehicles were excluded from the feebate analysis. Only changes in the sales mix affect those vehicles, and only those makes and models offered in 2003 are included in the market analysis.

A number of assumptions from Greene et al. were also carried over to the NAFAM. In particular, both models use the same market share price elasticities:

- −10 for make and model choices within a class at a market share of 1.5 percent
- −5 for the choice among classes at a market share of 10 percent

The overall price elasticity of LDV sales was assumed to be −1.0. These elasticities are believed to be on the high side, and this affects the estimated impacts of feebates, as discussed following.

Finally, in order to make the NAFAM a true North American model, a number of assumptions were reviewed or modified in order to properly reflect Canadian and U.S. markets. An important assumption in order to combine data from the United States and Canada is the currency exchange rate. This study assumed an exchange rate of C$1.20 per US$1.00. Table 7-2 shows other country-specific assumptions on fuel price and vehicle operations that were made by the authors of the study.

Nature of Analysis and Major Assumptions

The NAFAM is a static model, and as such, the model results presented in this chapter represent long-run equilibrium solutions 10 to 15 years in the future, when all manufacturers have had the time to retool their facilities in answer to the feebate. The results presented do not consider the transition period between the implementation of the feebate and this end point.

The analysis focused on three feebate rates, which were C$250, C$500, and C$1,000 per L/100 km. These rates were equivalent to US$208, US$417, and US$833 per L/100 km, respectively. Furthermore, the study focused on feebate designs either using a single pivot point or separate pivot points for cars and light trucks. Pivot points, defined in terms of L/100 km fuel consumption, are the boundaries that divide vehicles charged fees from those receiving rebates. By design, the NAFAM also ensured that all feebate cases were revenue neutral by making the fees collected by the government equal to the rebates given. No consideration was given to the administrative cost of the feebate program or to the potential tax revenue loss for the government caused by lower fuel sales.

Fuel consumption rates, consumer's surplus, manufacturers' revenues, and total government transactions were used to measure the impact of the various policies considered. In the case of consumer's surplus, the estimates were based on the assumption that consumers consider only the first three years of fuel savings when buying a vehicle. This is a strong assumption, as there are likely to be fuel savings well after three years. Thus, the analysis of changes in consumer's surplus will tend to overstate the negative impact of policies on the consumers.

The price elasticities used in this analysis are known to be toward the high end of what the published literature will support, and a sensitivity analysis, presented in the next section, showed that the impacts on manufacturers scale almost linearly with price elasticity. For this reason, the analysis probably overestimates the impacts of feebates on manufacturers, and thus these results should be interpreted with caution.

Results

A no-policy scenario was developed in order to estimate the incremental impacts of the various cases considered in this analysis. Consumers are assumed to purchase and manufacturers to implement fuel-saving technologies in the no-policy case up to the point where the marginal cost of the technologies equals the resulting marginal fuel savings. In the following sections of this paper, the impact of all scenarios will be measured by their incremental impacts compared to the no-policy case.

This base case follows Greene et al. by assuming that consumers take into account only the first three years of undiscounted fuel savings when making their purchase decisions. Under this case, a small but significant amount of fuel economy technology is adopted in both countries. The fuel consumption of cars would improve from 7.6 to 7.1 L/100 km in Canada, while the improvement in the United States would be from 8.2 to 7.6 L/100 km. The improvement is relatively more important for light trucks, which see their fuel consumption diminish from 10.7 to 9.8 L/100 km in

116 *Driving Climate Change*

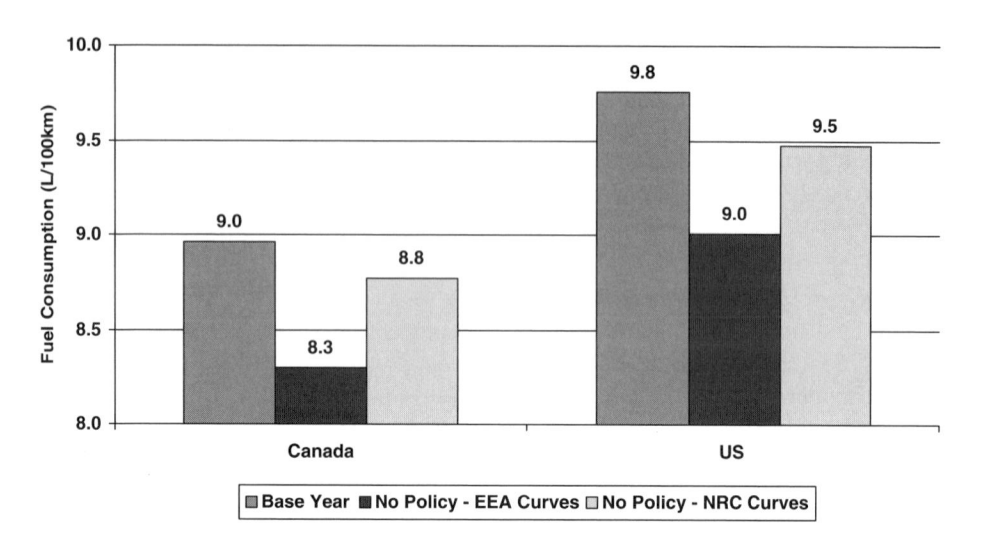

FIGURE 7-2. Effect of updating the fuel economy cost curves on no policy case.

Canada and from 11.4 to 10.4 L/100 km in the United States. Thus, the overall improvement for the new vehicle fleet is 0.7 L/100 km in Canada and 0.8 L/100 km in the United States.

On the other hand, if consumers are assumed to recognize fuel savings over the full life of the vehicle when making their purchase decisions, results are noticeably different. In Canada, passenger car fuel consumption dips down to 6.3 L/100 km, while trucks go down to 8.8 L/100 km, for a fleet average of 7.5 L/100 km. In the United States, the figures are 7.0, 9.5, and 8.3 L/100 km, respectively.

Impact of NRC Fuel Economy Technology Supply/Cost Curves

The base case uses the cost curves developed from the EEA, Inc., 2003 fuel economy technology cost data (EEA, 2005). A separate no-policy case was developed using the average fuel cost and miles per gallon (mpg) cost curves developed by NRC (2002). The results are shown in Figure 7-2. In both countries, the fuel consumption of LDVs decreases less when the NRC curves are used. In fact, the fuel consumption of passenger cars remains constant in both countries, at 7.6 L/100 km in Canada and 8.2 L/100 km in the United States. In the case of light trucks, their fuel consumption goes down to 10.3 L/100 km in Canada from 10.7 L/100 km and 10.8 L/100 km from 11.4 L/100 km in the United States. Those improvements lead to an overall improvement in average fleet fuel consumption of 2 percent in Canada and 3 percent in the United States.

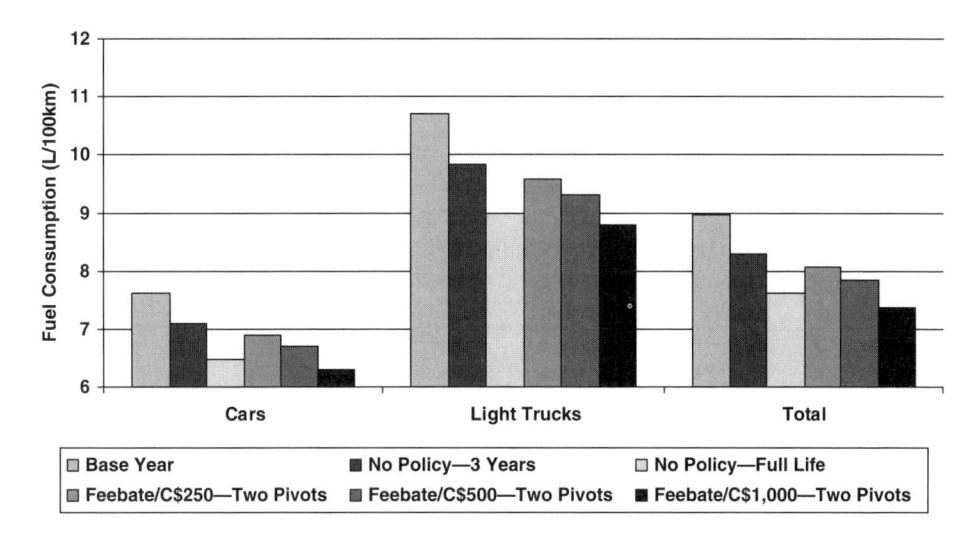

FIGURE 7-3. Impact of two pivot points feebates in Canada—Canada-only policies.

Canada-Only Instruments

The potential impacts of feebate systems implemented only in Canada and not in the United States are of great interest because, as a signatory of the Kyoto protocol, Canada is under obligation to reduce its GHG emissions to 6 percent below their 1990 level. To achieve this goal, many instruments, including economic instruments such as feebates, are being considered. How much can be achieved with different-sized feebates and the impacts on Canadian consumers and manufacturers can be estimated by the NAFAM. To date, the U.S. government is under no such pressure to reduce its GHG emissions, and has not expressed interest in feebates.

Two Pivot Points Instruments

Figure 7-3 shows the results of the various scenarios under two pivot point conditions. Introducing a feebate with a rate of C$250 per L/100 km with two pivot points, one for passenger cars and one for light trucks, in Canada has a small but measurable impact on the Canadian fleet of new vehicles. The fleet average fuel consumption was estimated to decrease from 8.3 to 8.1 L/100 km. Passenger car fuel consumption would decrease by 2.8 percent, while it would decline by 2.7 percent for light trucks. Marginally, consumers were worse off than under the no-policy case, as their net surplus decreases by US$15 million because they are forced into buying more fuel-efficiency technology than they would have wished.

For the Canadian government, the feebate results in total annual transactions of US$363 million. Total government transactions represent the sum of the absolute values of rebates and fees. For Canadian manufacturers, the feebate program leads to a decrease in revenue of US$734 million, or 2.1 percent but not in LDV sales. The fact that revenues decline while sales more or less stay constant is due to consumers buying proportionally smaller, more efficient, and less expensive cars. Because of these market shifts, the major North American manufacturers—Ford, General Motors, and DaimlerChrysler—are even more adversely affected by the feebate system. They see their revenues decline by US$872 million, while the other manufacturers collectively experience an increase in revenues of US$138 million. These results are highly dependent on the price elasticities used to conduct the analysis, and they should be treated with caution.

In the United States, the Canadian-only feebate has a small impact on fuel consumption as the fleet average fuel economy improves by 0.03 L/100 km, the same change as under the no-policy option. The Canada-only feebate also has only a limited impact on other key variables such as consumer's surplus, vehicle sales, and manufacturers' revenues.

Increasing the Canada-only feebate rate to C$500 per L/100 km with two pivot points more than doubles the impact of the feebate on fuel consumption and total government transactions, which increase to US$717 million per year. Under this scenario, the consumer's surplus decreases by US$84 million, because consumers feel more pressure to buy more energy efficient technologies. Manufacturers see their sales decrease by 0.2 percent, but because they sell more lower-priced vehicles, their revenues decrease by 4.1 percent. As a group, the major North American manufacturers face a net fee, which results when the total fees imposed on their vehicle sales are larger than the total rebates received from the sales of eligible vehicles. Other manufacturers are beneficiaries of a net rebate. In the United States, the Canada-only C$500 case has a marginal impact on fuel consumption, sales, and revenues when compared to the no-policy case.

Introducing a feebate with a rate of C$1,000 per L/100 km with two pivot points has roughly the same impact as if consumers considered the full value of fuel savings that can be realized when purchasing more fuel-efficient vehicles. The fleet average fleet economy would improve to 7.4 L/100 km. Passenger car fuel consumption would decrease by 17 percent, while it would decline by 18 percent for light trucks. However, such a feebate rate represents a carbon premium of approximately $175 per metric ton of carbon dioxide if amortized over 200,000 km, which is much higher than other energy consuming sectors of the Canadian economy are currently expected to bear in Canada's efforts to reach its Kyoto target.

Consumers would feel significantly worse off under such a steep feebate rate as their surplus actually decreases by US$388 million because they are induced to buy more fuel-efficient technology than they otherwise would have. In addition, the implementation of a feebate of that scale results in significant transactions, as much as US$1.4 billion a year. For Canadian manufacturers, the feebate program leads to reduced revenues of

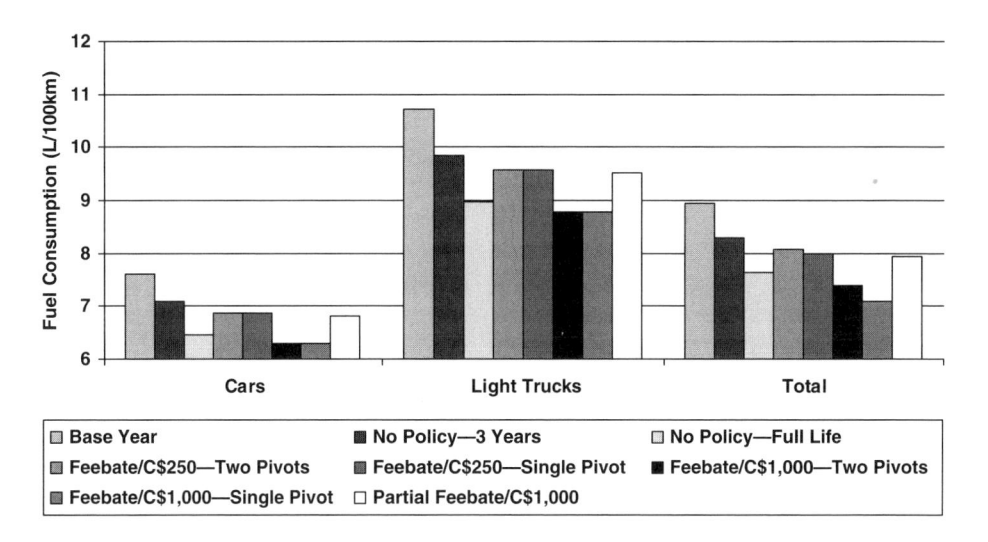

FIGURE 7-4. Impact of single and two pivot points feebates in Canada—Canada-only policies.

US$2.8 billion, or 8.0 percent, and a 1.1 percent drop in sales as consumers turn to more efficient but less expensive vehicles in response to the significant price signals provided by the feebate policy.

In this scenario, the major North American manufacturers are even more adversely affected by the feebate system. They see their revenues decline by US$3.5 billion, while the other manufacturers collectively experience an increase of US$648 million. In the United States, the feebate has a limited impact on fuel consumption. Passenger cars, light trucks, and the fleet average fuel consumption decrease by 0.1 L/100 km. Similarly, the Canada-only feebate has only a limited impact on other key variables such as consumer's surplus, vehicle sales, and revenues.

Single Pivot Points Instruments

Figure 7-4 shows the results of the various scenarios under single point conditions compared to two pivot point conditions. Keeping the feebate rate constant but using only one pivot point for all LDVs has no impact on the technology response that will be observed for each vehicle. For the consumer, a dollar received in the form of a rebate is equal to a dollar avoided in a fee. So manufacturers will put the same amount of technology in the vehicles as long as the feebate rate remains the same, regardless of where the pivot points are set. This explains why the average fuel consumption of cars and light trucks does not change when going from two to one pivot point.

Moving the pivot point will definitively have an impact on the rebate or fee that each vehicle will face, however. The change in the relative price of cars and light trucks explains the great impact that can be observed in the market shares of cars and light trucks when introducing a single pivot point. For instance, the market share of passenger cars increases by 3.3 percent with a rate of C$250 per L/100 km, and by 12.1 percent when the rate increases to C$1,000. These market shifts are responsible for the lower average fleet fuel consumption. The single pivot point also leads to slightly larger total government transactions and to a significantly larger impact on vehicle manufacturers. As in the two pivot points case, the major North American manufacturers are net fee payers, while the other manufacturers receive net rebates. The consumer's surplus, for its part, decreases by US$299 million, almost double the decline when compared to the two pivot points case. As is the case in the other Canada-only scenarios, the impact on the U.S. market is very limited.

One of the practical obstacles to the implementation of a feebate is the large number of transactions that a full feebate system would generate. One way to circumvent this problem would be to implement a "partial feebate," where only the most and least efficient vehicles would be affected. In effect, such a program could be seen as a combination of a gas-guzzler tax and a rebate on highly efficient vehicles. A new scenario was constructed to test the impact of such a program. It analyzed a partial feebate with a common pivot point for cars and light trucks and a rate of C$1,000 per L/100 km.

Under these conditions, LDVs with fuel consumption below 6.0 L/100 km were eligible for a rebate, while vehicles with a fuel consumption above 10.2 L/100 km faced a fee. These levels were chosen to make the partial feebate revenue neutral. Under such a program, the average fuel consumption of the fleet goes down to 7.9 L/100 km, at a relatively small cost—a decrease of US$79 million in consumer's surplus—to consumers. The partial feebate also leads to a significantly smaller loss of revenues for manufacturers, given the high feebate rate, especially for the major North American manufacturers. The partial feebate also limits the amount of total government transactions to US$352 million, the lowest of all the feebate options tested.

Rebates and Fees

The impact of rebate programs, using rates of C$250 and C$1,000 per L/100 km, were also considered. The rebate used two pivot points: for cars and light trucks. Rebates provide incentives for vehicles with fuel consumption below the pivot point but levy no fees on vehicles with higher fuel consumption rates. The pivot points are 6.5 L/100 km for cars and 8.6 L/100 km for light trucks. These pivot points were 25 percent below the current Canadian Company Average Fuel Consumption (CAFC) standards.

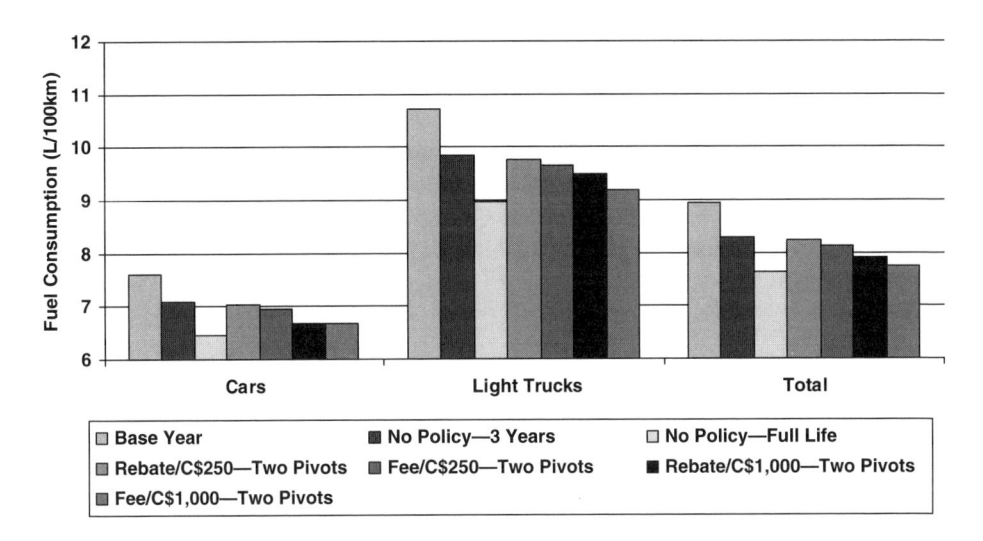

FIGURE 7-5. Impact of rebates and fees in Canada—Canada-only policies.

The results of these analyses are summarized in Figure 7-5. Using these relatively severe pivot points, the C$1,000 rebate system nevertheless results in fuel consumption averages almost as low as they would under the C$500 feebate scenario. The actual values were 6.7 L/100 km for cars and 9.5 L/100 km for light trucks, resulting in a 7.9 L/100 km fleet average.

A C$250 rebate policy would cost significantly less, US$69 million per year in rebates, but would be environmentally ineffective. With a C$250 rebate, sales increase by 0.2 percent, while revenues decrease by 0.1 percent. Only when the rebate rate is quadrupled do consumers move into cheaper, more efficient cars in a significant way as manufacturers see their revenues decrease by 0.6 percent even though sales increase by 1.3 percent. Once again, this program would have a very minor impact on the U.S. market.

If a fee was charged instead of a rebate for each LDV with a fuel consumption above the pivot points used in the rebate case, the resulting fuel consumption improvements would be larger than with the rebate. Of course, the fees would have a large impact on consumer's surplus, which would decrease compared to the no-policy case, and on manufacturers, who would see their revenues decrease.

United States-Only Instruments

The NAFAM analysis next switched to investigating the impacts of feebate scenarios implemented only in the United States. These analyses used two pivot points, for cars and light trucks, and assessed feebate policies that applied either to all of the U.S. market, or alternatively to only 25 percent of it. This proportion represents roughly the population of California and of

122 *Driving Climate Change*

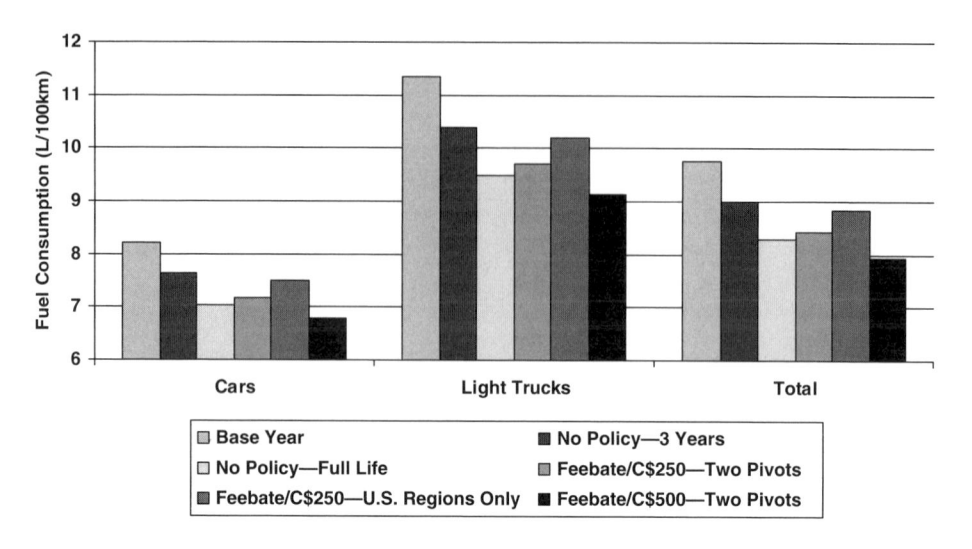

FIGURE 7-6. Impact of two pivot points feebates in the United States—U.S.-only policies.

eight northeastern states, including Connecticut, Maine, Massachusetts, New Hampshire, New Jersey, New York, Rhode Island, and Vermont. The results of these analyses are shown in Figure 7-6.

A C$250 feebate was first considered. Passenger cars and light trucks in the United States would see fuel consumption drop to 7.2 and 9.7 L/100 km, respectively, while the fleet average would fall to 8.4 L/100 km. U.S. consumers see their surplus diminish by US$460 million and total government transactions rise to US$3.7 billion per year. The market composition is only marginally affected by the feebate, and sales and revenues decline by 0.3 and 2.0 percent, respectively. The impact of the U.S. feebate policy on the Canadian market is similar to what it would be if Canada implemented a similar feebate in its market but at a smaller cost to Canadians.

Doubling the U.S. feebate rate to C$500 per L/100 km has a significant impact on the fuel consumption of new vehicles. The U.S. fleet average drops to 7.9 L/100 km. This significant improvement in fuel consumption is accompanied by a decline in consumer's surplus, an increase of annual government transactions to almost double what they were with the C$250 feebate rate, and a decrease of manufacturers' vehicle sales and revenues by 0.9 and 4.0 percent, respectively. In Canada, the impact of the U.S.-only policy is significant, dramatically improving fuel consumption. Canadian manufacturers would benefit by seeing their revenue increase by 1.1 percent, mostly because of the increase in vehicle price due to the new fuel-saving technologies being installed in vehicles.

A scenario applying a C$250 feebate to 25 percent of the U.S. market was analyzed to illustrate the effect of a state-level feebate. In this case, there are significant impacts on fuel consumption in the overall U.S.

North American Feebate Analysis Model 123

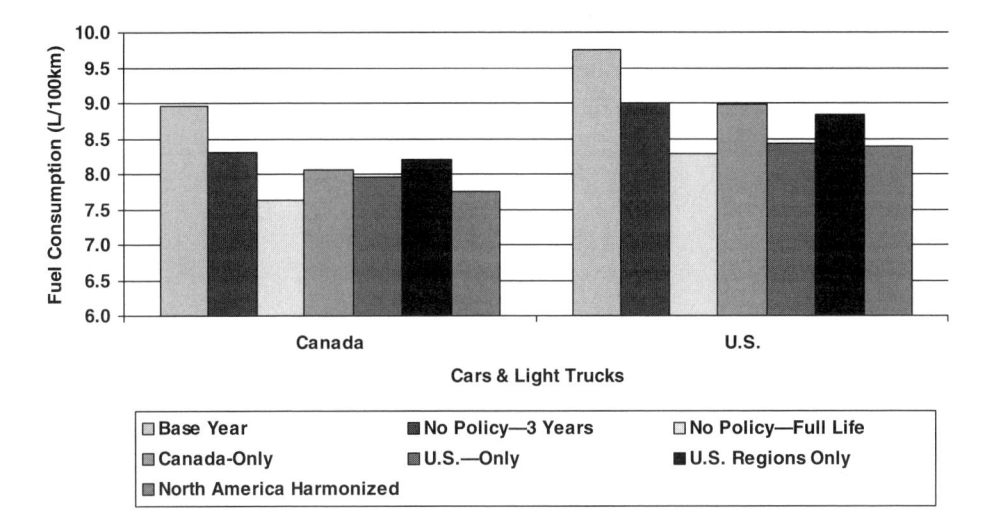

FIGURE 7-7. Impact of a two pivot points C$250 feebate on the average fleet fuel consumption—various market coverage policies.

market. Passenger cars and light trucks see fuel consumption decrease to 7.5 and 10.2 L/100 km, while the average fleet fuel consumption drops to 8.8 L/100 km. Consumers see their surplus decrease by US$119 million, while manufacturers see their sales remain roughly constant, while revenues decrease by 0.5 percent. Total government transactions are also much lower, at US$942 million. The regional feebate also has an impact on the Canadian vehicle fleet, where the average fuel consumption goes down to 8.2 L/100 km, which is only 0.2 L/100 km higher than the result that is achieved if Canada implemented a C$250 feebate on its own. Canadian consumers and manufacturers are also positively affected by the regional U.S. feebate.

North American Instruments

It is clear from the Canada- and U.S.-only scenarios just presented that the United States has a much more important impact on the Canadian market than Canada has on the United States. Figures 7-7 and 7-8 present the effect of various feebate policies on the fleet average fuel consumption for both countries with different market coverage such as Canada-only, U.S.-only, U.S. region-only, and finally a harmonized North American feebate policy. Figure 7-7 represents the impact of a C$250 feebate, while Figure 7-8 represents the impact of a C$500 feebate. Both figures show that the impact on the United States of Canada implementing a feebate on its own is small in all scenarios. Canada, however, would see the fuel consumption of it fleet improve dramatically if the United States implemented a feebate, whether the feebate policy is harmonized or not.

In the NAFAM, the adoption of fuel economy technologies in response to price signals is highly dependent on the fuel economy technology

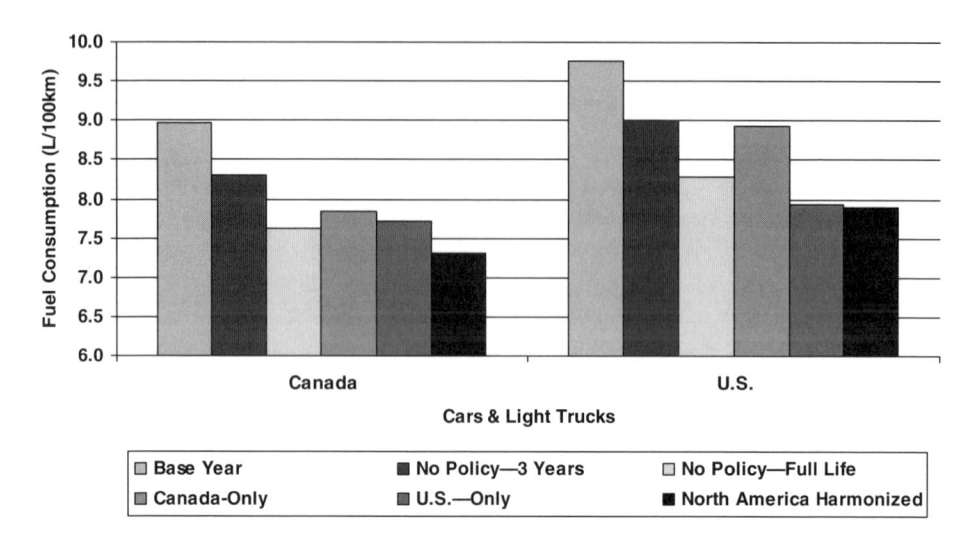

FIGURE 7-8. Impact of a two pivot points C$500 feebate on the average fleet fuel consumption—various market coverage policies.

supply/cost curves. As Canada represents a small proportion of the North American market, adoption of fuel economy technologies is always more important in the United States than in Canada when comparable policies are implemented in each country. Figure 7-9 illustrates the distribution of North American vehicle sales by fuel economy for various policies. Introducing a harmonized North American feebate reduces significantly the number of vehicles consuming more than 10 L/100 km.

Impacts of Changing the Vehicle Price Elasticities

This study used the same vehicle choice elasticities employed by Greene et al., 2000. However, those elasticities were deliberately chosen to be at the upper end of values appearing in the published literature and so might also be high for the Canadian market conditions. For this reason, a few scenarios were run with vehicle choice price elasticities at −5 and −2.5 for choice within a class or among classes, respectively. These elasticities are consistent with results reported by Bordley (1999), for example, for the United States. The overall price elasticity was maintained at −1.0 for all scenarios.

Figure 7-10 illustrates the impact of the alternative elasticities on the average fleet fuel consumption. Clearly, manufacturer revenues are much less affected by feebates when the lower elasticities are used. Revenue losses are roughly half as large. For this reason, the results of our analysis, especially the impacts on manufacturers, should be interpreted with caution because they are strongly dependent on the assumed price elasticities of vehicle choice.

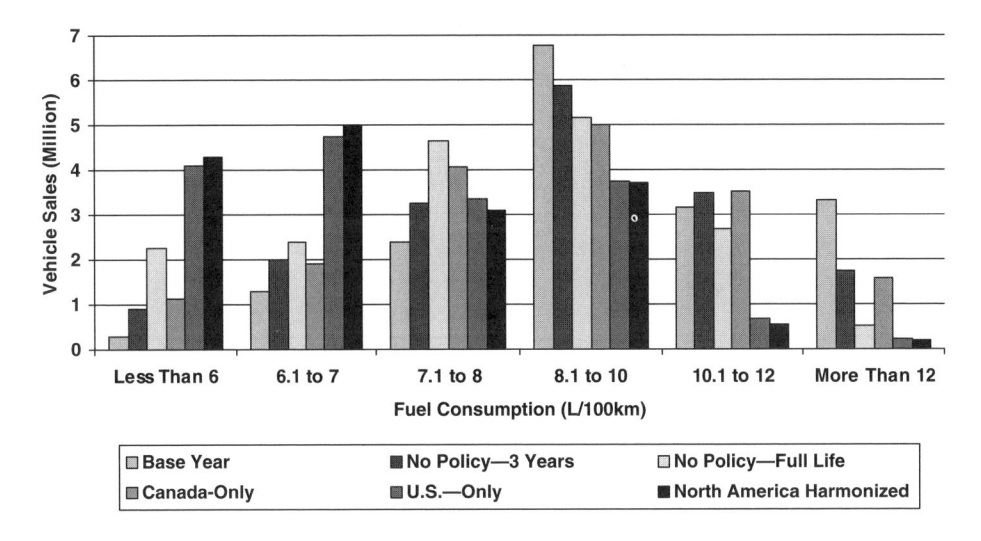

FIGURE 7-9. Effect of two pivot points C$500 feebates on distribution of vehicle sales by fuel consumption in North America—various market coverage policies.

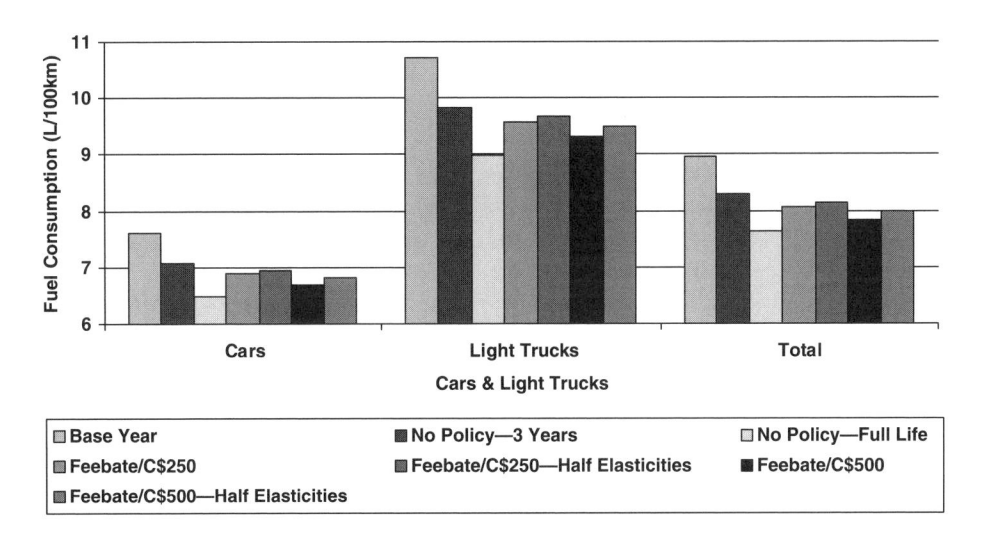

FIGURE 7-10. Impact of halving elasticities on the average fleet fuel consumption.

Conclusion

This chapter has presented the results of an analysis of the impacts of Canada-only, U.S.-only, and harmonized North American feebate policies on the North American vehicle market and on the Canadian and U.S. markets individually. The results presented here are dependent on two key assumptions:

- When modifying vehicles in response to a feebate policy, manufacturers will keep all vehicle characteristics but fuel economy constant
- The results apply to a single year, 10 to 15 years in the future, when all manufacturers have had the time to retool their facilities in answer to the feebate

If Canadian consumers do not fully value the lifetime fuel savings due to fuel economy improvements, then this market failure could be countered almost completely with a Canada-only C$500 feebate with two pivot points. If a North American harmonized feebate were introduced, the same results could be achieved with half this rate. This would result in a much smaller adverse impact on Canadian consumers and vehicle manufacturers.

This study presents for the first time the effect of Canada-only feebate policies as a change in the overall North American market demand for fuel economy. No country has ever implemented a large-scale feebate program. Consequently, some of the effects estimated in this study might be significantly different in a real world situation, thus calling for a cautious implementation of such an economic instrument.

There are still a number of issues related to feebates that would warrant further study, both in terms of data and model improvements. Although there is some evidence that there is a market failure for fuel economy technology in Canada, no research has been conducted to determine the extent to which such a market failure exists. Second, new and used vehicle price elasticities for the Canadian market have not been estimated. Although the fuel economy improvement estimates from the model are not very responsive to changes in the elasticities, the impacts on manufacturers and consumers are highly dependent on the elasticity values used. Third, the fuel economy technology cost curves used in the analysis are class averages and do not include diesel and hybrid electric technologies. Fourth, the solution provided by the model represents a long-run market equilibrium, after manufacturers have modified all their vehicles and fully retooled their plants in reaction to the feebate policy. The costs and benefits, and optimal feebate policy, as well as the impact of the policy on the overall LDV market during the transition, should also be investigated.

Author's Note

The views and opinions of authors expressed in this paper do not reflect those of the Canadian government.

References

Bordley, R. "An Overlapping Choice Set Model of Automotive Price Elasticities." Transportation Research B, vol. 28B, no. 6, 1994. pp. 401–408.

Davis, W. B., M. D. Levine, K. Train, and K. G. Duleep. *Effects of Feebates on Vehicle Fuel Economy, Carbon Dioxide Emissions, and Consumer Surplus.* DOE/PO-0031. Washington, D.C.: Office of Policy, U.S. Department of Energy, February 1995.

Energy and Environmental Analysis (EEA), Inc. *Report Automotive Technology Costs and Benefit Estimates.* Prepared for Transport Canada. Ottawa, Canada, March 2005.

Environment Canada. Canada's 2003 Greenhouse Gas Inventory.

Greene, D. L., and J. DeCicco. "Engineering-Economic Analysis of Automotive Fuel Economy Potential in the United States." Annual Review of Energy and the Environment, vol. 25, 2000. pp. 477–536.

Greene, D. L., P. D. Patterson, M. Singh, and J. Li. "Feebates, Rebates and Gas-Guzzler Taxes: A Study of Incentives for Increased Fuel Economy." Energy Policy, vol. 33, no. 6, 2005. pp. 757–775.

HLB Decision Economics, Inc. "Assessment of a Feebate Scheme for Canada." Project Number 6591 for Natural Resources Canada, National Climate Change Process. Ottawa, Canada, June 25, 1999.

National Research Council (NRC). *Effectiveness and Impact of Corporate Average Fuel Economy (CAFE) Standards.* Washington, D.C.: National Academy Press, 2002.

Natural Resources Canada. Canada's Emissions Outlook, an Update. 1999.

Small, K. A., and H. S. Rosen. "Applied Welfare Economics with Discrete Choice Models." Econometrica, vol. 49, 1981. pp. 105–130.

Train, K. *Qualitative Choice Analysis*, 1st ed. Cambridge, MA: MIT Press, 1986.

Turrentine, T., and K. Kurani. "Automotive Fuel Economy in the Purchase Decisions of Households." Presented at the 84th Annual Meeting of the Transportation Research Board. Washington, D.C., January 9–13, 2005.

CHAPTER 8

Reducing Growth in Vehicle Miles Traveled: Can We Really Pull It Off?

Gary Toth

For many years, international efforts to reduce greenhouse gas (GHG) emissions have focused on supply side fuel and vehicle technology strategies. This approach has reflected a belief that travel and land use strategies, the key alternatives to fuel and technology approaches, would be largely ineffective due to an intractable desire for ever increasing automobile use, particularly in the United States.

Work at the New Jersey Department of Transportation (NJDOT) suggests that this is not true and that there is strong public support for programs designed to reduce vehicle miles traveled (VMT). Moreover, the problem of excessive driving, fuel consumption, and GHG emissions is spreading around the world, particularly in the rapidly developing Asian nations. With 500 million new people moving into developing Asian cities over the next 20 years and rapid growth projected for parts of South America and Africa, it is questionable whether fuel supply and vehicle technology will be enough to reduce GHG emissions. Drawn on real-world experiences in New Jersey, this chapter examines the potential for restraining growth in VMT, and thus GHG emissions, by influencing the land use patterns which have characterized suburban development in the United States since World War II.

Evolving Transportation Approach to Solving Congestion

Transportation agencies in the United States have historically been reluctant to address transportation planning using strategies that influence land use. Most shared the belief that the love affair that exists in the United States between drivers and their automobiles would undermine any attempt to limit VMT. The result has been programs oriented almost exclusively to building more roads in the belief that road construction could keep ahead of traffic congestion.

Recent data suggests that outrunning traffic congestion is a Sisyphean task. For example, the 2005 Urban Mobility Report, completed by the Texas Transportation Institute (TTI), reveals that in spite of one of the biggest road building campaigns in the history of the world, congestion around the United States is worsening (Lomax and Shrank, 2005). In the 83 metropolitan areas studied by TTI:

- The time lost due to congestion annually jumped from 16 hours in 1982 to 62 hours per peak period traveler in 2000.
- The percentage of the major roadway system that is congested rose from 34 percent in 1982 to 59 percent in 2003.
- The number of hours when congestion occurs increased from 4.5 hours in 1982 to 7.1 hours in 2003.
- Sixty-seven percent of the peak period travel was congested in 2003, compared to 32 percent in 1982.
- Traffic congestion cost $63 billion in the United States in 2003 and 3.7 billion wasted hours each year.

Not only are drivers spending more time in congestion, but the ability to avoid it has virtually disappeared. When the post–World War II development and highway boom began, some commuters faced congestion on a few big highways during a relatively few peak hours. Today, in spite of major efforts at building the interstate highway system and many state and local highways, the number of congested major highways has increased, and congestion has spread to the back roads—once used to bypass the congestion—and the congestion has spread to off-peak hours. Over the last two decades, the average length of commute, miles driven, and time spent in traffic has increased at rates well above population growth. Walking is down dramatically.

The deterioration of public health has accompanied the rise in VMT and traffic congestion. Figure 8-1 shows the alarming increase in obesity rates in just 12 years from 1989 to 2001, as measured by the National Centers for Disease Control (CDC) (Jackson and Kochtitsky, 2001). The CDC has classified this rapid deterioration of public health as an inactivity epidemic and is warning that the health impacts of this inactivity include obesity, diabetes, cardiovascular disease, colon cancer, increased symptoms of depression and anxiety, and poorer development and maintenance of bones and muscles.

FIGURE 8-1. Obesity rates in the United States in 1989 compared to 2001. *Source:* From *Creating a Healthy Enviroment, The Impact of the Built Environment on Health*, Richard J. Jackson, MD, MPH; and Chris Kochtitsky, MSP; 2001.

Reasons for the Growing Traffic Congestion in the United States

These trends are the inevitable result of allowing transportation and land use systems to evolve separately over the last four to five decades. While interaction between transportation and land use is extremely complex, most state agencies are looking at a few basic reasons to explain the surge in VMT and the decline of walking:

- Induced demand
- The suburban sprawl factor
- Separated and spreading land use patterns
- The disappearance of the connected network
- Context-insensitive street designs

Figure 8-2 describes the induced demand phenomena. When paved road construction first started early in the twentieth century, it made land further out from the urban core more accessible. Cheap farmland in rural areas was more attractive to development, and people moved further out to take advantage of the newfound affordability and quiet rural lifestyle, initiating suburban sprawl. As more and more people moved out, the rural atmosphere changed to suburban, and ease of travel gave way to return of congestion. More roads were built, more cheap land was made accessible, more people moved; soon jobs and commercial uses followed. This led to more congestion, more loss of quiet atmosphere, more development, and more roads.

132 *Driving Climate Change*

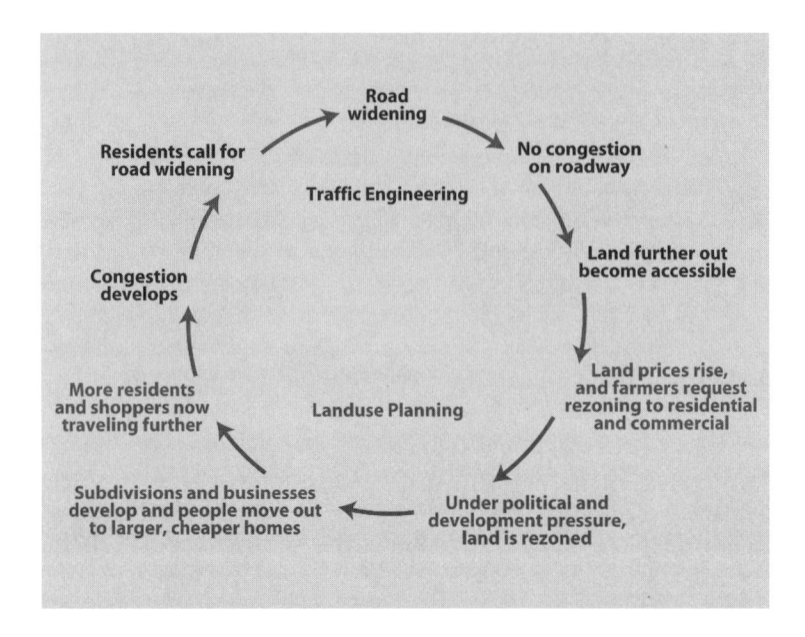

FIGURE 8-2. The endless cycle of capacity and sprawl. *Source:* Courtesy of Ian Lockwood of Glatting Jackson.

Compounding the sprawl factor is the post–World War II trend of separated and spread land uses. Prior to the rise of the automobile as the common form of transportation, land uses were, out of necessity, mixed and compact. Although many land use planners warned of the toll that this outward migration was beginning to take, few decision makers listened.

Ignoring the advice of planners and reacting to the desire to escape unattractive land uses, community planning shifted to separating land uses into specific and separate categories. This was a radical change from the traditional development scenario typical of communities created prior to World War II, which promoted land uses close to each other and connected by local streets. Convenience stores, compatible commercial uses, and neighborhood schools were located in the midst of residential areas. Not only could trips be made without using the big highway, some could actually be made on foot or bike, eliminating the demand for motor vehicles for these uses altogether.

Modern planning now, with limited exceptions, rejects the traditional integrated approach and intentionally separates and spreads different uses. Subdivisions are devoid of convenience uses and community schools, and these everyday uses are intentionally distanced from residential areas. Generally, these conveniences are placed on state and county roads, forcing these roads to bear not only through traffic but driving to local destinations as well. Road networks are intentionally disconnected and curvilinear. The

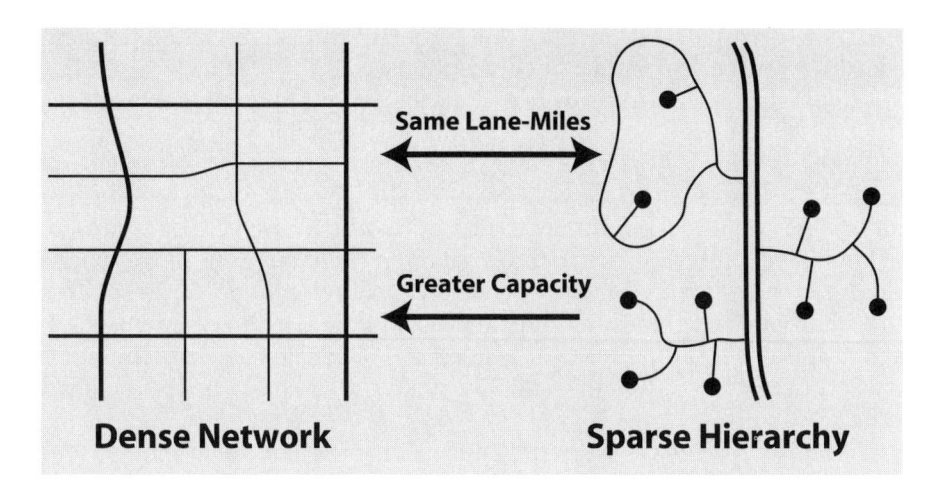

Dense Network — Same Lane-Miles — **Sparse Hierarchy**

Greater Capacity

FIGURE 8-3. Capacity of a dense and connected network versus the modern sparse hierarchal roadway system. *Source:* Courtesy of Ian Lockwood of Glatting Jackson.

desired result of lightly traveled residential areas is accomplished with the unintended consequence of runaway congestion on county and state highways.

Modern planning and circulation plans have another and perhaps more pervasive role in congestion. Traffic from isolated residential points of origin, called pods, in modern development designs can access big highways only at a few points. As residential communities grow, traffic engineers are forced to add traffic signals to more and more of these intersections, adding to the access time from the side roads. Moreover, once stopped, cars on the main roads do not instantly return to highway operating speeds when the light turns green. As a result, adding new signals to intersections, and later more access time to the side streets, cripples the capacity of the major highways.

Figure 8-3 shows that a dense and connected network, mile for mile, has more through carrying capacity than a sparse one. The latter forces out all traffic to the big highway as soon as possible and concentrates left turns at a few spots. Also, the main roads are almost always the taxpayers' responsibility, while the lesser, unconnected streets are generally built by private developers. In an era where government funds are shrinking, this failure to recapture private sector investment through sound infrastructure development is crippling efforts to reduce traffic congestion.

History of Traffic Development Patterns in the United States

Early development in America occurred around water bodies and rail lines. Without cars, the emerging urban centers needed to be compact and pedestrian oriented. The commercial, social, and recreational exchanges needed for prosperity could not occur otherwise.

In urban areas, street designs sought to improve channels of communication, drainage, sanitation, commercial success, and aesthetics. The connected network was of huge importance because it afforded good communication between various parts of the city. Drainage requirements for disposal and sanitation forced street networks to respect the contours of the land. Street frontage, block size, and the ratio of street frontage to buildable area were all important to commercial development. Because streets were front yards of properties and pathways for pedestrians, street trees and streetscaping were important for beauty and shade. Trees were encouraged between the back of the curb and the sidewalk to shield pedestrians from traffic.

When the freeways were introduced into the landscape, the automobile freed people from the need to locate near water or rail and it undercut the ability of cities to maintain connected networks. The grid pattern began to break down. Homes, which used to be located in close proximity to schools, jobs, and stores, were increasingly being built miles away.

The paradigm for street design and planning also became oriented toward the automobile. A new philosophy of road design shifted the focus toward the car, and design standards were based on streets classified according to functionality for the automobile. This new system treated the effect of roads on adjacent land uses as secondary, whether those uses were communities, open space, or ecosystems. New street design philosophy totally ignored the value of the adjacent community, with predictable results. Two examples of the incompatibility of modern land use planning, with its emphasis on the car, and pedestrian access to urban services are shown in Figure 8-4.

Can you spot the pedestrian?   *Could you cross here?*

FIGURE 8-4. Modern pedestrian unfriendly streets. *Source:* Sharon Roerty, Voorhees Transportation Center.

Changing the Paradigm

The adverse consequences of car-oriented growth on public health, the environment, traffic congestion, and social isolation are becoming clearer. The original public support for car-oriented land use decisions was motivated by a desire for freedom of travel, escape from congestion, and affordable housing. Although it worked well for a while after World War II, the policies to achieve these goals are now causing, not solving, problems.

The body of work that supports the connection between poor land use decisions and traffic congestion is growing. Studies done by Reid Ewing of the National Center for Smart Growth at the University of Maryland, for example, demonstrate that the per capita VMT in the least sprawling communities is 25 percent less than in the most sprawling (Ewing et al., 2003; Ewing, 2005).

Similarly, the body of work that supports the connection between modern land use and health is growing. A 2003 article in the *American Journal of Health Promotion* revealed that people who live in sprawling communities have higher body mass indexes, are more likely to be obese, and are more likely to have high blood pressure (Ewing et al., 2003). The *American Journal of Preventative Medicine* reported in 2004 that people who live in neighborhoods with a mix of shops and businesses within easy walking distance have a 35 percent lower risk of obesity (Frank, 2004). The *Annals of Behavioral Medicine* reported in 2003 that walkable neighborhoods encourage 15 to 30 extra minutes of walking per week, enough to lose a pound a year (Saelens, 2003).

Surveys conducted for Brooke Warrick's American Lives, a consumer market research firm, reveal that respondents expressed a greater desire for the conveniences of neighborhood life than for the amenities of middle class suburbia. Nearly four times as many respondents expressed a desire for a small cluster of convenience stores nearby or a neighborhood library than for clubhouses or dramatic entrances to their housing developments (Warrick, 1995).

A New Approach to Congestion Relief and VMT Reduction in New Jersey

A conservative estimate of the investment that it would take to relieve congestion on New Jersey's state highways indicates that it would cost over $20 billion to address congestion by the traditional method of adding lanes or building bypasses. Since the NJDOT can invest no more than $100 million per year on congestion relief, it would take 200 years to eliminate congestion by building new capacity, even if no new congestion emerges in the meantime.

With no other way to turn, the NJDOT has accepted the challenge of changing land use patterns. NJDOT's Smart Corridors Program, now called the New Jersey Future in Transportation (NJFIT) program, seeks to form partnerships with other New Jersey state agencies and local governments. The goal of the program is to build alternatives, increase transportation choices, lower design speeds, and provide more pedestrian friendly streetscapes. It also works with local jurisdictions to identify improvements to existing county and municipal roads to improve mobility. NJFIT partners with communities and developers to help revise internal circulation plans to provide connectivity between adjacent developments to facilitate movement without entering the state highway system. Finally, the program will provide planning assistance and consultant resources to local jurisdictions to help them develop land use planning alternatives that shift trips to nonautomobile modes and make better use of the local road infrastructure. This approach achieves coordination of all levels of planning activities to leverage the full value of infrastructure investments made in New Jersey transportation corridors.

The NJFIT program is based on four key principles:

- Downsizing new investment in new capacity for state highways
- Working with communities to create a connected network of local streets
- Helping communities with land use design
- Implementing context sensitive street designs (CSDs) on highways

Downsizing, or "right sizing," as Secretary Allen Biehler of Pennsylvania DOT calls it, is necessary because the huge costs of eliminating congestion at dozens of locations in New Jersey will allow only a few congestion hot spots to be fixed each decade. Causing most communities to wait decades if not a century for a solution is unacceptable, if not absurd. So in each Smart Corridor where a study is underway, NJDOT engineers are working to identify key locations where choke points are responsible for disproportionate amounts of congestion. Sometimes segments of new state highway may have to be added to support or complete the travel network, but these segments will be smaller in size and less ambitious in design approach.

Connectivity is best described in a 1990 paper entitled *Hierarchical and Connected Road Systems*. The paper indicates that a well-connected road or path network has many short links, numerous intersections, and cul-de-sacs. As connectivity increases, travel distances decrease and route options increase, allowing more direct travel between destinations, creating a more accessible and resilient system (Kulash, Anglin, and Marks, 1990).

From the 1960s through the 1990s, roadway design practices favored a poorly connected, hierarchical network, with numerous cul-de-sacs. This increases the amount of travel required to reach destinations, concentrates traffic onto fewer roads, and creates barriers to nonmotorized travel. A connected road network emphasizes accessibility by accommodating more direct travel with traffic dispersed over more roads, while a hierarchical road network emphasizes mobility by accommodating higher traffic volumes and speeds on fewer roads.

NJFIT land use policies support improved connectivity as a way to increase land use accessibility. For a particular development or neighborhood, connectivity applies both internally for streets within that area and externally for connections with arterials and other neighborhoods. Increased street connectivity can reduce vehicle travel by reducing travel distances between destinations and by supporting alternative modes. Increased connectivity tends to improve walking and cycling conditions, particularly where paths provide shortcuts, so walking and cycling are relatively faster than driving. This also supports transit use.

Traffic modeling by Kulash, Anglin, and Marks predicts that a connected road network reduces VMT within a neighborhood by 57 percent compared with conventional designs. A U.S. Environmental Protection Agency study found that increased street connectivity, a more pedestrian-friendly environment, and shorter route options have a positive impact on per-capita vehicle travel, congestion delays, traffic accidents, and pollution emissions (EPA, 2004).

Providing a well-connected street network and a backbone of strategic investment in the state highway system can go a long way to addressing congestion in a study area. However, to fully reap the benefits of that work, the land use must be arrayed in an intelligent manner to minimize unnecessary trips or trips that are unnecessarily lengthy. As part of the NJFIT program, the NJDOT, in collaboration with the New Jersey Office of Smart Growth, will provide planning assistance and consultant resources to local jurisdictions. The assistance is to help them develop land use planning alternatives that shift trips to nonautomobile modes and make better use of the local road infrastructure. Assistance will be provided in the form of in-house expertise and consultant services provided at state expense.

CSD is an important part of the NJFIT program, due to the increased reliance that the program places on local and county roads. On these roads, street design should lead drivers to adopt driving behavior appropriate to local conditions. Since vehicular speeds affect local context as surely as the physical dimensions of the street, roadway designers should carefully consider the appropriate target speed for a roadway section. This consideration must be based on land use conditions, building densities, the environment, and the disparate needs of the residents and the users of the facility. Streets not only serve transportation-related functions but are also places of commercial and social encounter. Therefore, designers should also consider the nonvehicular uses of a roadway and seek consistency between all aspects of the roadway, its environment, and the chosen design speed.

There is a wide range of options available to the designer to do so, including some that fall under the umbrella "traffic calming." These include neckdowns, rotaries, and speed humps; however, these could also include narrow lanes and shoulders, and curvilinear alignments.

The tendency to develop streets that are wide, flat, and straight in search of safety will sometimes lead to inappropriate vehicle operating speeds, particularly in downtown or "Main Street" environments. In these locations, where the true intent may have been for a slower pace of traffic, street design needs to support pedestrian safety and acknowledge the importance of pedestrian quality of life, and related socioeconomic factors. Additionally, there is evidence emerging that wider roadways and faster speeds during traffic yielding are not always safer, even when viewed strictly from a motorist's perspective.

Therefore, when working with local road networks, designers need to consider the adjacent land use and desired function of the road. The streets need to be designed to be sensitive to these contexts and encourage the intended operating speeds. Modern roadway design, particularly as it relates to secondary and tertiary streets, needs to carefully weigh whether the use of these elements creates a desirable balance between the competing interests of adjacent land use, nonmotorized transportation, and motor vehicles.

The NJDOT expected a poor reception to its new direction for congestion relief and VMT reduction. It assumed that the public would balk at the deemphasis of roadway expansion as the prime solution to congestion. It also expected local jurisdictions to oppose the efforts of any state agency attempting to influence their local land use planning. Instead, the NJFIT received an unexpected welcome, and officials in most cities have embraced the effort. It appears that local communities are being overwhelmed with development applications and are working under tight budgets that in most cases do not allow them to plan effectively.

Project Examples from New Jersey

Some examples of local programs growing out of the NJFIT statewide initiative include the following success stories.

For Route 31 in Flemington and Raritan, an extensive and connected local road network has been planned, and a slower-moving, two-lane rural parkway has been proposed to replace the abandoned freeway bypass concept. This network is not only less costly, but it also spreads its investment in congestion relief over 10 to 15 years. The original bypass solution would have not only cost the New Jersey taxpayers almost double, $150 million as opposed to $70 million for the network, it would have concentrated the investment over two fiscal years. Furthermore, the bypass solution would have squandered the opportunity to recapture the developer infrastructure for public use.

Similarly, in Trenton, a new network has been proposed for the Trenton Waterfront Redevelopment area. The existing land use is a series of huge parking lots and state offices that, combined with the Route 29 Freeway, currently severs Trenton from its waterfront. Conversion of Route

29 into a boulevard and the addition of a local street network will diffuse traffic.

For the 12 communities in the Route 9, Ocean County, corridor, some jurisdictions were skeptical at first and held back on cooperation and participation for several months. In other places, NJDOT and OSG support was immediately welcomed. One local planner told the NJDOT, "It's about time someone from the state came here to help us." After a few months, the NJDOT lead engineer was getting so many invitations to attend local planning board meetings that his other projects had to be reassigned.

On Route 31 in Hopewell, local officials were initially incensed at the suggestion that the solution to their traffic problems might lie in their land use planning. Six months later, local officials were working side by side with NJDOT engineers and OSG planners to resketch the future of their community.

In the Route 33 Smart Corridor project, local officials in Hamilton agreed to work with the NJDOT from the beginning and the lead engineer now gets calls from the Hamilton mayor almost weekly, asking for help with another land use development somewhere in the town.

In Manalapan, the NJDOT and OSG have helped local officials and developers reshape new development adjacent to the Monmouth Battlefield. The original plan for the area called for almost 2,000 new homes to be built in four unconnected pods. All travel to and from each pod would have to be made on the adjacent county roads and highways. No street connections would have been made to a new "lifestyle shopping center" planned to be located adjacent to the new residential areas. Although the two land uses would literally be within shouting distance of each other, the street plan would force everyone into their cars. The new plan for the Monmouth Battlefield area provides for multiple and walkable connections between all residential areas and the new commercial development. Furthermore, the commercial area has been replanned to create a town center for Manalapan, a feature that is currently missing from and desired by the community.

Other State Initiatives

Other progressive transportation agencies have embarked on programs similar to the NJFIT program. For example, over the past few years, the Pennsylvania DOT has collaborated with the Delaware Valley Regional Planning Commission to undertake several cutting edge integrated transportation and land use strategies studies. They have collaborated in support of new street connections, sought context sensitive solutions to reinforce historical main streets, and provided land use planning expertise to communities. Their work has sought to change densities, mix uses, and foster development that inspires nonautomotive modes of transportation.

The New Hampshire DOT (NHDOT) has also stepped out of traditional roles and helped prepare a manual on smart growth in New Hampshire. It also began to make the transportation and land use connections on projects such as Interstate 93 (I-93) and State Route 16. On I-93, NHDOT provided communities with $2 million for integrated transportation and land use planning.

More recently, the Portland Area Comprehensive Transportation Study (PACTS) enacted bold policy and priorities regarding transportation investments. PACTS, which is the Metropolitan Planning Organization for the Portland, Maine, region, now requires any arterial corridor roadway project that will reduce commuter travel times between an urbanized and a nonurbanized area be accompanied by a land use plan that preserves the arterial's capacity, protects its mobility function and the public investment, and minimizes sprawl.

The Vermont Agency of Transportation recently had a major circumferential highway project stopped by the courts after several decades of study. The Chittenden County Circumferential Highway was remanded back to the state for a reexamination, due to the failure of the proponents to adequately assess the induced and cumulative growth impacts of the highway. This is the first instance that a state agency was found to violate federal environmental laws primarily based on land use and secondary impact assessment considerations.

Conclusion

The NJDOT followed the traditional approach of trying to widen existing roads or build new roads in response to congestion for many years. It has recently faced the reality that this solution has not been working. Congestion continues to increase on the state highway system and has spread to secondary and tertiary roads. This is not unique to New Jersey but is manifesting itself all around the country.

With most transportation agencies facing the need to focus on aging infrastructure programs, the option of building more roads faster is not only unlikely but impossible. This is borne out by the 2005 Urban Mobility Report, which indicates that from 1982 to 2001, states have been able to provide only 41 percent of the new capacity needed to abate congestion (Lomax and Shrank, 2005).

The NJDOT turned to a new approach for dealing with congestion and reducing vehicle miles traveled, an approach based on proactive integration of transportation and land use planning. Attempts to influence the demand for automobile usage are not being met by the public resistance many predicted. Public willingness to accept the new paradigm supports the conclusions that sound planning and integration of transportation and land use can in fact reduce VMT.

Acknowledgments

I would like to thank a number of people. Fred Kent, cofounder and president of the Project for Public Spaces, taught me about the importance of public spaces and helped me learn how to scope and design transportation projects to enhance and not detract from the public realm. Ian Lockwood, formerly the transportation engineer for the city of West Palm Beach and currently with the firm of Glatting Jackson, et al., has been my mentor in understanding the relationship of good street and road design to helping improve our communities.

David Burwell, formerly of the Surface Transportation Project and currently with the Project for Public Spaces, provided me with consistent encouragement and inspiration as we continue to move forward and push toward the "Tipping Point." Robin Murray, formerly of the New Jersey Office of Smart Growth and currently with the School of Architecture at the New Jersey Institute of Technology, has been my primary mentor in the field of Urban and Land Use Planning. Walter Kulash of Glatting Jackson et al., in addition to much inspiration and education, also allowed me to use verbatim a paper on Street Connectivity entitled "Hierarchical and Connected Road Systems" (Kulash, Anglin, and Marks, 1990). Ansel Sanborn, director of Planning for the New Hampshire DOT, provided graphics and inspiration.

Thanks also to Reid Ewing, formerly with the Voorhees Transportation Center at Rutgers University in New Jersey and currently with the National Center on Smart Growth, for much inspiration and information sharing in reports, articles, and presentations.

Thanks to the following for editorial assistance: Dominic Critelli of NJDOT, David Burwell and Andy Wiley-Schwartz of the Project for Public Spaces, and Ian Lockwood of Glatting Jackson. Thanks to Carol Murray, commissioner, New Hampshire DOT, and Allen Biehler, secretary, Pennsylvania DOT, for their leadership, support, and inspiration in changing the paradigm for transportation planning in the United States. And thanks to Assistant Commissioner for Planning at the New Jersey Department of Transportation (NJDOT) Dennis Keck, who has supported me throughout my 32-year career at NJDOT.

Last, but not least, thanks to my current commissioner at the New Jersey Department of Transportation, Jack Lettiere. Jack has not only provided leadership for a new way of business at NJDOT, but he has also fostered the concepts while president of the American Association of State Highway and Transportation Officials. Without Jack's encouragement and, more important, his support, every step of the process would have been difficult if not impossible.

References

Duany, A., E. Plater-Zyberk, and J. Speck. *Suburban Nation: The Rise of Sprawl and the Decline of the American Dream.* North Point Press, 2000.

Ewing, R., T. Schmid, R. Killingsworth, A. Zlot, and S. Raudenbush. "Relationship Between Urban Sprawl and Physical Activity, Obesity, and Morbidity." *American Journal of Health Promotion*, 18(1): 47–57, September/October 2003.

Ewing, R. "Generalizing from Sacramento: What Is Really Possible." Presentation made at the 2005 Asilomar conference (www.its.ucdavis.edu/events/outreachevents/asilomar2005/presentations/Ewing.pdf), August 2005.

Frank, L. D. "Obesity Relationships with Community Design, Physical Activity, and Time Spent in Cars." *American Journal of Preventive Medicine*, vol. 27, issue 2, August 2004.

Garry, G. "Potential Travel Reductions with Managed Growth: A Sacramento Case Study." Presentation made at the 2005 Asilomar conference (www.its.ucdavis.edu/events/outreachevents/asilomar2005/presentations/Garry.pdf), 2005.

Jackson, R. J., and C. Kochtitsky. *Creating a Healthy Environment: The Impact of the Built Environment on Health.* Washington, D.C.: Sprawl Watch Clearinghouse, 2001.

Kulash, W., J. Anglin, and D. Marks. "Traditional Neighborhood Development: Will the Traffic Work?" *Development* 21, July/August 1990.

Lomax, T. J., and D. Schrank. "The 2005 Annual Urban Mobility Report." College Station, Texas: Texas Transportation Institute (mobility.tamu.edu/ums/report), May 2005.

Saelens, B. E., J. F. Sallis, and L. D. Frank. "Environmental Correlates of Walking and Cycling: Findings from the Transportation, Urban Design, and Planning Literatures." *Annals of Behavioral Medicine*, 25(2): 80–91, Spring 2003.

USEPA. "Characteristics and Performance of Regional Transportation Systems." Document Number EPA 213-R-04-001. 2004.

Warrick, B. *Survey of Surveys.* American Lives, 1995.

CHAPTER 9

International Comparison of Policies to Reduce Greenhouse Gas Emissions from Passenger Vehicles

Feng An

Recent world events in the oil market, natural disasters, and Mideast conflicts bring renewed attention in the United States to energy security and climate changes. Securing energy from developing countries such as China and India, coupled with hurricanes on the Gulf Coast, generated a "perfect storm" in late 2005 that pushed oil prices over $70 per barrel and retail gasoline prices spiked over $3 per gallon in the United States.

Oil demands have been steadily increasing not just in rapidly motorizing developing countries but also in the developed world. Oil demand growth is primarily driven by the growth in passenger vehicle population and total vehicle miles of travel in all regions of the world. Greenhouse gas (GHG) emissions associated with passenger vehicle uses not only are soaring in non-Kyoto countries such as the United States and developing countries, but also threaten the commitments to the Kyoto treaty by the European Union (EU) nations, Japan, and Canada.

How to control energy demand and GHG emissions from personal use vehicles becomes a major challenge faced by today's world. Clearly, curbing vehicle growth, reducing travel demand, and improving vehicle fuel efficiency are three key elements to reducing oil demand. Indeed, a wide variety of approaches to address these three areas have been introduced in different parts of the world.

Nearly every major country in the world has established a program to address climate change resulting from transportation emissions. Most of these programs are more ambitious than the program underway in the United States. Fuel economy programs and GHG emission targets, either mandatory or voluntary, have proven to be among the most effective tools in controlling oil demand and GHG emissions from the transportation sector. While fuel economy standards for passenger vehicles have been largely stagnant in the United States over the past two decades, the rest of world—especially EU nations, Japan, and recently China and the U.S. state of California—has moved forward, establishing or tightening GHG or fuel economy standards. This chapter reviews and compares the programs under way around the world.

In a broader sense, fuel economy programs include both numeric standards and fiscal incentives to improve energy efficiency of individual vehicles per unit of travel distance. In today's technology-driven world, new technologies offer great promise to drastically improve vehicle fuel economy. However, realizing such technological promise has proven to be a big challenge. Historical trends in United States have clearly demonstrated that technological advancement tended to be used to boost vehicle size and performance over fuel economy, given a lack of regulatory pressure, as trends demonstrated from the mid-1980s to today. However, technology development has also been capable of responding to regulatory requirements to improve vehicle fuel economy, when such requirements were in place from mid-1970s and mid-1980s (An and DeCicco, 2005).

Fiscal incentive programs have improved fuel economy or reduced fuel use, especially in combination with standards. Incentives can be directed at improving the efficiency of the vehicle fleet, through variable registration fees or taxes, or at limiting vehicle use, through fuel taxes and road use fees. Many European countries have established vehicle tax systems either based on engine size, fuel efficiency, or carbon dioxide (CO_2) emission rates, in support of mandatory standards. Higher fuel taxes in the EU reinforce efforts on the part of automakers to meet voluntary GHG emission targets. Taxes are a major factor in the predominance of smaller and more fuel-efficient vehicle models and the limited growth in vehicle miles traveled (VMT) in Europe. Table 9-1 summarizes major approaches for the purpose of reducing automobile fuel consumption and GHG emissions.

Comparison of Vehicle Standards around the World

Research at Energy and Transportation Technologies, LLC, indicates that at least nine countries and regions around the world have established or proposed their own motor vehicle fuel economy or GHG emission standards, as shown in Table 9-2. Motor vehicle fuel economy standards have been established for most of the developed world, including the United States, EU nations, Japan, Canada, and Australia. The EU has also negotiated voluntary vehicle CO_2 emission rate targets as a means to control GHG

TABLE 9-1. Measures to Promote Fuel-Efficient Vehicles Around the World

Approach	Measures/Forms		Country/Region
Standards	Fuel economy	Numeric standard averaged over fleets or based on vehicle subclasses	U.S., Japan, Canada, Australia, China, Taiwan, South Korea
	GHG emissions	Grams/km or grams/mile	EU, California
Fiscal Incentives	High fuel taxes	Fuel taxes at least 50% greater than crude price	EU, Japan
	Differential vehicle fees and taxes	Tax or registration fee based on engine size, efficiency & CO_2 emissions	EU, Japan
	Economic penalties	Gas guzzler tax	U.S.
Support for new technologies	R&D programs	Funding for advanced technology research	U.S., Japan, EU
	Technology mandates and targets	Sales requirement for ZEVs	California
Traffic control measures	Incentives	Allowing hybrids to use HOV lanes	California, Virginia, and others states in the U.S.
	Disincentives	Banning SUVs on city streets	Paris

Source: Based on Table 1 in An & Sauer, 2004.

TABLE 9-2. Fuel Economy and GHG Standards for Vehicles Around the World

Country/Region	Type	Measure	Structure	Test Method	Implementation
United States	Fuel	mpg	Cars and light trucks	U.S. CAFE	Mandatory
European Union	CO_2	g/km	Overall light-duty fleet	EU NEDC	Voluntary
Japan	Fuel	km/L	Weight-based	Japan 10–15	Mandatory
China	Fuel	L/100-km	Weight-based	EU NEDC	Mandatory
California	GHG	g/mile	Car/LDT1 and LDT2	U.S. CAFE	Mandatory
Canada	Fuel	L/100-km	Cars and light trucks	U.S. CAFE	Voluntary
Australia	Fuel	L/100-km	Overall light-duty fleet	EU NEDC	Voluntary
Taiwan, South Korea	Fuel	km/L	Engine size	U.S. CAFE	Mandatory

Source: Energy and Transportation Technologies, LLC.

emissions. The state of California in the United States has also recently proposed its own GHG emission standards for vehicles. China and South Korea have their own recently adopted new vehicle fuel efficiency standards, while Taiwan has had its own fuel economy standards for more than a decade.

Directly comparing vehicle standards among different regions and countries is challenging. Different countries and regions have chosen to adopt different fuel economy or GHG standards for various historic, cultural, and political reasons. These standards differ in stringency—by their apparent forms and structures and by how the vehicle fuel economy or GHG emission levels are measured—that is, by testing methods. They also differ by implementation requirements, such as mandatory versus voluntary approaches.

Automobile fuel economy standards can take many forms, including numeric standards based on vehicle fuel consumption, such as liters of gasoline per hundred kilometers of travel (L/100-km) or fuel economy, such as miles per gallon (mpg), or kilometers per liter (km/L). Automobile GHG emission standards are usually expressed as grams per kilometer (g/km) or grams per mile (g/mile). Test methods include the U.S. Corporate Average Fuel Economy (CAFE) test, New European Drive Cycle (NEDC) test, and the Japan 10-15 Cycle test.

Comparison of Countries and Regions

Recently announced fuel economy regulations by the Chinese government have inspired new interest in analyzing and understanding fuel economy and GHG programs around the world. An and Sauer recently wrote a report published by the Pew Center called "Global Climate Changes: Comparison of Passenger Fuel Economy and GHG Emissions Standards around the World" (An and Sauer, 2004). In the report, they proposed a methodology to directly compare fleet average fuel economy of passenger vehicle fleets in different regions and countries. The significance of the report is that, prior to the study, fuel economy programs in different countries and regions had largely been isolated issues. These international comparisons have put these programs in the spotlight and put pressures on countries that either are lagging behind or lack the standards of the rest of the world.

The three largest automobile markets—the United States, the EU, and Japan—approach the regulation of fuel economy quite differently. The United States uses the CAFE standards, which require each manufacturer to meet specified fleet average fuel economy levels for cars and light trucks. Canada's automobile industry has voluntarily agreed to follow the U.S. CAFE standards in Canada.

In Japan and China, fuel economy standards are based on a weight classification system, where vehicles must comply with the standard for their weight class. Similarly, the fuel economy standards in Taiwan and South Korea are based on an engine size classification system. However, China is following testing procedures developed by the EU, and Taiwan and Korea are following testing methods that are similar to U.S. CAFE procedures. Japan maintains its own test procedures.

In the EU and Australia, the automobile industry has signed a voluntary agreement with the government to reach an overall fleet average fuel economy or CO_2 emissions level by a specific date. The entire industry must meet one target. This contrasts with the U.S. CAFE approach where each company must individually meet standards for cars and light trucks. Tracking of compliance in EU nations is left up to the Association des Constructeurs Européens d'Automobiles (ACEA) and the other automaker associations.

In order to create comparisons among the programs underway in different countries, the vehicle fuel economy or GHG standards must first be converted into fleet averages, using the methodology developed by An and Sauer. For standards already designed as fleet averages, including those in the United States, EU, and Australia, this step is not necessary. For regions with standards designed by categories—such as vehicle type, weight, or engine size—this analysis assumes that the vehicle fleet mix in each country stays constant from 2002 throughout the time period analyzed. In other words, the comparisons do not address the implications of changing the vehicle size or weight composition of the current fleet. Next, the U.S. CAFE equivalent mpg and EU NEDC equivalent standard measuring grams of CO_2 per kilometer (km) are selected as the reference standards. Finally, conversion factors to convert local standards to the reference standards are developed and applied where necessary.

Figures 9-1 and 9-2 show comparisons of fuel economy and GHG emission standards normalized around metrics and vehicle test cycles as described in the preceding procedure. These figures show that the EU and Japan have the most stringent standards and that the United States and Canada have the weakest standards in terms of fleet-average fuel economy rating. These figures also show that the United States and Canada also have the highest CO_2 emission levels based on EU testing procedures. If the California GHG standards go into effect, they would narrow the gap between U.S. and EU standards, but the California standards would still be less stringent than the EU standards.

Figure 9-3 shows that the EU, China, Canada, and California all will have fleet average fuel economy improvements within the next decade equal to or greater than 25 percent over their corresponding 2002 baseline cases. Figure 9-4 shows the fleet average GHG and fuel reduction over 2002 baseline year for these countries and regions.

148 *Driving Climate Change*

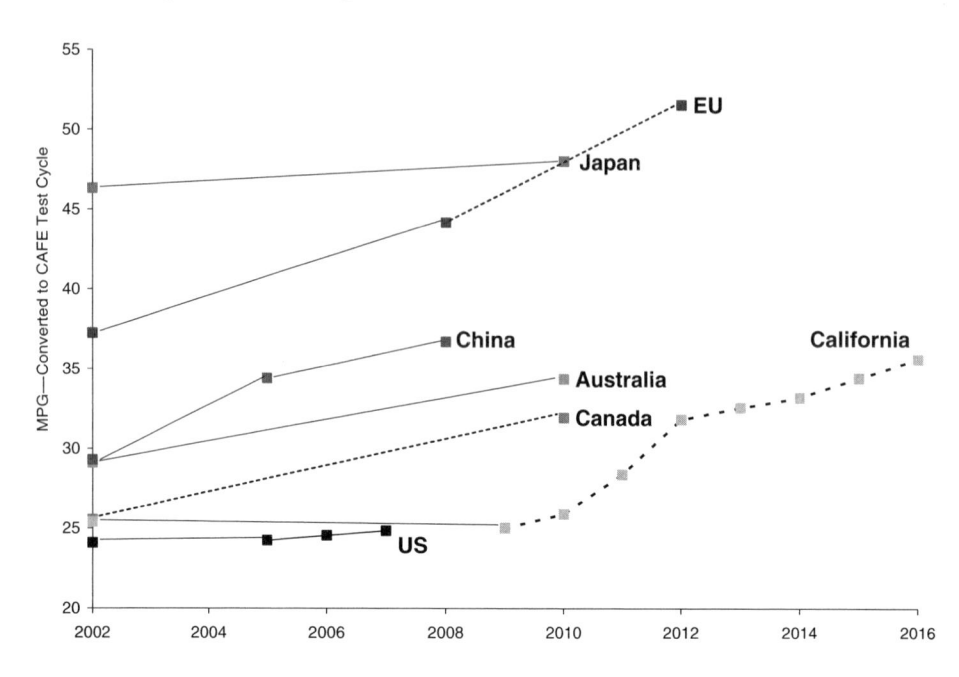

FIGURE 9-1. Comparison of fuel economy and GHG emission standards normalized by CAFE-converted mpg. *Source:* An and Sauer, 2004. *Note:* Dotted lines denote proposed standards.

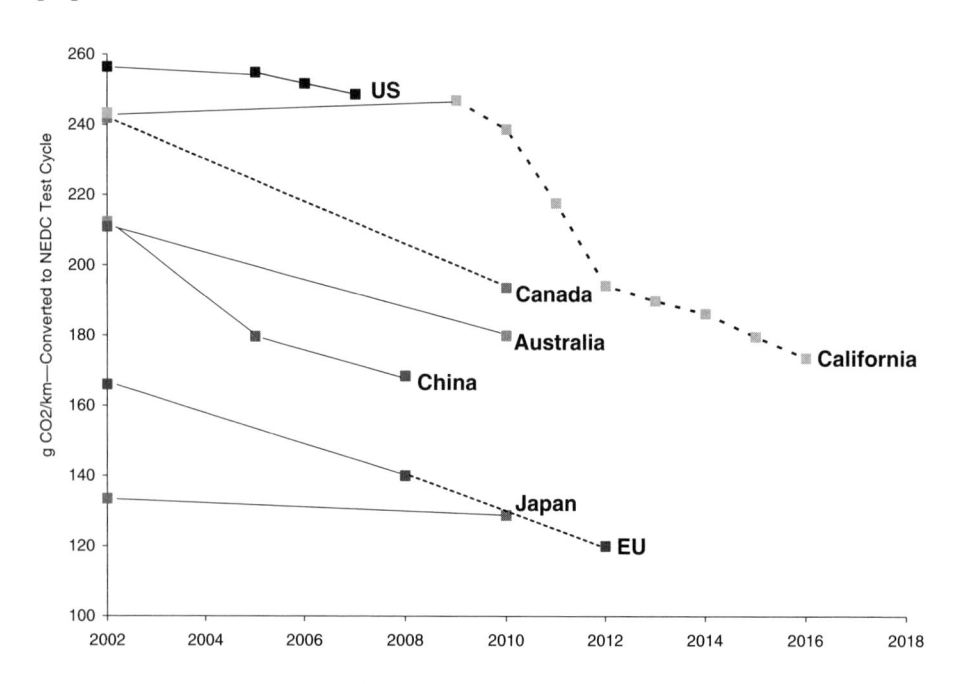

FIGURE 9-2. Comparison of fuel economy and GHG emission standards normalized by NEDC-converted g CO_2/km. *Source:* An and Sauer, 2004. *Note:* Dotted lines denote proposed standards.

International Comparison of Policies 149

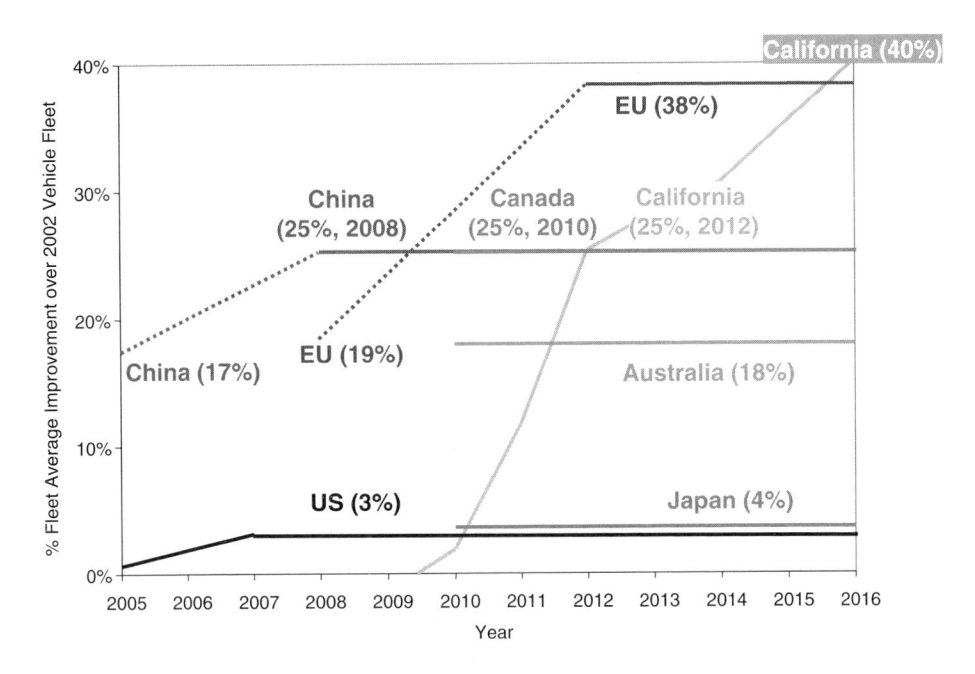

FIGURE 9-3. Fleet average fuel economy improvements over the 2002 level. *Source:* Feng An, Energy and Transportation Technologies LLC.

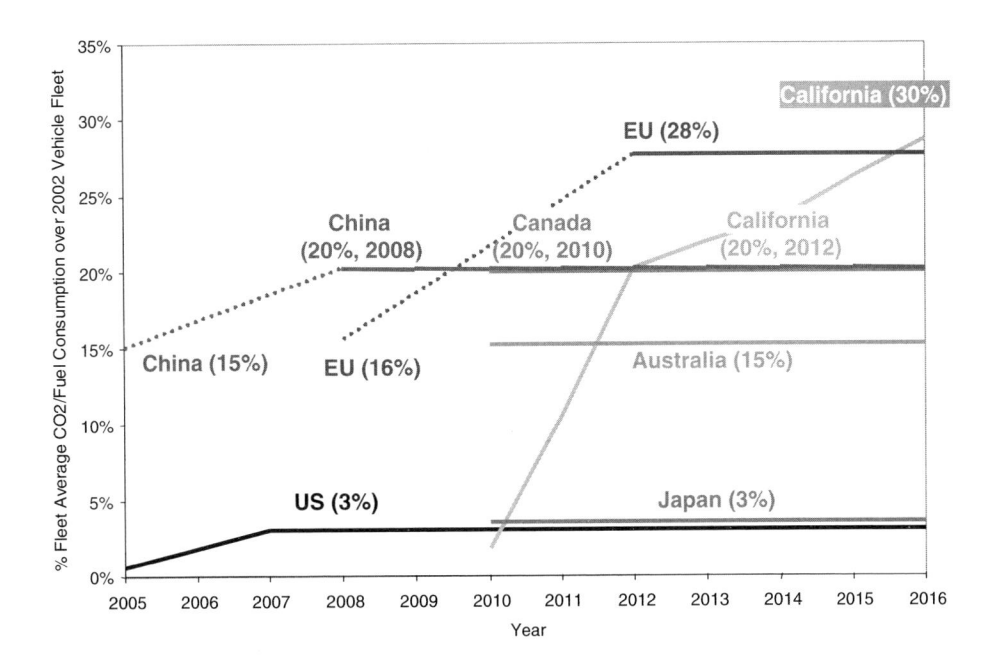

FIGURE 9-4. Fleet average GHG/fuel reduction over the 2002 level. *Source:* Feng An, Energy and Transportation Technologies LLC.

The international comparison clearly highlighted the fact that the fuel economy and GHG emission performance of the U.S. automobile fleet—both historically and projected based on current policies—lag behind most other nations. The United States not only has the lowest standards in terms of fleet-average fuel economy rating and the highest GHG emission rates based on the EU testing procedure but also has the lowest percentage improvement targets in the foreseeable future.

Country and Regional Profiles

More detailed profiles of the programs in effect in the countries and regions that have established or proposed, in the case of California, vehicle GHG emission or fuel economy standards are included in this section.

The United States

The United States was the first country to establish vehicle fuel economy standards. In the wake of the 1973 oil crisis, the U.S. Congress passed the Energy Policy and Conservation Act of 1975 with the goal of reducing the country's dependence on foreign oil. Among other things, the act established the CAFE program, which maintains an important distinction between passenger cars and light trucks, with each having their own standard. Under the regulations, passenger cars are classified as any four-wheeled vehicle not designed for off-road use that transports ten people or fewer. Light trucks, on the other hand, include four-wheeled vehicles that are designed for off-road operation or vehicles that weigh between 6,000 and 8,500 pounds and have physical features consistent with those of a truck.

The distinction between cars and light trucks was originally included in the CAFE legislation when light trucks were a small percentage of the vehicle fleet, with the most common light truck being pickups, used primarily for business and agricultural purposes. Since that time, however, the distinction between passenger cars and light trucks has become increasingly fuzzy, in part because automakers have introduced crossover vehicles that combine features of both cars and light trucks. Meanwhile, light-duty vehicles classified as trucks, such as minivans and sport utility vehicles (SUVs), are used primarily as personal transport vehicles. The result has been a 7 percent decrease in the overall light-duty fleet fuel economy since 1988, associated with the rapid growth of light trucks used as passenger vehicles beginning in the mid-1980s (EPA, 2004).

The CAFE standard for passenger cars has remained unchanged since 1985 at 27.5 mpg, although the standard was rolled back for several years in the late 1980s in response to petitions filed by several automakers (Union of Concerned Scientists, 2003). The standard for light trucks has recently been increased from the existing standard of 20.7 mpg in 2004 to 21.0 mpg for 2005, 21.6 mpg for 2006, and 22.2 mpg for 2007 (Federal Register, 2003).

TABLE 9-3. Examples of Proposed Size-based Fuel Economy Target

Footprint	Fuel Economy Target			
	2008	2009	2010	2011
20	28.5	30.0	29.9	30.4
30	28.2	29.5	29.6	30.2
40	26.7	27.6	27.9	28.6
50	23.3	23.9	24.3	24.4
60	20.8	21.6	21.9	22.2
70	20.1	21.0	21.3	21.8
80	20.0	20.9	21.2	21.8
100	20.0	20.9	21.2	21.8

Source: National Highway Traffic Safety Administration 49 CFR Parts 523, 533 and 537 [Docket No. 2006-24306] RIN 2127-AJ61 Average Fuel Economy Standards for Light Trucks Model Years 2008–2011.

"In April 2006, the National Highway Traffic and Safety Administration (NHTSA) adopted a reformed CAFÉ scheme that is based vehicle size defined by light-truck footprints (areas between four wheels). A complicated formula correlate fuel economy targets with vehicle sizes would be applied. Example of the new schemes is shown in Table 9-3 [Federal Register, 2006]". For the first three years, from 2008 through 2010, however, manufacturers can choose between size-based targets and truck-fleet average targets of 22.5, 23.1, and 23.5 mpg, respectively.

The short- and long-term impacts of the newly proposed rules are still unclear. However, an analysis by the NHTSA shows that, as a result of the different compositions of automakers' truck fleets, each company would have its own fuel economy targets, as shown by Table 9-4. The table shows that the major U.S. automakers—DaimlerChrysler (DCX), General Motors (GM), and Ford—and Nissan, the Japanese automaker, would have the lowest fuel economy targets among all automakers.

California

California has long been a world leader in imposing stringent vehicle tailpipe criteria pollutions. Frustrated by the lack of efforts and substantial progress toward tightening CAFE standards at the federal level, in 2002, California enacted legislation directing the California Air Resources Board (CARB) to achieve the maximum feasible and cost-effective reduction of GHGs from California's motor vehicles. The standard will take effect with the 2009 model year passenger vehicles. The states of New York, Massachusetts, New Jersey, Maine, Connecticut, Rhode Island, Vermont, and Washington have all recently approved adopting the California regulation for their use (Bernton, 2004). Canada has also expressed its intention to follow California's lead.

TABLE 9-4. Estimates of Required Fuel Economy Levels and Gains Based on the Proposed Target Levels and Current Information

	Fuel Economy Targets (MPG)				MPG Gains over 2008		
	2008	2009	2010	2011	2009	2010	2011
Hyundai	24.2	25.9	25.7	26.3	7.0%	6.2%	8.7%
BMW	23.8	24.8	25.1	25.7	4.2%	5.5%	8.0%
Toyota	23.2	24.1	24.5	25.0	3.9%	5.6%	7.8%
VW	22.7	23.9	24.3	24.8	5.3%	7.0%	9.3%
Honda	23.1	24.0	24.2	24.8	3.9%	4.8%	7.4%
DCX	22.8	23.5	23.7	24.2	3.1%	3.9%	6.1%
GM	22.2	22.8	23.2	23.7	2.7%	4.5%	6.8%
Nissan	22.1	22.8	23.2	23.7	3.2%	5.0%	7.2%
Ford	22.4	22.9	23.1	23.6	2.2%	3.1%	5.4%

Source: Federal Register, 29 CER Part 533, Table 7, Light Trucks, Average Fuel Economy; Model Years 2008–2011; Proposed Rules, August 2005.

Calculations suggest that these states, including California, and Canada represent approximately 30 percent of all cars sold in North America, excluding Mexico (Ward's Vehicle Facts & Figures, 2003).

CARB has proposed near-term standards to be phased in from 2009 through 2012, and midterm standards to be phased in from 2013 through 2016. The GHG emission standards will be incorporated directly into the current low-emission vehicle (LEV) program, along with other light- and medium-duty automotive emission standards. The LEV program applies to passenger cars, light-duty trucks, and medium-duty vehicles weighing 8,500 to 10,000 pounds, and it establishes exhaust emission standards. Accordingly, there would be a GHG emission fleet-average requirement for the passenger car/light-duty truck 1 (PC/LDT1) category, which includes all passenger cars regardless of weight and light-duty trucks weighing less than 3,750 pounds equivalent test weight (ETW). The second category is light-duty truck 2 (LDT2) for light trucks weighing between 3,751 pounds ETW and 8,500 pounds gross vehicle weight (GVW). ETW includes the vehicle curb weight plus passenger weight of 300 pounds and rounded by every 250 pounds. GVW is mostly used for Class 2b to Class 8 trucks, including vehicle curb weight plus rated vehicle load. Furthermore, vehicles weighing 8,500 to 10,000 pounds that are classified as medium-duty passenger vehicles (MDPVs) will be included in the LDT2 category for GHG emission standards.

The legislation will be phased in for both the near-term and medium-term standards. Table 9-5 outlines the GHG emission standards approved by CARB.

The California legislation also authorizes the granting of emission reduction credits for any reductions in GHG emissions achieved in model

TABLE 9-5. California Air Resources Board Approved Standards

Time Frame	Year	GHG Emission Standard (g/mi)		CAFE-Equivalent Standard (mpg)	
		PC/LDT1	LDT2	PC/LDT1	LDT2
Near-term	2009	323	439	27.6	20.3
	2010	301	420	29.6	21.2
	2011	267	390	33.3	22.8
	2012	233	361	38.2	24.7
Medium-term	2013	227	355	39.2	25.1
	2014	222	350	40.1	25.4
	2015	213	341	41.8	26.1
	2016	205	332	43.4	26.8

Source: California Environmental Protection Agency Air Resources Board, August 2004.

year 2000 through 2008 vehicles built prior to the date the regulations take effect. Under the early credit proposal, manufacturer fleet average emissions for model years 2000 to 2008 will be compared to the near-term standard on a cumulative basis. Manufacturers that had cumulative emissions below the near-term standards would earn credits. Similarly, credits can be accumulated during the phase-in years and used to offset compliance shortfalls up to one year after the end of the phase-in at full value or at a discounted rate in the second and third years after the end of the phase-in.

CARB estimates that the proposed GHG emission standards will reduce projected GHG emissions from the light-duty vehicle fleet by 17 percent in 2020 and by 25 percent in 2030 (CARB, 2004). In absolute terms, however, total GHG emission reductions due to the legislation would be more than offset by growth in vehicle population and travel by 2020, and they would stabilize at today's GHG emission level by 2030.

In December 2004, the automobile industry filed a lawsuit to challenge the CARB rules in court on the basis that GHG emissions are closely related to fuel economy and that only the federal government has the authority to regulate fuel economy under the CAFE legislation. California officials, including the governor, remain committed to seeing these regulations come into force, arguing that they regulate greenhouse gases, not fuel economy, and that the state is permitted to do so under the Clean Air Act. Because California state regulations preceded the enactment of the Clean Air Act (CAA), California has a special status under the CAA that allows the state to design its own air pollution regulations for vehicles. Other states are mandated to follow either federal regulations or California regulations.

Canada

Canada's Company Average Fuel Consumption (CAFC) goal was introduced in 1976 for the new passenger vehicle fleet. This voluntary goal is equivalent to the targets set in the U.S. CAFE program but measured in terms of L/100-km of driving. Legislation was introduced in 1982 to make the fuel efficiency program mandatory instead of voluntary, with penalties for noncompliance. This legislation closely matched key provisions in the CAFE program, including a credit system and the use of the CAFE test driving cycle to determine fuel consumption. Although the legislation was passed by Parliament, it did not go into effect because the motor vehicle industry agreed to comply voluntarily with the requirements of the act.

One difference between the U.S. CAFE system and Canada's CAFC goal is that the Canadians do not distinguish between domestic and import fleets as they do in the United States. Canadian goals have continued to match the U.S. standards each year for new passenger car and new light-duty truck fleets, with the Canadian vehicle fleet outperforming the U.S. fleet overall for average fuel economy by about 3 percent. This is due in part to different tax provisions for fuels, vehicles, and income, and also to the different sales mix of vehicles in the two countries. Overall, Canadians purchase slightly fewer pickups and SUVs and more minivans than do their U.S. counterparts. Canadians also exhibit a lower vehicle ownership level than U.S. car owners. In 2004, 70 percent of the driving age population owned cars in Canada. Car ownership in the United States is nearly universal. Also, the split in Canada between passenger cars and light trucks has been relatively steady since 1997—at about 55 percent versus 45 percent—while the market share of light trucks in the United States continues to increase, and for the first time in model year 2003, light trucks outsold cars (Automotive News, 2005).

As part of Canada's plan to meet its CO_2 obligations under the Kyoto Protocol, the Canadian government recently reached a voluntary agreement with industry for a reduction of GHG emissions from light-duty vehicles through 2010. Nineteen automakers signed the agreement to collectively reduce GHG emissions in 2010, plus interim targets. The Canadian government estimates that this target is consistent with the reduction of the average fuel consumption of the new vehicle fleet by 25 percent in 2010.

European Union

The European automotive industry is currently committed to reducing passenger vehicle CO_2 emissions through a voluntary agreement with the European Commission. Signed in March 1998, the "ACEA Agreement" is a collective undertaking by the European automobile manufacturers association and its members to reduce voluntarily the CO_2 emission rates of vehicles sold in the EU. The ACEA agreement covers all vehicles produced or imported into the EU by member companies—including BMW, DaimlerChrysler, Fiat, Ford, GM, Porsche, PSA Peugeot Citroën, Renault, and the VW Group.

As part of the agreement with ACEA, the European Commission initiated similar negotiations in 1998 with the Korean and Japanese manufacturers. The Korean Automobile Manufacturers Association (KAMA) includes Daewoo, Hyundai, Kia, and Ssangyong. The Japanese Automobile Manufacturers Association (JAMA) includes Daihatsu, Honda, Isuzu, Mazda, Mitsubishi, Nissan, Subaru, Suzuki, and Toyota. Altogether, vehicles sold by companies under the ACEA voluntary agreement, including the Korean and Japanese components, make up nearly 90 percent of total EU vehicle sales.

Specifically, the ACEA agreement establishes industry-wide targets for average vehicle emissions from new vehicles sold in Europe of 140 grams of CO_2 per kilometer (gCO_2/km) by 2008, with the possibility of tightening the target to 120 gCO_2/km by 2012. Furthermore, there is an intermediate target range in 2003 of between 165 and 170 gCO_2/km. A recent estimate by an EU source predicted that European automakers' CO_2 emissions would be in the range of 145 g/km to 148 g/km in 2008, missing the 140 g/km target. The last monitoring report indicates that the European and Japanese auto companies are on track to meet this target, while the Korean companies lag behind (Commission of the European Communities, 2004).

JAMA and KAMA agreed to similar commitments to those of ACEA, with the following modifications. KAMA has until 2004 to achieve the 2003 intermediate target. At 165 to 175 gCO_2/km, JAMA's 2003 intermediate target range is wider. Both JAMA and KAMA have an extra year to achieve the final 140 gCO_2/km target.

According to EU member states data, in 2002, the average CO_2 emissions from ACEA's new vehicle fleet was 165 gCO_2/km. Gasoline-fueled cars showed an average emission rate of 172 g/km. Diesel-fueled cars had a lower average emission rate of 155 g/km. Emissions from alternative-fueled cars were highest of all, at 177 g/km. These emissions are in line with the 2003 intermediate target range. Compared with 2001, the 2002 levels represent a reduction of 1.2 percent in new vehicle emissions. Despite the progress, companies will need to accelerate their efforts in the years ahead. Figure 9-5 charts ACEA's, JAMA's, and KAMA's progress under the ACEA agreement compared to future targets.

The growth in sales of diesel vehicles made it easier for companies to meet their intermediate 2003 target and is likely to contribute greatly toward reaching the 2008 final target. Diesel has grown from 14 percent of European vehicles in 1990 to 44 percent in 2003, and it is expected to grow to 52 percent of market share by 2007. The reasons for strong diesel demand are mainly tax incentives that lowered taxes on diesel fuel and imported diesel cars in some EU countries, high fuel prices that encourage purchase of lower-cost diesel, and the superior driving capabilities of diesel engines.

Despite reluctance on the part of industry to extend the ACEA Agreement to the 120 gCO_2/km target in 2012, the European Commission has

156   *Driving Climate Change*

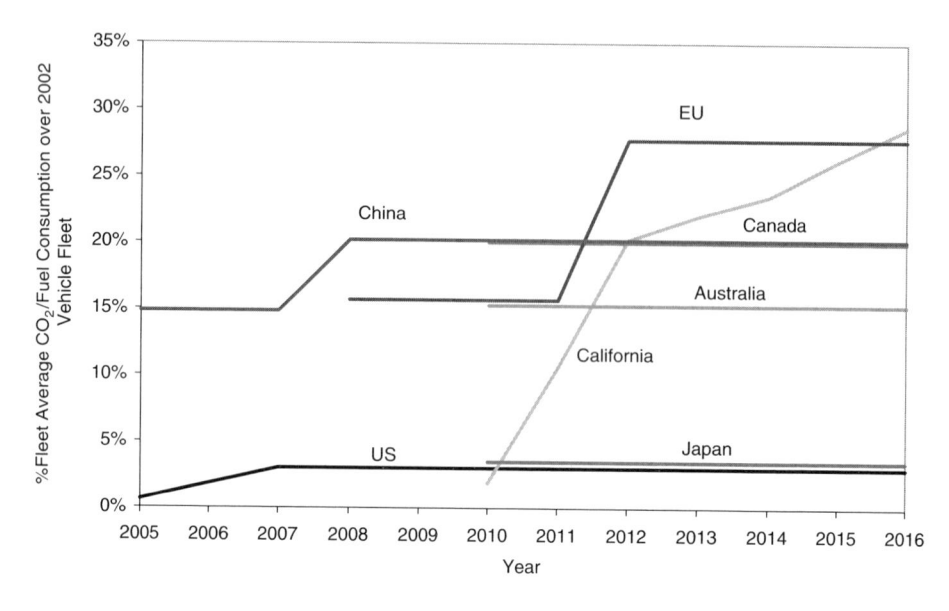

FIGURE 9-5. Progress and targets under the ACEA agreement. *Source:* European Commission.

reaffirmed its objective to reduce average per-car CO_2 emissions to this goal (Thisdell and Weernimk, 2004). The 2012 commitment is likely to be based on a broader set of incentives, including tax incentives, greener driving initiatives, and alternative fuels. Natural gas-based fuels and biofuels are the likely candidates for alternative fuels, given their beneficial well-to-wheels lifecycle CO_2 emission characteristics.

Japan

The Japanese government has established a set of fuel economy standards for gasoline and diesel powered light-duty passenger and commercial vehicles, with fuel economy targets based on average vehicle fuel economy by weight class. The targets for gasoline vehicles are to be met by 2010, while 2005 was the target year for diesel vehicles. The regulations were revised in 2001 to allow automakers to accumulate credits in one weight class and use them in another weight class, with some limitations. Table 9-6 illustrates the improvements required by fuel economy standards for gasoline vehicles.

Assuming no change in the vehicle mix, these targets imply a 23 percent improvement in 2010 in gasoline passenger vehicle fuel economy and a 14 percent improvement in diesel fuel economy compared with the 1995 fleet average of 14.6 km/L. According to the Japanese government, this

TABLE 9-6. Japanese Weight Class Fuel Economy Standards for Gasoline Passenger Vehicles

Vehicle Classes by Maximum Vehicle Curb Weight		Fuel Economy Fleet Average Target by Class	
kg	lbs	km/L	mpg
<702	<1,548	21.2	49.8
703–827	1,550–1,824	18.8	44.2
828–1,015	1,826–2,238	17.9	42.1
1,016–1,265	2,240–2,789	16.0	37.6
1,266–1,515	2,791–3,341	13.0	30.6
1,516–1,765	3,343–3,892	10.5	24.7
1,766–2,015	3,894–4,443	8.9	20.9
2,016–2,265	4,445–4,994	7.8	18.3
>2,266	>4,997	6.4	15.0

Source: Ministry of Transportation, Japan.

improvement will result in an average fleet fuel economy of Japanese vehicles of 35.5 mpg by 2010. The regulations include penalties if the targets are not met, but these penalties are very small. Furthermore, the majority of vehicles sold in Japan in 2002 were already in compliance with the 2010 standards.

China

Mindful of its rapidly growing passenger vehicle fleet and increasing oil demand, China recently approved regulations for new fuel economy standards for its passenger vehicle fleet. These standards are primarily designed to mitigate China's increasing dependence on foreign oil, but another objective is to encourage foreign automakers to bring more fuel-efficient vehicle technologies to the Chinese market.

The new standards will be implemented in two phases. Phase 1 took effect on July 1, 2005, for new vehicle models and will take effect on July 1, 2006, for continued vehicle models. In the Chinese regulations, "continued vehicle models" refers to existing vehicle models that continue to be produced at the effective date of the regulation. Phase 2 will take effect on January 1, 2008, for new models and on January 1, 2009, for all vehicle models.

The standards will be classified into 16 weight classes, ranging from vehicles weighing less than 750 kg, or approximately 1,500 pounds, to vehicles weighing more than 2,500 kg, or approximately 5,500 pounds. The standards cover passenger cars, SUVs, and multipurpose vans (MPVs), collectively defined as M1-type vehicles under the EU definition, with separate standards for passenger cars with manual and automatic

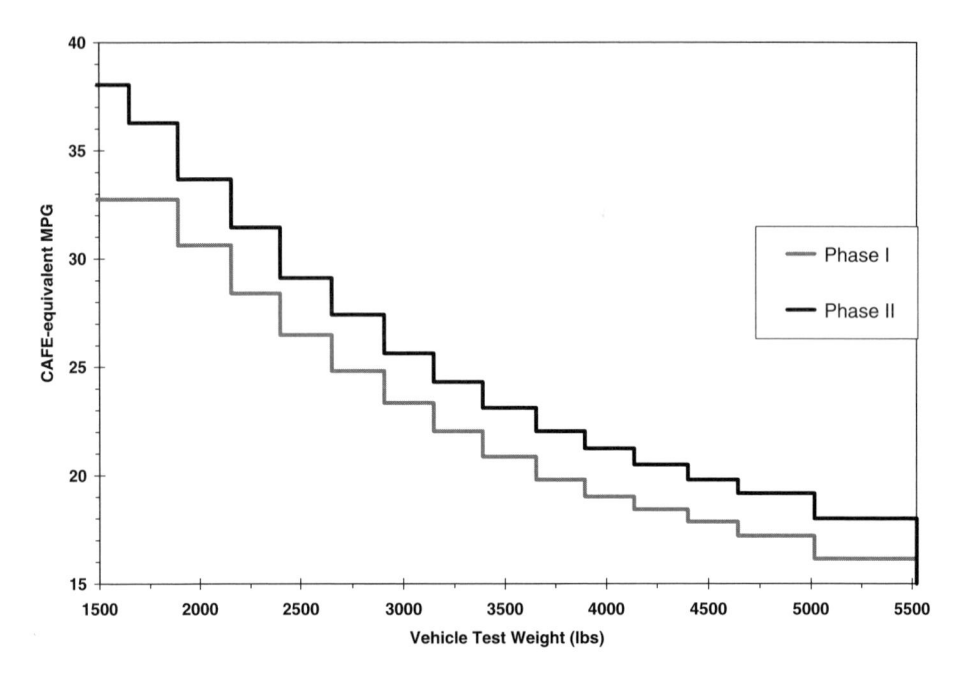

FIGURE 9-6. China's automotive fuel economy standards for passenger vehicles with automatic transmissions and for SUVs/MPVs (CAFE-equivalent mpg). *Source:* An and Sauer, 2004.

transmissions. SUVs and MPVs, regardless of their transmission types, share the same standards as passenger cars with automatic transmissions. Commercial vehicles and pickup trucks are not regulated under the standards.

Table 9-7 summarizes the new Chinese standards, with maximum limits for fuel consumption (L/100-km) or minimum CAFE-equivalent mpg limits. Figure 9-6 shows minimum CAFE-equivalent mpg limits of Chinese standards for vehicles with automatic transmissions and SUVs/MPVs.

One distinctive feature of the Chinese standards is that they set up maximum allowable fuel consumption limits by weight category, rather than being based on fleet averages. Every individual vehicle model sold in China will be required to meet the standard for its weight class. The system does not include a credit system to allow vehicles that exceed compliance to offset those that do not.

The current level of fuel economy of the Chinese vehicle fleet is not well known, as the data have not become publicly available, and thus the relative stringency and effect of these standards is not well understood. However, the standards were designed to be bottom heavy, meaning that they become relatively more stringent in the heavier vehicle classes than

TABLE 9-7. Maximum Limits for Fuel Consumption (L/100-km) and Minimum CAFE-Equivalent mpg Limits, for Passenger Vehicles in China (Excluding Taiwan)

Weight (lbs)	Maximum Fuel Consumption Limits, Based on NEDC Cycle (L/100-km)				Minimum Fuel Economy Limits, Based on U.S. CAFE-Equivalent (mpg)			
	Phase I [2005]		Phase II [2008]		Phase I [2005]		Phase II [2008]	
	Manual	Auto/SUV	Manual	Auto/SUV	Manual	Auto/SUV	Manual	Auto/SUV
≤1,667	7.2	7.6	6.2	6.6	36.9	35.0	42.9	40.3
≤1,922	7.2	7.6	6.5	6.9	36.9	35.0	40.9	38.5
≤2,178	7.7	8.2	7.0	7.4	34.5	32.4	38.0	35.9
≤2,422	8.3	8.8	7.5	8.0	32.0	30.2	35.4	33.2
≤2,678	8.9	9.4	8.1	8.6	29.9	28.3	32.8	30.9
≤2,933	9.5	10.1	8.6	9.1	28.0	26.3	30.9	29.2
≤3,178	10.1	10.7	9.2	9.8	26.3	24.8	28.9	27.1
≤3,422	10.7	11.3	9.7	10.3	24.8	23.5	27.4	25.8
≤3,689	11.3	12.0	10.2	10.8	23.5	22.2	26.1	24.6
≤3,933	11.9	12.6	10.7	11.3	22.3	21.1	24.8	23.5
≤4,178	12.4	13.1	11.1	11.8	21.4	20.3	23.9	22.5
≤4,444	12.8	13.6	11.5	12.2	20.8	19.5	23.1	21.8
≤4,689	13.2	14.0	11.9	12.6	20.1	19.0	22.3	21.1
≤5,066	13.7	14.5	12.3	13.0	19.4	18.3	21.6	20.4
≤5,578	14.6	15.5	13.1	13.9	18.2	17.1	20.3	19.1
>5,578	15.5	16.4	13.9	14.7	17.1	16.2	19.1	18.1

Source: China Automotive Industry Information Website: http://www.autoinfo.gov.cn/zfwj/040330fg.htm.

in the lighter weight classes. For example, a World Resources Institute analysis shows that 66 percent of cars currently sold in the United States would meet the Chinese standards, while only 4 percent of light trucks would comply (Sauer and Wellington, 2004). This will help to create incentives for manufacturers to produce lighter vehicles for the Chinese market.

Issues and Methodologies Involved with Comparing Vehicle Standards Around the World

The previous sections described various fuel economy and GHG standards around the world. Because these standards differ greatly in structure, form, and underlying testing methods, it is challenging to compare them directly. This section identifies key issues involved with comparing diverse standards, and it proposes a generic methodology with which to compare them.

Differences in Test Driving Cycles

Several countries have developed their own testing protocols to measure vehicle emission and fuel economy levels. These test protocols have been variously adopted by other countries. One key element of the testing protocol is the selection of a driving cycle, which ideally is designed to represent on-road vehicle driving patterns in a given country. However, in reality, these driving cycles could be far different from how the vehicles are actually driven, resulting in gaps or shortfalls between certified fuel economy levels and real-world fuel economy levels. This poses a special challenge when comparing vehicle standards and performance around the world.

Countries and regions use essentially three different test cycles to determine fuel economy and GHG emission levels: The NEDC, the Japan 10-15 cycle, and the U.S.-based CAFE cycle. The U.S. CAFE cycle has two test cycle components: city driving and highway driving. The combined CAFE cycle is composed of 55 percent city driving and 45 percent highway driving. These test cycles are very different in terms of average speed, duration, distance, acceleration and deceleration characteristics, and frequencies of starts and stops. All these factors significantly affect fuel economy ratings. In general, average speeds of the test cycles and associated fuel economy ratings are positively correlated.

The U.S. combined CAFE cycle has a highest average speed of close to 30 mph and a highest fuel economy rating of about 31 mpg for the sample vehicle. The average speed of the NEDC is about 21 mph, with the fuel economy rating of the same vehicle about 27 mpg. The average speed of the Japanese cycle is about 15 mph, with a fuel economy rating of 23 mpg.

The variations in fuel economy ratings among these cycles may change somewhat from vehicle model to model. On average, analysts at the U.S. Argonne National Laboratory estimate that the CAFE cycle values are about 13 percent higher than NEDC cycle values, and CAFE cycle values are about 35 percent higher than Japan 10-15 cycle values. In other words, to roughly convert fuel economy rating based on the EU cycle to the rating based on the U.S. CAFE cycle, one multiplies by a factor of 1.13. Similarly, to roughly convert a fuel economy rating based on the Japanese cycle to one based on the U.S. CAFE cycle, one multiplies by 1.35.

Among the countries and regions that have vehicle standards, the United States, California, Canada, Taiwan, and South Korea use the U.S. CAFE cycle. The EU, China, and Australia use NEDC. Japan's fuel economy ratings are based on Japan 10-15 cycle.

Fuel Economy Versus Fuel Consumption Versus GHG Emissions

The relationship between GHG emissions and fuel consumption is important because CO_2 is the dominant source of GHG emissions from an automobile and the level of CO_2 emissions from automobiles is directly linked to vehicle fuel consumption. California's proposed rule would regulate all GHG emissions in terms of CO_2-equivalent emissions, and the EU regulates CO_2 emissions only. Because the vast majority of automobiles consumes petroleum-based fuels such as gasoline and diesel, the conversion factors from CO_2 to gasoline and diesel fuels were treated in this analysis as constants among most countries and regions, even though small variations do exist due to differences in fuel quality and additives. However, these differences are likely to remain relatively minor unless use of alternative fuels that are not petroleum based becomes widespread.

Table 9-8 provides conversion factors from measures associated with different regions to U.S. CAFE-equivalent mpg ratings, EU-equivalent CO_2 emission rates (in g/km), and California-equivalent CO_2 emission rates (in g/mi). Because diesel fuel has a different heat content and density from gasoline fuel, a gasoline-equivalent fuel economy (MPGge) measure was developed to convert diesel fuel into a comparable gasoline equivalent.

Regulatory Versus Voluntary Approaches

There is a clear difference between a regulatory and voluntary approach to fuel economy and GHG emission standards. While a regulatory target with sufficient enforcement and penalties for noncompliance can be more or less guaranteed in the future, a voluntary target is less certain. However, this analysis compares both regulatory and voluntary targets, assuming that voluntary targets will be met in future years.

TABLE 9-8. Conversion Factors to CAFE-Equivalent mpg, EU-Equivalent CO$_2$ (in g/km), and California-Equivalent CO$_2$ Emission Rate (in g/mi)

Country	Cycle	Type	Measure (Y)	Converted to CAFE-Equivalent mpg	Converted to EU-Equivalent CO$_2$ (g/km)	Converted to CA-Equivalent CO$_2$ (g/mi)
United States	U.S. CAFE	Fuel	mpg	Y*	1/(Y) *	6,180	1/(Y) *	8,900
Taiwan	U.S. CAFE	Fuel	Km/L	Y*	2.35	1/(Y) *	2,627	1/(Y) *	3,783
South Korea	U.S. CAFE	Fuel	Km/L	Y*	2.78	1/(Y) *	2,226	1/(Y) *	3,206
Canada	U.S. CAFE	Fuel	L/100-km	1/(Y) *	235.2	Y*	26.2	Y*	37.8
California	U.S. CAFE	CO$_2$	g/mi	1/(Y) *	8,900	Y*	0.69	Y*	1.00
European Union (gasoline)	NEDC	CO$_2$	g/km	1/(Y) *	6,180	Y*	1.00	Y*	1.44
European Union (diesel)	NEDC	CO$_2$	g/km	1/(Y) *	7,259	Y*	1.00	Y*	1.44
Japan	Japan	Fuel	km/L	Y*	3.18	1/(Y) *	1,946	1/(Y) *	2,803
China, Australia	NEDC	Fuel	L/100-km	1/(Y) *	265.8	Y*	23.2	Y*	33.5

Source: Table 11, An and Sauer, 2004.

Corporate Fleet Averages Versus Minimum Requirements

Among all the standards, only the Chinese standards are based on minimum fuel economy requirements that are applicable to individual vehicle models. All other existing or proposed standards throughout the world are based on sales-weighted averages either by whole vehicle fleet or by vehicle class/weight categories. The Chinese standards pose a special challenge to cross-country comparisons, because a number of assumptions must be made to translate the minimum requirements into a fleet average.

The minimum requirement simply provides a floor for all the vehicle models. The fleet average fuel economy level should be above the minimum requirement. This analysis assumes that all vehicle models will at least meet the floor requirements. For vehicle models that are already performing better than the standards, this analysis assumes that they will maintain their current fuel economy levels in the future years.

Vehicle Categories and Weight Classes

Different standards around the world are structured with significant differences in definitions of vehicle categories and weight classes. It is difficult to compare one standard against another because of these differences. This analysis, therefore, compares them on an entire fleet average basis. Such a comparison requires vehicle databases by these countries and regions that provide sales figures and fuel economy ratings for individual vehicle models, which are difficult to obtain for some countries. Data were available for all the countries and regions studied with the exception of the Taiwan and South Korea markets.

Another challenge is to project future fleet average fuel economy figures for different regions. Fuel economy projection efforts usually require a projection into future years of sales breakdowns by vehicle weight classes and categories defined by the standards themselves. Historical data in the United States and Japan have shown significant shifts in sales from one category to another, mostly from lighter vehicle groups to heavier ones. However, it's beyond the scope of this analysis to make such projections. The analysis assumes that the current sales composition of vehicle categories will be maintained, and future fleet average fuel economy was projected under such assumptions.

Conclusions

Fuel economy programs or GHG targets, either mandatory or voluntary, have proven to be among the most effective tools in controlling oil demand and GHG emissions from the transportation sector. Nine major regions around the world have implemented or proposed various fuel economy and GHG emission standards. Yet, these standards are not easily comparable, due to differences in policy approaches, test drive cycles, and units of measurement. This chapter discusses a methodology to compare these programs to better understand their relative stringency. Key findings include the following:

- The EU and Japan have the most stringent standards in the world.
- In the next ten years or so, the EU, China, Canada, and California all will have fleet average GHG reduction greater than 20 percent compared to a 2002 baseline case.
- The fuel economy and GHG emission performance of the U.S. automobile fleet—both historically and projected based on current policies—lag behind most other nations. The United States and Canada have the lowest standards in terms of fleet-average fuel economy rating, and they have the highest GHG emission rates based on the EU testing procedure.
- The new Chinese standards are more stringent than those in Australia, Canada, California, and the United States, but they are less stringent than those in the EU and Japan.
- If the California GHG standards go into effect, they would narrow the gap between U.S. and EU standards, but the California standards would still be less stringent than the EU standards.

References

An, F., and A. Sauer. *Comparison of Passenger Fuel Economy and GHG Emissions Standards around the World*. Washington, D.C.: Pew Center on Global Climate Change, 2004.

An, F., and J. DeCicco. "Indices of Technical Efficiency." A draft working paper. 2005.

Automotive News' Data Center. www.autonews.com/datacenter.cms.

Bernton, Hal. "Tighter Vehicle Emission Standards Proposed for State." *Seattle Times*, December 2, 2004.

California Air Resources Board (CARB). "Staff Proposal Regarding the Maximum Feasible and Cost-Effective Reduction of Greenhouse Gas Emissions from Motor Vehicles." Sacramento, CA: CARB, August 2004.

Commission of the European Communities. *Implementing the Community Strategy to Reduce CO_2 Emissions from Cars: Fourth Annual Report on the Effectiveness of the Strategy (Reporting Year 2002)*. Brussels, Belgium, 2004.

Federal Register, 29 CER Part 533: Light Trucks, Average Fuel Economy; Model Years 2008–2011; Proposed Rules. Washington, D.C., August 2005.

Federal Register. Docket Number 68 FR 16867. Washington, D.C., April 7, 2003.

Sauer, A., and F. Wellington. "Taking the High (Fuel Economy) Road." Washington, D.C.: World Resources Institute, November 2004.

Thisdell, Dan, and Wim Oude Weernink. "Brussels Readies for CO_2 Fight." *Automotive News Europe*, November 15, 2004.

Union of Concerned Scientists. *Life in the Slow Lane*. Washington, D.C., 2003.

U.S. Environmental Protection Agency. *Light-Duty Automotive Technology and Fuel Economy Trends: 1975 through 2004*. Washington, D.C., 2004.

Ward's Vehicle Facts & Figures, 2003. Southfield, MI.

CHAPTER 10

Reducing Transport-Related Greenhouse Gas Emissions in Developing Countries: The Role of the Global Environmental Facility

Walter Hook

Transportation is the fastest-growing sector of greenhouse gas (GHG) emissions globally. It is also the sector where the least progress has been made in addressing cost-effective GHG reductions. According to the Paris-based International Energy Agency (IEA), over the next 20 years, the growth of transportation sector energy consumption and related GHG emissions will be greater than for any other sector. Transportation's share of total energy use is projected to increase from 28 percent in 1997 to 31 percent in 2020. While the developed world will continue to be the main source of the problem, the growth of GHG emissions in transportation will increasingly come from developing countries over the next 20 years (IEA, 2002).

Though much depends on future oil prices, most experts believe that without significant intervention to reverse these trends, growth in motor vehicle use will overwhelm any efficiency gains from new fuels and technologies. While oil use in industrialized countries is growing by only 1 percent per year, it is growing by 6 percent annually in Africa, Asia, and Latin America. From 1995 to 2020, worldwide vehicle ownership is expected to grow by 75 percent to over 1.3 billion vehicles (OECD/ECMT, 1995), with the greatest rate of growth to occur in Latin America and Asia.

The Global Environmental Facility (GEF)—managed jointly by the World Bank, the United Nations Development Programme (UNDP), and the

UN Environment Programme (UNEP)—is increasingly important as a source of financing sustainable transportation projects around the world. After some initial missteps, the GEF is playing an increasingly constructive role in bringing about the sort of dramatic shift in transport paradigm that will be required to avert significant global warming.

Despite its rapidly increasing importance, transportation was one of the last major sectors that contribute significantly to global warming to be considered for GEF funding. The GEF was for many years reluctant to become involved in the transport sector, fearing that the cost of interventions would inhibit a productive role for the GEF, and fearing the GHG emissions impacts of transport sector projects would be difficult to quantify.

In its early years, the GEF funded two transportation-focused projects. One in Tehran, Iran, funded a number of studies of emission monitoring systems, inventories of pollution sources, and proposed new policy initiatives. A second project, approved in 1996, was a $7 million project in Pakistan, focused on establishing vehicle inspection and maintenance centers.

Creation of Operational Program #11

The GEF felt that these projects lacked focus, and the benefits were difficult to quantify. In the late 1990s, it decided to make transportation a specific operational program and began to draft Operational Program #11 (OP #11). The GEF Standing Technical Advisory Panel (STAP), which advises the GEF on technical matters, issued a set of draft recommendations, but the GEF ignored the STAP recommendations and hired a fuel cell researcher to write OP #11 as an outside consultant. Not surprisingly, the initial draft of OP #11 focused exclusively on hydrogen fuel cell vehicle technology. The GEF embraced the hydrogen strategy, and, for many years, hydrogen and fuel cell programs dominated the GEF portfolio of transportation project funding.

This first draft OP #11 provoked significant criticism among some government officials, nongovernment organizations (NGOs), and transport experts. Some government departments, including the U.S. Environmental Protection Agency (EPA) and Germany's Umwelt Bundesamt, complained, as did the UN Habitat. The initial draft was also viewed with skepticism by some within the World Bank. Intervention by these groups managed to get the mandate broadened to include the following funding priorities (GEF, 2001):

- Modal shifts toward more efficient and less polluting forms of public and freight transport through measures such as traffic management and avoidance and increased use of cleaner fuels
- Nonmotorized transport (NMT)
- Fuel cell or battery operated two- or three-wheeled vehicles designed to carry more than one person

- Hydrogen-powered fuel cell or battery-operated vehicles for public transport and goods delivery
- Hybrid electric buses equipped with internal combustion engines
- Advanced technologies for converting biomass feedstock to liquid fuels

The First Years of the GEF Transportation Program: Hydrogen Fuel Cells

During the first several years of implementing OP #11, some $36 million in GEF funds were approved for single-initiative hydrogen fuel cell bus demonstration projects, all of them sponsored by UNDP. For a project to be financed by the GEF, it had to meet one of the preceding criteria, and be endorsed both by the GEF focal point within the country and by an implementing agency. In the first phase of the GEF OP #11, the implementing agency could be the World Bank, UNDP, or UNEP. In the second phase of the effort, regional development banks also became eligible as implementing agencies for GEF projects. Project sponsors also had to provide 50 percent matching funds from non-GEF sources. This cumbersome and often difficult approval process constituted a fairly significant barrier to entry to many good projects.

The GEF-UNDP Fuel Cell Bus Program initially supported the commercial implementation of fuel cell bus and associated refueling systems in six of the largest bus markets in the developing world:

- Beijing, China
- Shanghai, China
- Sao Paulo, Brazil
- Cairo, Egypt
- Mexico City, Mexico
- New Delhi, India

As of the fall of 2005, however, only Beijing had received any fuel cell buses. The projects in Shanghai and Brazil are still moving ahead, with fuel cell bus procurement processes underway. Mexico has moved its program away from a focus on hydrogen toward various hybrid electric bus technologies. The projects in India and Egypt have been delayed indefinitely, awaiting the outcome of the other projects.

Criticism of Fuel Cell Bus Effort

Critics of the hydrogen fuel cell technology approach of the first GEF projects focused on several areas of concern. They pointed out, for example, that new and unproven technologies are not generally first introduced in developing countries but rather are brought to scale in developed economies and

TABLE 10-1. Greenhouse Gas Emissions by Mode

Mode	CO_2-Equivalent Emissions (grams/ vehicle-km)	Maximum Capacity (passenger)	Average Capacity (passenger)	CO_2-Equivalent Emissions (grams/ passenger-km)
Pedestrian	0	1	1	0
Bicycle	0	2	1.1	0
Gasoline motor scooter (2-stroke)	118	2	1.2	98
Gasoline motor scooter (4-stroke)	70	2	1.2	64
Gasoline car	293	5	1.2	244
Gasoline taxi car	293	5	0.5	586
Diesel car	172	1.2	1.2	143
Diesel minibus	750	20	15	50
Diesel bus	963	80	65	15
CNG bus	1,050	80	65	16
Diesel articulated bus	1,000	160	130	7

Source: Sperling and Salon, 2002.

then exported to developing economies only after the production costs have dropped significantly and the technology has become more mature. The other major criticism was that focusing on bus technology in isolation from the specific public transit markets where they will operate is likely to have perverse emissions effects.

In all of the target countries, the income of bus passengers is quite low, with India being the most extreme case, where bus passengers often have incomes as low as one or two dollars a day. With such low incomes, bus price elasticity of demand is very high, so a small increase in the bus fare leads to a fairly rapid shift to motorized two-wheeler and three-wheeler shared taxi trips. Each shift of this type leads to an increase in emissions. While the impact of this modal shifting on GHG emissions will be case specific, some typical numbers are listed in Table 10-1.

Given that the shift of a single passenger from a bus to a motor scooter is likely to increase the per trip GHG emissions by nine times, even the smallest increase in the bus price that diverts bus passengers to private motorized modes is likely to have huge adverse impacts on aggregate GHG emissions. The purchase prices of various transportation vehicles appear in Table 10-2. At a projected market price of $1.5 million for each hydrogen fuel cell bus, these buses are 150 times the price of some buses currently being used in developing countries.

TABLE 10-2. Bus Vehicle Costs

Vehicle Type	Purchase Cost (US$)
Small, new, or secondhand bus seating 20–40 passengers, often with truck chassis	$10,000–$40,000
Large, modern-style diesel bus that can carry up to 100 passengers, produced by indigenous companies or low-cost import	$40,000–$75,000
Diesel bus meeting Euro II standard, produced for (or in) developing countries by international bus companies	$100,000–$150,000
Standard OECD Euro II diesel bus sold in Europe or United States	$175,000–$350,000
Diesel with advanced emissions controls meeting Euro III or better	$5,000 to $10,000 more than a comparable standard diesel bus
CNG, LPG buses	$25,000 to $50,000 more than a comparable standard diesel bus (less in developing countries)
Hybrid-electric buses	$75,000 to $150,000 more than a comparable standard diesel bus
Fuel-cell buses	$875,000–$1,200,000 more than a comparable standard diesel bus

Source: Adapted by IEA, 2002, p. 120; Wright, 2006.

Out of this experience, some important lessons should be learned about GEF involvement in the transit vehicle sector. First, it is probably inappropriate for the GEF to be picking technological winners and imposing them on developing countries. Prepicking technological winners runs the great risk that the technology will not actually reach expectations. Often, less publicized technological improvements on existing technologies achieve the same objective faster at a lower, more commercially viable price.

It is highly probable that far greater GHG emission-reductions could have been achieved if bus operators were given direct incentives to find the best solution to reducing their GHG emissions. The decision to switch bus technology ultimately has to be made by the bus operator, and if this switch is to be made in a way that does not disrupt bus services, the bus operators and bus system regulators need to be involved directly in the process.

Current GEF Transport Priorities

In the second phase of the OP #11 effort, the GEF shifted decidedly away from hydrogen fuel cell bus projects and refocused on projects with a demonstrated impact on shifting trips to less energy intensive modes. The turning

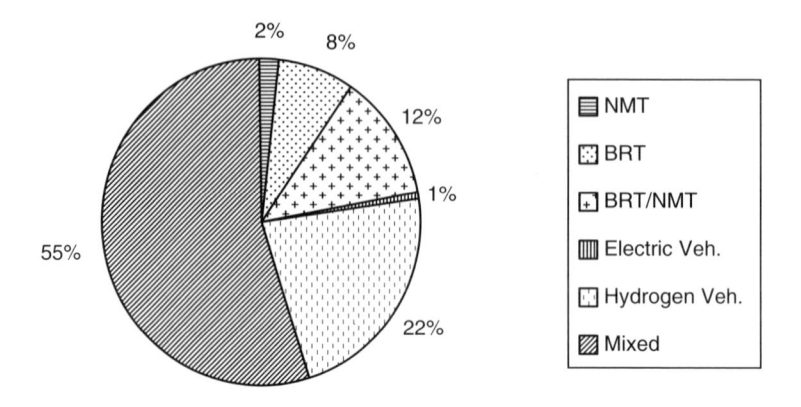

FIGURE 10-1. GEF resources by project type.

point occurred at a STAP meeting held in Nairobi in 2002. The conclusions of this meeting were summarized in a World Bank report (Karekezi, Majoro, and Johnson, 2002). This document asserted that the existing OP #11 was consistent with the World Bank's own urban transportation policies (World Bank, 2002) in the following areas:

- Promotion of low cost public transport modes, such as BRT
- NMT, including bikeways and pedestrian walkways
- Transport and urban planning to facilitate efficient and low GHG modes of transportation
- Transport demand management (TDM) measures the favor or enables public transport and NMT

By 2005, the hydrogen fuel cell part of the portfolio has fallen to less than a quarter of anticipated expenditures. It was replaced by projects which lump together several interrelated interventions, usually involving NMT, BRT, some TDM measures, and often a host of other measures. The distribution of 2005 funding by program area is shown in Figure 10-1. The shift also represented a move from funds granted by the UNDP toward World Bank funding. As shown in Figure 10-2, the World Bank now contributes 60 percent of OP #11 transportation funding.

This growing domination of the World Bank in GEF funding has some positive and negative impacts. On the positive side, the World Bank has an extensive network of field offices and a staff with long experience in transportation. The projects developed under the World Bank are also better grounded in the ongoing transportation decision-making process in the beneficiary countries than were the hydrogen fuel cell bus projects. The UNDP, by contrast, decided not to make transportation a priority area of their technical assistance in the late 1990s. However, UNDP continued to play an ad

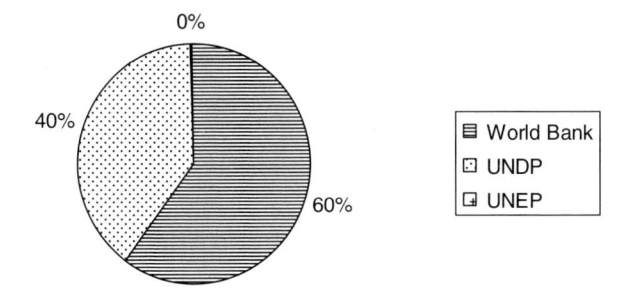

FIGURE 10-2. GEF transport funds by agency.

hoc but often important role in providing technical support to transport projects, particularly in Latin America. The UNDP in fact financed some of the technical work on both Bogota and Quito BRT systems.

Ultimately, the World Bank has a significant advantage over the UNDP and UNEP in that it has the capacity to use its own loans to provide the matching funds that the GEF Council requires. In Ghana, for instance, the government decided to work with the World Bank rather than the UNDP mainly because the World Bank also promised to provide low interest loans to finance the implementation of the project. The UNDP has some technical assistance funds at its disposal, while UNEP has very few of its own resources that it can bring to the table.

A government is much more likely to get a GEF grant from the World Bank if they are also considering a loan from other sources within the bank. The World Bank's management prioritizes GEF funds for this purpose. What this means in practice is that projects where the governments themselves do not need World Bank funds to implement a good project, which surely is a sign of political commitment, will face greater difficulties than those simultaneously approaching the World Bank for loans.

The influence the World Bank has over the GEF relative to the other implementing agencies means that it is often able to get approval for projects with only nominal political commitment, and where the specifics of the project are vague and the GHG emissions impacts are unknown and often unknowable, while other implementing agencies frequently face a higher standard of assessment. Lack of clarity about the approval process and the approval criteria have led to frustration on the part of potential beneficiaries, many of whom have spent long hours preparing projects in response to sometimes contradictory guidance from the GEF Secretariat and the implementing agency. Despite these shortcomings, the quality of projects being funded by the GEF is improving significantly.

Both the World Bank and the UNDP face a similar institutional tension between being responsive to the requests and requirements of the beneficiary countries on the one hand, and implementing projects that successfully

172 *Driving Climate Change*

alleviate poverty and improve the environment on the other. An unpublished study of the UNDP's technical assistance in the transportation sector showed that 60 percent was targeted to civil aviation, which had no direct poverty alleviation benefit, despite the fact that poverty alleviation is the primary mandate of the UNDP.

The World Bank's transportation portfolio has in the past faced criticism from NGOs for doing little to alleviate poverty or improve the environment, while many large road loans had significant adverse air quality and involuntary resettlement impacts. While the World Bank has made significant changes in response to these criticisms, nonetheless, the staff is ultimately rewarded for making loans, and if a developing country government wants to borrow money for a problematic project, it is difficult for World Bank staff to refuse. The fundamental problem faced, therefore, by both UNDP and World Bank staff is how to generate large projects that will actually reduce poverty and improve air pollution, when borrowing governments frequently lack creativity in this regard, or have other motives.

GEF funds are increasingly playing a vital role in helping both the UNDP and the World Bank be more proactive in generating good projects, rather than just waiting for good projects to come to them. As shown in Figure 10-3, more than half of all GEF transportation grants are awarded to projects in Latin America. The Latin America division of the World Bank has been the most entrepreneurial in this regard, and the GEF projects for Mexico City and Santiago played a critical role in initiating BRT and NMT projects in those cities. The Asia and Africa divisions are increasingly using the GEF in this way as well, with new World Bank GEF projects underway in China, Vietnam, and Ghana. These projects are some of the most exciting initiatives in the transportation sector. The potential exists for truly historical change in transport sector system development being played by the

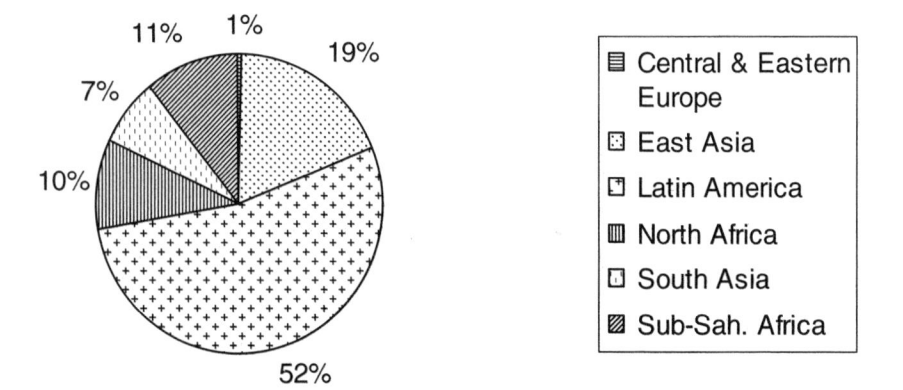

FIGURE 10-3. GEF Transport Funds by Region.

GEF. However, most of these projects are still only in the planning stages, and few concrete successes have actually been implemented through this mechanism to date.

The UNDP has also shifted the focus of its GEF transport program, and increasingly it is focused also on BRT and NMT projects. When India dropped out of the hydrogen fuel cell bus program, the money was reprogrammed to a project preparation grant for a large scale GEF project. This project now rests with the Urban Development Ministry, which is likely to focus on NMT improvements in secondary cities. The UNDP in cooperation with an NGO, the Institute for Transportation and Development Policy (ITDP), also helped initiate BRT projects in Accra, Ghana, and Dakar, Senegal, both of which received project preparation grants. While the project for Dakar is moving forward, the project in Accra has been taken over by the World Bank.

The third player in transportation GEF projects has been UNEP. Until recently, most of UNEP's work in this area was focused on multicountry studies and information sharing. Starting in 2002, the UNEP has become much more active in creating GEF transportation projects. The UNEP and the ITDP have a medium-sized GEF grant to develop the BRT and NMT project in Dar es Salaam, Tanzania, the BRT and NMT system in Cartagena, Colombia, and a BRT Planning Guide. A project preparation grant funded delegations from these and several other countries to participate in workshops on BRT and NMT in Bogota, Colombia. While not all the participating countries decided to move forward, the project played a key role in securing political commitment to these measures in Dar es Salaam. The UNEP is currently developing several GEF projects for different projects around the world, such as in Jakarta, Indonesia, and is playing a particularly important role where the government is not interested in World Bank loans. The regional development banks, particularly the Inter-American Development Bank and the Asian Development Bank (ADB), also have some GEF transport projects being developed. The former has a BRT project under development in Managua, Nicaragua, and the latter has a BRT project under development in Manila, the Philippines.

Ultimately, the overworked and understaffed GEF is trying its best and doing a good job, but in the end, decisions frequently reflect simply the degree of trust the GEF has in the implementing agency to implement the project. While this has led to domination by the World Bank, it is imperative that UNEP and UNDP be kept actively involved, if for no other reason than to keep the World Bank marginally accountable, and to provide an alternative mechanism for financing good projects where municipalities are ready to implement great projects with their own funds but do not want, for whatever reason, to involve the World Bank.

World Bank dominance over the GEF has also influenced the allocation of GEF funds by region. The predominance of Latin America in the OP #11 GEF portfolio stems largely from the fact that the World Bank's Latin

America division was the first to rely heavily on this funding mechanism. It also reflected the fact that two megacities in Latin America—Mexico City and Sao Paulo—were included in the hydrogen fuel cell bus demonstration project. The World Bank's Africa division, on the other hand, did not originally feel that urban transportation was a priority, though with rapid motorization in some cities this view has been modified somewhat. The low prevalence of GEF funds in South Asia is linked to the lack of World Bank urban transportation lending in the region and complicated and nontransparent procedures for focal point approval.

In conclusion, after a period of institutional learning, the GEF has reoriented its focus onto projects that are much more likely to lead to profound GHG emission reductions. It has also grounded itself much better in what the development banks are doing in the transportation sector and thus has the potential to profoundly influence, not only the use of the GEF money, but also to leverage multilateral development bank loans toward more sustainable projects. As such, the potential exists for the GEF transportation portfolio to play a historical role in reorienting global transportation systems.

The devil, however, is in the details. Setting the basic programmatic direction of the GEF transportation program on tasks that will truly reduce GHG emissions has been largely accomplished. However, turning these programmatic priorities into successful projects is extremely difficult, and the track record to date is not that impressive. Getting up-to-date information on GEF projects is extremely difficult and requires querying the parties responsible for project implementation. The next three sections in this chapter discuss projects where the ITDP has had some sort of involvement or familiarity. They identify the range of difficulties currently being encountered in implementing GEF transportation projects.

NMT Projects Financed under the GEF

Support for using GEF funds for NMT infrastructure projects has been strong, although total dollar value of the NMT projects has accounted for just 12 percent of all GEF transportation project grants. The bigger problem has been finding governments that want to make significant improvements in NMT facilities.

The idea behind NMT projects is to promote bicycling and walking to reduce GHG emissions. If their modal share could be retained or increased, this would make a significant contribution to reducing GHG emissions at a very modest cost. There is considerable evidence from cities like Bogota, Colombia, that a municipal investment in bicycling facilities could result in a significant increase in NMT. The mode share of cycling in Bogota increased from less than 0.5 percent of all daily trips to over 4 percent of daily trips in less than five years with the construction of some 300 kilometers (km) of new bicycle facilities.

Bogota's success was implemented without any GEF funds, but it seemed certain that the use of GEF funds would help to induce other cities to follow Bogota's lead. Early project experience has been a mixed success, however, indicating that getting GHG emissions reductions from bicycle infrastructure projects is not guaranteed but requires political will, proper planning, and complimentary measures.

There were two early GEF projects that focused on NMT: Marakina, a district of Manila in the Philippines, and Gdansk, Poland. The first project was funded by the World Bank and the second by the UNDP. There were also several projects that have already been implemented where NMT infrastructure was a component of the project, including World Bank–sponsored GEF projects in Lima, Peru, and Santiago, Chile. Many others are in development.

The Marakina project proposed to spend about $1.27 million for pilot bikeways and bike promotion in a district of Manila where some bikeways already existed and that appeared relatively sympathetic to bikes. This project began in 1996 and was approved for World Bank funding in 2001. As of 2004, only about $400,000 had actually been spent. The tensions surrounding the project are typical of what has happened in many NMT projects. The main interest of the district mayor was to build a largely recreational facility along a watershed that was used as a park.

Advocacy groups in Manila complained that this focus had relatively little importance for expanded use of bicycles by the local population and that the planning for the project was done with no involvement from actual cyclists. Cyclists wanted bike lanes on the major arterials, where they could use them to reach shops and centers of employment safely, but the mayor was reluctant to implement these strategies for fear of antagonizing motorists. Some money was spent on promoting cycling in the area, with some positive effects. Most recent information, however, indicates that only the recreational elements of the bike network have been built to date, and the GHG emissions benefits have been minimal.

More recently, the Metro Manila Development Authority, which is responsible for transportation management in the metropolitan region, together with President Arroyo have launched a pilot bicycle plan for downtown Manila. It appears, however, that the facilities will be built on secondary streets where their need is less.

The Gdansk project was implemented somewhat better. The Municipality of Gdansk, a local organization, the Polish Ecological Club, and the UNDP cosponsored the project. The project, financed with $1.0 million from the GEF and $1.5 million in cofinancing, constructed a core network of cycling facilities and paths. To date, 12 km of bicycle paths have been completed or are under construction, with another 20 km planned. Other activities include measures to reduce traffic speed using speed bumps, strict speed limits, and public outreach and information campaigns to encourage cycling. According to the UNDP, the number of people in Gdansk

FIGURE 10-4. Lima bike lane financed with a World Bank loan.

that are now cycling as a result of the project has doubled so far, though it was from a very low baseline.

More recently, new GEF projects approved for Lima and Santiago de Chile designed in conjunction with World Bank urban transportation loans propose to build a number of bike lanes and to promote cycling. Lima had already built some cycling facilities under a previous World Bank loan (see Figure 10-4). These facilities were physically separated bike lanes both curbside and in a median, serving an industrial area where it was hoped that industrial workers would begin to cycle to and from work. The design of these initial bike lanes was not entirely successful, and utilization is fairly modest. They are in the process of being rehabilitated under another World Bank urban transportation loan not using GEF funds. There are GEF funds to implement some new bike facilities that were identified under a new bicycle master plan, and those that will be funded under the GEF have just been selected but not yet implemented.

Santiago de Chile has built a much more extensive network of 53 km of bike lanes in a few pilot districts in the city center using GEF funds under the World Bank project. The system is now being expanded by 16 additional km. GEF funds also finance support for bike promotion, which is being handled by Cuidad Viva, a well-qualified local NGO, and a 900-person bicycle caravan was held. Other districts are also building bike lanes inspired by the Santiago de Chile experience but not implemented with GEF funds. The districts of Bella Vista, Plaza Nunoa, Brasil, and Lastarria now have bike lanes built or under construction. Some are high-caliber,

FIGURE 10-5. Shared bicycle and pedestrian facilities in Santiago de Chile.

grade-separated facilities, while others are on sidewalks, with some conflicts with pedestrians.

In most cases, there was considerable struggle over the placement of the facilities, with struggles around whether space would be taken away from motor vehicles or pedestrians. The ultimate compromise is inevitably a political decision, and the GHG emissions impacts will be in part the result of this political discussion. Shared bicycle and pedestrian facilities, as shown in Figure 10-5 from Santiago de Chile, do not work well if there are any significant pedestrian volumes. In such cases, modal shift impacts are likely to be marginal.

In summary, the potential of GEF-funded bicycle infrastructure projects has only partially been realized to date. Some new facilities have increased bicycle mode share, but the bike facilities have done little to improve bicycle mode share in other cases. Nowhere have GHG emission reductions been very impressive. However, the cost of these projects was also fairly modest, the projects have inspired municipalities to do more on their own, and, given the amount invested, the result was reasonable.

Ultimately, designing proper bicycle facilities requires the strong will of a mayor who is truly committed to cycling as a means of transportation. Involving cycling advocacy groups in the design process along with road engineers will also help, as many road engineers have never ridden a bicycle in their lives. Standard road designs include detailed parameters developed over many years and codified in highway design manuals and, until recently, with little concern for NMT. By contrast, NMT infrastructure design is still an area very much in development, with even basic principles open to debate.

In summary, it is insufficient for the GEF to fund projects supporting NMT infrastructure in a categorical manner. Ultimately, both the specifics of the proposal and the level of commitment of the local government officials need to be carefully reviewed. Systems of accreditation have to be established to determine the competency of technical people chosen to plan and design NMT facilities. Without careful scrutiny, bad NMT projects with more powerful promoters may take precedence over good ones with promoters who understand less well how the GEF system operates.

The GEF and Bus Rapid Transit

BRT systems include various integrated improvements that increase the speed, capacity, and quality of bus-based transit services. The main point of BRT is to create an integrated mass transit system that has the quality of service and performance that can be achieved only at a much higher price by rail-based systems. The most complete examples of BRT systems are the URBS system in Curitiba, Brazil, and TransMilenio in Bogota, Colombia. These systems include physically separated lanes that take buses out of mixed traffic congestion, prepaid enclosed boarding and alighting platforms that significantly increase average bus speeds, larger buses that increase the capacity of the system, bus priority at intersections, and a clear marketing image.

Because BRT systems are much less expensive to build and operate than rail-based systems and can be built much more rapidly, they are the only mass transit system investment that has proven that it can halt and reverse a downward trend in public transit ridership on a citywide basis. While metro rail systems have increased transit mode share in specific corridors, they have never been able to reverse a citywide trend toward declining transit mode share on their own.

Furthermore, because BRT systems often require the reconstruction of central urban road corridors, they also create the opportunity to build complementary facilities for pedestrians and cycling. Most of Bogota's TransMilenio corridors include bicycle lanes and wide sidewalks that play an important role in the GHG emissions reductions benefits.

Mayors have embraced BRT projects in unprecedented numbers in recent years because they are visible, attractive projects that offer a significant political payoff within a single term of office without costing much to implement. Stand-alone bicycle and pedestrian improvements are generally harder to sell politically. Thus, BRT promotion has created a

TABLE 10-3. Public Transit Modal Split before and after BRT and Metro Construction

City	% of Trips Before	% of Trips After
Metro Systems		
Mexico City	80	72
Buenos Aires	49	33
Bangkok	39	35
Kuala Lumpur	34	19
Santiago	56	33
Warsaw	80	53
Sao Paulo	46	33
Tokyo	65	48
Seoul	81	63
BRT Systems		
Bogotá	53	56
Curitiba	74	76
Quito	76	77

Source: Compiled by ITDP.

TABLE 10-4. Select BRT Projects Funded by the GEF OP #11

City	Implementing Agency	Status
Mexico City	World Bank/WRI	Implemented
Santiago	World Bank	Partially implemented
Lima	World Bank	Advanced planning stage
Dar es Salaam	UNEP/ITDP	Advanced planning stage
Colombian Cities	World Bank UNEP/ITDP (Cartagena only)	Pereira and Cali under construction
Accra	World Bank	Early planning stage
Dakar	UNDP/ITDP	Planning stage
Hanoi	World Bank	Planning stage

Source: Compiled by ITDP.

successful wedge issue to begin a process of change in the approach to dealing with urban transportation that can be easily expanded to include measures promoting public space, cycling and walking, and transit-oriented development.

Recently, the GEF has played an increasingly important role in financing the initiation and development of good BRT projects. Examples of modal shifts due to BRT systems appear in Table 10-3. Some projects are listed in Table 10-4. Several others are in the discussion stage. The most important factor in the development of most of these projects was not the possibility of GEF funding but rather high-profile meetings and presentations made by the dynamic former mayor of Bogota, Enrique Penalosa and former Curitiba mayor Jaime Lerner, and the powerful visible impact that Bogota's TransMilenio and Curitiba's BRT have had with political leaders around the world.

The GEF's involvement in BRT promotion and implementation has been particularly successful in Mexico City's new BRT system, though the GEF was only one of several actors involved. The World Bank and World Resources Institute's EMBARQ program, with money from the Shell Foundation, played the most important international role, with some modest ITDP involvement. Initially, the World Bank had hoped that the GEF money would be used for a BRT system in the state of Mexico that surrounds the Federal District of Mexico City that would serve as a feeder system to the existing metro, which faces declining ridership. The state of Mexico, however, was heavily in debt and unable to borrow additional funds to implement the project.

Meanwhile, Penalosa from Bogota and EMBARQ played a key role in convincing Mexico City Mayor Obrador to consider BRT on a major urban arterial where it would get political exposure, and the decision was made to put the first corridor along Avenida Insurgentes, largely because the bus routes were controlled by only one private bus concessionaire, making institutional conversion easier. The GEF paid the salary of the staff person based in the municipal government that led the technical development of the project. The system, despite issues with the ticketing system and construction, is now carrying 250,000 passengers daily. While the GHG emission benefits have yet to be quantified, they are likely to be quite positive.

Detailed planning has also been done on a BRT system in Lima, Peru. The new system is fully designed, but as of the end of 2005, contracts for the construction of the BRT corridors and bus operations had yet to be awarded. Initially, the Lima BRT system was to receive GEF funds to finance the scrapping of older, highly polluting buses by the new BRT bus operators, but the scrapping component of the project has itself been scrapped. The actual construction of the system is to be covered by a joint loan from the World Bank and the Inter-American Development Bank, and not by the GEF. The BRT system in Lima has been continually delayed by political struggle with a competing metro project and general political turmoil.

TransSantiago's BRT project in Santiago, Chile, is similar to the bus sector reforms in Sao Paulo's Interligado system rather than the Bogota or Curitiba systems. Roughly 26km of new exclusive bus lanes have been built, and bus routes have been restructured to reflect more trunk and feeder lines, increasing bus system profitability. Some new articulated buses have been procured, but the new bus lanes continue to have old buses operating on them. There is no prepaid platform-level boarding and alighting, and the system does not have a clear marketing identity. The few new buses procured operate both within exclusive bus corridors and also in mixed traffic.

Of the two sections of exclusive bus corridors, the first section was open for bus traffic in the fall of 2005.

The Dar es Salaam project is the farthest along of the GEF projects being funded for BRT in Africa. In this case, the GEF money is routed through the UNEP to the ITDP. This money is covering the business plan, the institutional development and capacity building of the BRT agency, which will be called DART. The physical design and operational plan, costing roughly $1 million, is being financed by a World Bank loan to the national government, but the authority for the planning has been vested in a project management unit under the Dar City Council. In this case, the World Bank is involved in the design and is a likely source of funds for implementation, but not with GEF funds. Though other implementation financing options exist, they come with more strings attached and are less desirable as a result. While the detailed designs should be completed in early 2006, it will probably be 2008 before implementation because of the time it is likely to take to put together the financing. A similar project in Dakar may be done under the UNDP, if the national government gives the project its approval. This second Africa project would probably also rely on the World Bank to finance the infrastructure.

The BRT projects being financed through the GEF by and large are good projects, and the institutions sponsoring them are quite competent. As with the NMT projects, however, the BRT projects are not categorically going to reduce GHG emissions. Several new BRT systems have been developed in recent years that have had adverse GHG impacts at least in their initial phases, such as the trial phase of the Beijing BRT system. The Beijing system initially carried less than 2,000 passengers per day, largely because there is no feeder system, the busway is only separated from mixed traffic where there is no congestion, and the busway enters mixed traffic at the most congested part of the corridor. In 2006, however, the line was extended to 15 km and ridership has increased to over 70,000 daily trips.

A project by TransJakarta in Indonesia had technical support from the ITDP using a U.S. Agency for International Development (AID) grant, but it financed the infrastructure and buses using general funds from the DKI Jakarta government budget. It is not yet a complete success, but it is politically successful, and the second and third corridors are already under implementation. The GHG emissions benefits have been significant.

The problems faced by the Indonesian project have been largely the result of underestimating the technical complexity of designing a successful BRT system, and an unwillingness to involve foreign experts directly in the design process. At first, the governor and members of his staff who had never even seen a BRT system began to design a system with curb level bus lanes and no enclosed stations. After technical staff visited BRT systems in Bogota and Quito under the U.S. AID project, the designs were changed to median bus lanes with prepaid boarding stations, but the stations were extremely small. After flying the governor of Jakarta personally to Bogota, he ordered the stations redesigned. The system still uses buses with only a single door, which is creating passenger boarding and alighting bottlenecks.

A second problem stems from the lack of a feeder bus system and from allowing most of the existing bus lines to continue to operate in mixed, and congested, traffic lanes. The result of this decision was that ridership on the busway was lower than it could have been, and the continuation of many buses in the mixed traffic lanes significantly contributed to congestion in these lanes. As a result, travel speeds in the mixed traffic lanes dropped much more than anticipated.

These problems have cut the capacity of the Jakarta system to a third or less of what it could be. This has had an adverse effect on congestion and the concentrations of some transportation air pollutants. Evidence suggests, however, that the BRT has been successful in significantly increasing modal shifts. Some 19 percent of the TransJakarta passengers had previously been using taxis, private cars, or motorcycles. As of the fall of 2005, daily ridership has risen to a reasonable 75,000, and it has been increasing by about 10,000 daily passengers each year. Now that more parallel bus lines are being eliminated, passengers are being forced into two fare zones, with some adverse equity impacts, but this is also helping to decongest the mixed traffic lanes.

Ideally, BRT systems are designed to reduce mixed traffic congestion. This is not ideal in terms of maximizing short-term GHG emission reductions, but in the initial phase of a BRT project, political acceptance of the new system and building a political coalition for expanding the system is most important to the system's long-term impact on modal split. It is normally easier to win political acceptance for the system if even private motorists benefit from the new system. This is generally achievable in BRT projects by removing high volumes of existing buses from the mixed traffic lanes.

Other Areas for Future GEF Transport Sector Involvement

There are four other transportation project categories which are not currently receiving funding from the GEF but should be eligible under OP#11 enabling guidelines. They could play an important role in reducing transport sector GHG emissions. The project categories include TDM, traffic avoidance, nonmotorized vehicle improvements, and travel blending or social marketing. Brief descriptions of recent program activity in each of these areas follow.

Traffic Demand Management (TDM)

The potential GHG emission reduction benefits from TDM projects should in principle be far greater than through any other mechanism. As a result, the GEF should actively pursue projects in this area. As TDM is one of the surest ways of inducing modal shift, such projects should be eligible for GEF funding under existing guidelines.

The success of the BRT schemes in Bogota, Curitiba, and other cities was in part made possible by the simultaneous implementation of a number of TDM measures. These measures reduce the total number of trips by a particular mode, normally private motor vehicle driving. It is important to distinguish such measures from traffic management measures generally aimed to increase the capacity of the existing road system, which tend to stimulate rather than suppress traffic. Assessing the separate impact of the BRT versus the TDM measure in this context is thus quite difficult. However, some countries have implemented only TDM measures, and these efforts provide some indications of the relative importance of this approach.

Congestion pricing is the holy grail of TDM measures, as it could, in theory, fully internalize the marginal social cost of operating a private motor vehicle into the individual's cost of making the trip. The recent political success of London's congestion charging scheme has proven that it is politically possible to implement congestion charging in a democracy. Singapore's area licensing scheme, which has recently been upgraded to include electronic road pricing, provides another example of congestion charging that has been successful.

Another fundamental obstacle to implementing TDM measures is political will. TDM measures are technologically, contractually, and institutionally complex, sapping enthusiasm of political leaders looking for quick fix solutions. Willingness on the part of the World Bank GEF to help Sao Paulo's former mayor with a pilot congestion charging project played an important role in overcoming the reluctance of political leaders to secure preliminary political commitment. The project stalled, however, when the mayor lost the election and the new mayor did not want to continue.

Finally, most cities indirectly subsidize on-street and sometimes off-street municipal parking by charging far less than market value for parking. In the developing world, undercharging for parking is an indirect subsidy for the wealthiest sector of the population. Many of the benefits of congestion charging could be achieved by internalizing parking charges.

Traffic Avoidance

The existing GEF guidelines mention traffic avoidance as a desirable goal of transportation programs. Traffic avoidance generally refers to programs that seek to eliminate motorized trips, mainly by individual passenger cars, in the long run through colocation of economic, commercial, and residential activity. Transit use and walking and cycling can be encouraged by increasing the density of development along corridors served by transit or provided with good walking and cycling facilities. Zoning regulations can be modified to encourage transit-oriented development. This has been successfully done in Curitiba and elsewhere. Such zoning changes do not really require the involvement of the GEF because they can be implemented at the local political level with little direct cost.

More complex and difficult is intervening in the land development process early on to proactively encourage transit-oriented development. In the United States, rapid motorization and suburbanization starting in the 1920s and continuing today has led to a massive problem of urban blight that drove people from high-density urban environments to outlying suburban regions. This dramatically escalated the U.S. vehicle miles traveled and GHG emissions. The scale of this phenomenon in the United States is unique, but urban blight is spreading to numerous other megacities as motorization and suburbanization accelerate. Downtown Sao Paulo, North Jakarta, and the downtown brownfields of the former Socialist block countries are just a few examples of urban blight and rapid suburbanization.

Africa faces a unique problem. With a handful of exceptions, African cities were always historically associated with colonialism, and never became vibrant centers of indigenous culture. Developing viable projects to revitalize these city centers, particularly around transit nodes, will require a new set of technical and institutional skills and relationships that are largely absent in many developing African cities.

There is fairly limited international development bank involvement in traffic avoidance projects. No GEF funds have been used to support traffic avoidance measures, but the World Bank has funded some interesting urban revitalization projects in northern Africa with non-GEF funds, particularly in Morocco. The European Bank for Reconstruction and Development (EBRD) is exploring an urban revitalization project in Lublin, Poland. The Inter-American Development Bank has an ambitious city center revitalization project for downtown Sao Paulo that merits monitoring. These experiences need to be systematically assessed, and an appropriate role for GEF funds, if any, should at least be considered.

Low income housing projects that site new facilities on transit corridors should also be considered. The major development banks largely pulled out of the housing sector in the 1980s, but they have renewed activity. Housing loans have increased in Mexico, for instance, although there is still no direct linkage of these loans to transit-oriented development.

Nonmotorized Vehicle Interventions

The OP #11 guidelines do not clearly specify that nonmotorized transport interventions be focused on bicycle lanes, though in practice this is how the guidelines have been interpreted. While the initial focus on hydrogen fuel cell buses may have soured the GEF on working with the vehicle sector, to the extent that GEF funds are going to be used to make vehicles cleaner, they should also be usable to promote the dissemination of vehicles that do not generate any pollution. To date, there has not been any GEF activity in this area, but there has been work done by NGOs in this area, with financing from a range of bilateral donors.

Nonmotorized vehicle interventions began in rural areas in the 1970s. Most were aimed primarily at increasing farmer productivity. The British-based Intermediate Technology Development Group (ITDG) and the consulting firm that grew out of this group, IT Transport, were early leaders in this area. In the 1980s, a large number of groups began exporting used bicycles from the United States, Japan, and Europe to Africa, mainly as a charitable activity. This has led to the emergence of a viable secondhand bicycle market in a few countries, especially Ghana. Studies indicate, however, that the vast majority of the bicycles are used in rural areas and few replace motorized trips; hence, their GHG emissions impact is marginal. Most did not lead to continuing viable commercial supply of nonmotorized vehicles or a viable bicycle industry. This industry has emerged independently in many countries.

Inspired by the successes of ITDG, which were primarily focused on rural poverty alleviation, the U.S.-based ITDP began in the 1990s to try to apply some of the lessons from these projects to urban areas. For years the Indian and Bangladeshi governments supported projects to improve cycle rickshaw technology, but none of them led to any significant commercial adoption as they remained based at university research departments.

In 1997, ITDP began a U.S. AID-funded Indian cycle rickshaw modernization project by working with existing cycle rickshaw manufacturers in the Agra and Delhi regions to make the vehicle more comfortable and to bring down the vehicle weight in order to attract more passengers. The cycle rickshaw had not substantially changed in design since its introduction in the 1940s. Being a technology used primarily by low income people, profit margins were low, and there was little capacity or incentive for the business community to modernize the vehicle.

The ITDP's technical experts worked with local industry to develop a rickshaw design that was more comfortable, 33 percent lighter weight, and carried more passengers and baggage, yet cost the same to produce. Thanks to extensive promotional work, this new design has caught on commercially, and there are an estimated 150,000 of these modernized cycle rickshaws on Indian roads, all being manufactured and sold with no subsidies. According to surveys by a local NGO called Lokayan, each vehicle makes on average nine trips per day, and their average trip length is 1 km. While most of their passengers were taken from traditional cycle rickshaws, 11 percent were diverted from bus trips, 6 percent from auto rickshaws, 19 percent from higher capacity auto rickshaws known as vikrams, and 2 percent from motor scooters. Using emission factors from Table 10-1 and the Urb-Air Study from Mumbai, an estimate of 3.2 tons of CO_2 emissions reductions per day is a reasonable estimate of the impact of this technological innovation. It is possible that the entire fleet will switch to the modern design over the next decade. If this happens, 1,980 tons of CO_2 emissions may be reduced per day.

The entire rickshaw project cost was only $350,000. Moreover, the incomes of the cycle rickshaw operators increased by 20 to 50 percent. These operators are among the lowest-income people in India.

This project is currently being replicated for the becak, a vehicle similar to the rickshaw, in Indonesia under the auspices of the ITDP, Gadjah Mada University, GTZ, the Toyota Foundation, and Instrans, a local NGO. To date, roughly 110 modern becaks have been built, but finding a solid market has proven difficult. Unlike in India, where in many cities the market for new cycle rickshaws was tens of thousands of units per year, in Yogyakarta competition from motorcycles has led to a declining use of becaks, and the willingness of fleet owners to invest in new vehicles has been weak.

The ITDP is also trying to modernize and popularize the urban bicycle in Africa with its California Bike project. Many of the reasons for the lack of cycling in African cities are related to road safety, but some of the reasons are related to the immaturity of the bicycle industry. For decades, the traditional English roadster has dominated the African bicycle industry. When first introduced among the very poor, these vehicles were high-status items. As years passed, however, these bicycles became associated more and more with the rural poor. Lack of modern high-quality bicycles with trusted brand names undermined bicycle use among young people. Rather than switching to modern bicycles, much of urban Africa switched instead to cars and paratransit vehicle use.

The ITDP developed the California Bike in cooperation with Trek. The bike is one of the more expensive bicycles available in Africa, but it is at least 25 percent cheaper than any bicycle of equivalent quality available in the African market. By combining the sales of bicycles to small, independent bicycle dealers, donor agencies, government agencies, and large employers who sell the bikes to their own staff through payroll deduction schemes, the ITDP has managed to sell all 1,920 bicycles sent to four countries—Ghana, Senegal, South Africa, and Tanzania—in 2004, and earned a 16 percent rate of return on its initial investment. The project has brought 35 independent private bicycle dealers into a distributors network and trained them in modern business techniques. A second shipment of an additional 1,920 California Bikes arrived in Africa in 2005 and the bicycles are selling well.

Travel Blending or Social Marketing

Travel behavior is in part cultural. People take more polluting modes mainly because they are cheaper, faster, and more convenient, but there is also a cultural element in the decision-making process that should not be overlooked. Most people find it difficult to switch all of their trips to nonmotorized and less-polluting modes, but recent efforts show that many people can easily be convinced to switch to less polluting modes for at least some of their trips with modest levels of effort. Recent studies show that cultural attitudes not only are important, but they are also subject to influence through social marketing and should not be seen as an excuse for doing nothing.

Bogota's Mayor Penalosa used the mayor's office as a bully pulpit to promote car-free days, car-free Sundays, and other high-profile public events that were able to create a culture of walking and cycling. The international Car Free Days movement has caught on in Europe and many cities in the developing world, from Surabaya, Indonesia, to Chengdu, China, to Paris, France. The ITDP and other NGOs have sponsored similar events around the world to great effect.

Several cities in Australia and Europe have developed a new technique for achieving dramatic changes in mode shares at very low costs through a form of social marketing known as "travel blending." The idea is to give people more information on their commuting options through a completely personalized process and then facilitating changes in travel behavior. While the focus to date has been in developed countries, recent successes in Santiago de Chile indicate that it may be applicable to higher-income developing economies.

The technique involves phone contact with all households in the area, identifying the proportion of respondents who would be interested in making some changes in travel behavior and supplying them with information, such as public transport timetables and maps of cycling routes. Household visits are conducted with interested respondents. Respondents are asked to complete seven-day travel diaries, which teams later analyze to devise suggestions on alternatives for the participant. In some cases the diaries and interviews lead to changes in the local transportation systems—such as better access to public transport services, new bus stops, provision of new timetables, and the extension of service hours.

The results to date have been remarkable. In the first trial in Perth, Australia, approximately $61,500 was expended to conduct the surveys and provide information. Of the 380 households targeted, the program produced a 6 percent decrease in auto use immediately and an additional 1 percent decrease after 12 months. Public transport trips rose from 6 percent of all trips to 7 percent, and cycling trips doubled from 2 to 4 percent. The results have held even two years after the assistance was delivered. The technique is now being applied throughout Australia and in some cities in Europe, where similarly impressive results are being achieved at extremely low costs.

The consulting firm Steer Davies Gleave implemented a travel blending program in Santiago, Chile. The Santiago results suggest that travel blending could become part of an effective, low-cost emissions reduction package for cities in developing nations. Steer Davies Gleave reports an astonishing 17 percent reduction in car driver trips as a proportion of combined participating and nonparticipating households, with a 23 percent reduction in kilometers driven and a 17 percent drop in the time spent traveling (Hutt, 2002).

Travel blending techniques may be well suited to an active role by NGOs, particularly in the collection of survey data and the development and dissemination of transport alternatives. In many communities, NGOs maintain a close dialog with residents and thus would be well suited to this sort of activity.

Conclusions

The GEF is an increasingly important source of financing for bringing about a fundamental transformation in travel behavior that could dramatically reduce the level of GHG emissions being generated from the transport sector. This benefit is disproportionate to the money it brings because it overcomes the critical structural difficulty faced by the development banks: proactively generating good projects worthy of bank financing. After a questionable start, the priorities of most recent GEF projects are well focused on reasonable projects with promising opportunities to bring about significant GHG emission reductions. Now the challenge is to ensure the implementation of successful projects. Everyone has a stake in their success.

References

ECMT/OECD. *Urban Travel and Sustainable Development*. Paris: 1995.
Global Environmental Facility (GEF). *Operational Program #11: Supporting Environmentally Sustainable Transport*. 2001.
Hutt, G. Una herramienta para el cambio de conducta. Tranvia, No. 18, 1 July 2002. www.revistatranvia.cl/revista/tv18.
International Energy Agency. *Sustainable Urban Transport Program (SUTP) Draft Final Report*. Paris: IEA, 2002.
Karekezi, S., L. Majoro, and T. Johnson. "Climate Change and Urban Transport: Priorities for the World Bank." Washington, D.C.: Environment Department, The World Bank, 2002.
Menckhoff, Gerhard. Interview by telephone to Washington, D.C. August 2005.
Sperling, D. and D. Salon. *Transportation in Developing Countries: An Overview of Greenhouse Gas Reduction Strategies*. Arlington, VA: Pew Center on Global Climate Change, 2002.
World Bank. *Cities on the Move: A World Bank Urban Transport Strategy Review*. Washington, D.C.: World Bank, 2002.
Wright, L., ed. *Bus Rapid Transit Planning Guide*, 3rd ed. New York: ITDP, 2006.

CHAPTER 11

What Multilateral Banks (and Other Donors) Can Do to Reduce Greenhouse Gas Emissions: A Case Study of Latin America and the Caribbean

Deborah Bleviss

Energy use and greenhouse gas (GHG) emissions from transportation in developing countries are increasing more rapidly than in the wealthier industrialized countries, as shown in Figure 11-1. Most strategies and policies under consideration to counter these trends, such as fuel economy standards, target the design of vehicles or the fuels they use. GHG reductions achieved by these types of strategies, however, are likely to be dwarfed by increased emissions from the expected flood of new vehicles in developing countries, most of which will inevitably be fueled by carbon-intensive petroleum fuels. The light-duty vehicle stock in developing countries is projected to equal that in the rest of the world by 2050, compared to only one-third of the rest of the world today.

Developing countries suffer from substantial infrastructure limitations that raise questions about whether or not the high rate of growth projected will actually be realized. The capacity of existing roads is very low, leading already to severe traffic congestion, not only in the megacities of the developing world, but in many of the secondary and tertiary cities as well. Most of these countries have limited access to funds to expand the existing

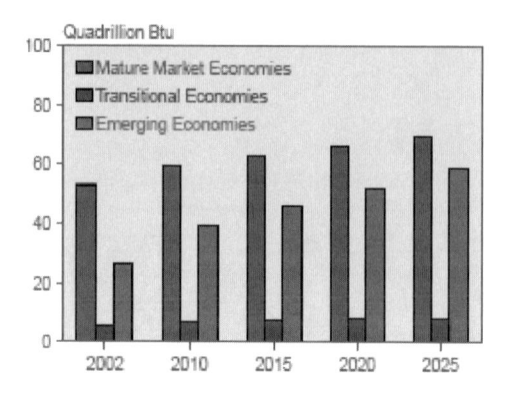

FIGURE 11-1. Projected transportation sector energy consumption by region, 2002–2025. Source: EIA, 2005.

infrastructure. Moreover, even with the current infrastructure and vehicles in these countries, their cities are already suffering from substantial—and increasingly intolerable—levels of air and noise pollution from transportation, levels that will only rise as the number of roads and vehicles grow.

With rising oil prices, oil import costs have also become unbearable. The existing levels of congestion only exacerbate pollution and increase oil consumption, since fuel consumption and pollution are the highest at the lower road speeds and stop-and-go conditions of congestion. Finally, traffic congestion has increasingly hampered the mobility of the poor, who comprise the majority of the population in these countries. They rely primarily on walking and public transportation to meet their mobility needs, both of which are compromised as private vehicles increasingly dominate the existing roads.

Given the challenges facing developing countries, a systemic approach is needed to slow the rate of energy growth for transportation and, therefore, the rate of growth of GHG emissions. This type of approach, referred to as "sustainable transportation," enables transportation and mobility needs to be met in a financially sustainable manner, while also minimizing local pollution, global greenhouse gas emissions, noise, accidents, congestion, and barriers to transportation access by the poor. In addition to the policies already mentioned, such an approach also consists of the following strategies:

- Emphasize high-quality, efficient, and clean public transportation for a substantial majority of the population, including the middle class
- Establish incentives and capacities for good nonmotorized transportation, including walking and biking, as well as other alternatives to motorized transportation such as telecommuting and electronic commerce

- Discourage the use of private vehicles when other modes are available
- Encourage good land use management to reduce congestion and promote demand for public transportation and nonmotorized transportation

The adoption of sustainable transportation approaches broadens the types of governments involved in transportation policymaking. Rather than a primarily national governmental approach that is necessary for implementation of policies such as fuel economy standards and adoption of alternative fuels, local governments must become increasingly involved as well, since they are responsible for oversight of public transportation and implementation of land use priorities.

Opportunities for Donor Agencies in Climate Change and Transportation

Multilateral development banks (MDBs) and other donor agencies have many opportunities to assist in the implementation of sustainable transportation strategies in developing countries to reduce the risk of climate change. These include tools directly available to limit carbon emissions from the transportation sectors of developing countries, as well as indirect tools that are oriented to other goals but that provide the cobenefit of reducing GHG emissions.

The most mature of the direct tools to address transportation and climate change in developing countries is within the Global Environmental Facility (GEF). The GEF was established in 1991 to help developing countries fund projects and programs that protect the global environment, including those that protect against climate change. Operational Program #11 within the climate change portion of the GEF's mandate specifically addresses environmentally sustainable transportation, including improved transportation systems. GEF projects in sustainable transportation, as in other areas, may be developed directly by the World Bank and the regional development banks, including the Asian Development Bank, the Inter-American Development Bank (IDB), the African Development Bank, and the European Bank for Reconstruction and Development. The United Nations (UN) may also develop GEF projects, primarily through the UN Development and Environment Programmes. Other donors may develop GEF projects by partnering with one of these agencies.

A new tool to address transportation and climate change in developing countries is the Clean Development Mechanism (CDM), defined by the Kyoto Protocol, which recently entered into force. The CDM enables public and private entities in developed countries to invest in projects or programs in developing countries that reduce projected GHGs. By doing so, these entities may get credits that can be applied against the GHG emissions targets agreed to by industrialized country signatories to the Kyoto Protocol. Many bilateral donors have been sponsoring demonstration programs using the CDM for several years in preparation for the full-scale implementation of the Kyoto Protocol. Moreover, several developed country governments, led by the Netherlands, are already planning to invest in CDM projects, in several cases developed by MDBs, to acquire carbon credits. All CDM projects must be approved by the CDM Executive Board. To date, the Board has not approved any transportation system projects, although one application is pending from the city of Bogota, Colombia.

There are other programs by donor agencies that can render benefits for GHG emissions in transportation. Programs to reduce local air pollution, if focused on transportation systems, often reduce GHG emissions as well. Programs in governance reform, particularly to support the decentralization of governance, can lead to the creation of strong and capable local agencies specializing in the planning for and regulation of local transportation systems. Assistance in urban development reform and investment can include strategies to encourage better land use management and creation of pedestrian-only commercial zones. Donor programs in transportation sector reform and investment can include the establishment of bus rapid transit systems to improve public transportation, sidewalks, and other pedestrian facilities, and strategies to reduce the excessive driving of private motorized traffic, including taxis.

While there are numerous opportunities for donor agencies to support activities in sustainable transportation that will reduce the risks of climate change, the challenges are also substantial. A viable sustainable transportation approach requires that donors work with national and subnational governments. To improve the chances of success, it is especially important to identify cities and city governments that are most prepared to move forward with sustainable transportation approaches.

The IDB undertook a study in 2004 to determine which cities in Latin America and the Caribbean are most prepared to advance to become candidates for GEF or CDM funding (Bleviss, 2004). The study had the following objectives:

- To identify medium-sized cities undertaking some sustainable transportation activities
- To define criteria in the context of successful experiences elsewhere for identifying the cities most prepared to advance
- To identify a first tier and a second tier of cities most prepared to advance, using the defined criteria
- To provide recommendations to the IDB on next steps based upon the analysis undertaken

Review of Cities and Development of Criteria

Over 50 cities in Latin America and the Caribbean were identified in the IDB study that are undertaking some types of sustainable transportation activities. Most involve efforts to reform urban public transportation systems. However, activities in such diverse areas as creating bicycle paths, developing pedestrian walkways, and establishing parking programs were also identified. Sources of information in identifying the active cities included existing literature, websites, and information from contacts throughout the region, including from the MDBs. Successful experiences from five cities around the world were documented to serve as a baseline for evaluation criteria. The cities were Curitiba, Brazil; Bogota, Colombia; Cuenca, Ecuador; London, England; and Singapore.

Curitiba was the first city to initiate a unique transportation path. Initial efforts began during the 1970s for this city with a population today of 1.6 million. The goal was to avoid a transportation path of intense motorization, exemplified by Los Angeles, California. Emphasis in Curitiba was placed instead on developing an efficient and cost-effective bus transportation system that would enable users to easily and quickly travel from their homes to work and other destinations. This system consists of high volume, exclusive lane trunk lines along major arteries of the city, which subsequently connect to smaller feeder lines and then to neighborhood lines. One fare is charged for entrance into the system, and all transfers to other lines are free.

Long-term concessions are awarded on a competitive basis by the city to private professional companies to operate specific public transportation lines. The municipal government retains the responsibility to plan future changes and expansions to the public transportation system and provides oversight and regulation of the concessionaires. Accompanying the design of this public transportation system in Curitiba was the development of substantial pedestrian-only walking areas, including shopping districts; land use code changes that required the most dense development to be near public transportation lines; and an extensive bicycle path system.

Bogota, Colombia, a much larger city than Curitiba with a population of over 7 million, adopted many of the principles of Curitiba when it began reforming its transportation sector in the mid-1990s. Today, it has a single-fare public transportation system, also consisting of trunk lines connecting to feeder lines. The process of creating this trunk-and-feeder-line system also included reform of its public transportation concessions process. New, exclusive concessions are now awarded to professional companies, in many cases consisting of groups of individual bus owners that had previously long dominated public transportation. In addition to public transportation changes, the city also enjoys pedestrian-only zones and an extensive bike system that is integrated with the public transportation system. Furthermore, to discourage the driving of private cars during peak traffic hours, the city has instituted a constraint on vehicle use. Vehicle owners are prohibited from using their vehicles at peak hours on specific days of the week, depending on the last number on the license plate.

Cuenca is a smaller city than Curitiba, with just over 250,000 inhabitants. Nevertheless, it, too, has adopted many of the features of larger cities.

They include an integrated public transportation system with some exclusive-lane trunk routes and new public transportation concessions agreements focused on professional companies, pedestrian-only areas, and bicycle lanes. The city has also developed a well-enforced parking program that has reduced traffic congestion substantially. Vehicle users buy prepaid parking cards and park their cars in spots where the allowable length of time for parking is clearly marked by the city.

Finally, London and Singapore have adopted congestion pricing to discourage use of private vehicles in their center cities. In these cities, private vehicles must pay a toll for entering designated areas. Furthermore, Singapore has invested in an efficient metro rail system. It also has other strategies for discouraging the ownership and driving of private vehicles, including a vehicle quota system in which the government determines how many new vehicles of different size classes may be registered each year. An auction is conducted on the Internet for certificates of entitlement allowing purchase of these vehicles by the highest bidders. In addition, Singapore has established a road tax that is assessed annually and increases with the size of the vehicle engine and the age of the vehicle.

Based on the experiences of these five cities, two categories of criteria were developed in the study for the IDB—those criteria most important in the short term and those most effective over the longer term. The identification of cities most prepared to advance tended to weight criteria aimed at the short term more heavily, with the thought that the criteria aimed at the longer term could be developed by the cities that were selected as they evolved. Criteria identified as most important for cities to advance in the short term included:

- Strong support and leadership from political leaders, especially at the local level. In the experiences of both Curitiba and Bogota, the mayors of these cities established the initial vision, worked with the public to gain its acceptance, created the local governmental capacity to plan for and regulate the transportation sector, and stuck with their vision even when problems occurred.
- Substantial progress in establishing and beginning to implement effective transportation master plans. Such plans need to address reform of the public transportation system in most cities, although the most successful cities have also added components to encourage nonmotorized transportation and to discourage the use of private vehicles.
- Strong local planning capability in transportation and urban planning, preferably in a local government institution. Curitiba and Bogota owe a good measure of their success in sustainable transportation to the establishment of such institutions. The most successful are characterized by their capacity to address a multitude of issues related to transportation, rather than just being constrained to planning for public transportation.

- Strong local regulatory authority to design public transportation concessions and oversee the sector. Local regulatory authorities in Curitiba and Bogota were critical in the reform of the public transportation sector, which included the renegotiation of concessions agreements with larger, more professional public transportation companies.
- Existence by the local government of the financial capacity to invest in the needed transportation infrastructure. Ultimately, the success of a sustainable transportation strategy depends on a municipality's capacity to invest in the necessary infrastructure, either by borrowing or through internal resources.

The following criteria were identified as most important for cities to advance in the longer term:

- Ample awareness of and support by the public of efforts to change the transportation system. The most successful cities have sought to build support from all major stakeholders, including existing bus owners and operators and their associated support industries such as mechanics and garage operators, public transportation users, and the general public.
- Substantial progress in the decentralization of governance. The most successful cities have established the capacity to create strong local planning and regulatory institutions without the interference of national institutions. In addition, they have had the local capacity to collect revenues to support these institutions.
- Significant advancement in establishing and implementing an Urban Development Master Plan. Such a plan has proved most successful when it is structured to complement a Transportation Master Plan, emphasizing mixed use development and denser development near public transportation corridors.
- Sufficient financial resources for the operating budgets of local transportation sector planning and regulatory agencies. The most successful cities have dedicated specific revenues, such as licensing fees or inspection fees, for the budgets of these agencies so that they can achieve greater autonomy from political pressures that inevitably would arise if such organizations had to rely solely on the annual municipal budgeting process.
- Progress in transforming public transportation concession owners into mature public transportation companies. In most Latin American and Caribbean cities, one of the greater challenges to successfully implementing public transportation reform is facilitating the transition between the present owners and operators of the system, who tend to own only one or two buses and have access to very limited financial credit, and the operators of the future, which will hopefully be professional companies with the expertise to run their companies well and the financial assets to invest in their buses and other equipment.

- A strategy developed for attracting financing into new public transportation companies. A major challenge for these companies is being able to attract both debt and equity financing. In several cities, public and private financing programs have been established to respond to this challenge.

Identification of Candidate Cities

Two tiers of cities were identified by the IDB: a first tier of those most prepared to advance and a second tier of those still prepared to advance but not as quickly as the first tier. For both of these tiers, the ranking of each city was determined by how they "rated" against each criterion. The identified first tier cities were Concepcion, Chile; Cordoba, Argentina; Cuenca, Ecuador; Fortaleza, Brazil; Guatemala City, Guatemala; Queretaro, Mexico; and Quito, Ecuador.

Table 11-1 shows the ranking of each of these cities by criterion; criteria identified as most important for the short term are shown in gray.

The identified second tier cities were Arequipa, Peru; Cali, Colombia; La Paz, Bolivia; Panama City, Panama; San Salvador, El Salvador; Sao Bernardo do Campo, Brazil; and Rosario, Argentina.

Table 11-2 presents the ranking of each of these cities by criterion; again, criteria identified as most important for the short term are shown in gray.

Conclusions and Recommendations

The IDB study demonstrated that MDBs and other donors have the potential to play a catalytic role in helping cities in Latin America and the Caribbean make progress toward sustainable transportation as part of a strategy to reduce carbon emissions growth from the transportation sector in the developing world. The process of identifying the cities most prepared to advance in sustainable transportation strategies enables cities to be targeted where the chances of success are greatest. Similar studies need to be done examining opportunities in Asian and African cities.

The study identified specific unmet needs facing MDBs and donors in the area of climate change and transportation. Chief among these is the lack of adequate methodologies and data to assess baseline carbon emissions and the carbon emissions savings resulting from transportation system improvements. The most efficient way to address this inadequacy is for MDBs and donors to work together to develop common methodologies and improve data so that projects may eventually qualify for CDM credits. Otherwise, needless delays will occur, as differing data sets and methodologies are likely to emerge. Additional funds will inevitably be required downstream to develop the common data sets and methodologies needed in the first place.

What Multilateral Banks Can Do to Reduce Greenhouse Gas Emissions   197

TABLE 11-1. Ranking of First-Tier Cities by Criterion

City	Political Commitment	Transportation Master Plan	Local Planning Capability	Local Regulatory Capability	Development of Public Transport Companies	Government Financial Investment Capability
Concepcion, Chile	√	√√	?	?	√√	√√
Cordoba, Argentina	√	√	√	√	—	?
Cuenca, Ecuador	√√	√	√√	√√	√√	√
Fortaleza, Brazil	√	√	√√	√√	√√	√
Guatemala City, Guatemala	√√	√	√	√	√	√
Queretaro, Mexico	√	√	√	√	√	√
Quito, Ecuador	√	√	√	√	√	?

City	Public Support	Decentralization of Government	Urban Development Master Plan	Sufficient Resources for Local Institutions	Financial Climate for Transport Companies
Concepcion, Chile	√	?	√√	?	√
Cordoba, Argentina	√	√	√	?	—
Cuenca, Ecuador	√	√	√	√√	?
Fortaleza, Brazil	?	√√	√	√√	√
Guatemala City, Guatemala	?	√	√	√	√
Queretaro, Mexico	√√	√	√	?	?
Quito, Ecuador	?	√√	?	—	?

√ Positive   √√ More positive
— Negative   √√ Most positive
? Not known
Source: Bleviss, 2004.

TABLE 11-2. Ranking of Second-Tier Cities by Criterion

City	Political Commitment	Transportation Master Plan	Local Planning Capability	Local Regulatory Capability	Government Financial Investment Capability
Arequipa, Peru	√√	—	√	√	—
Cali, Colombia	√√	√	?	?	√
La Paz, Bolivia	√	—	?	?	?
Panama City, Panama	?	√	?	?	√
Rosario, Argentina	?	√√	√	√	—
San Salvador, El Salvador	—	√	?	?	√
Sao Bernardo do Campo, Brazil	√√	—	√	√	√

City	Public Support	Decentralization of Government	Urban Development Master Plan	Sufficient Resources for Local Institutions	Development of Public Transport Companies	Financial Climate for Transport Companies
Arequipa, Peru	?	√	—	?	?	?
Cali, Colombia	?	√	?	?	?	?
La Paz, Bolivia	?	√	√	?	—	?
Panama City, Panama	?	—	?	?	√	√
Rosario, Argentina	?	√√	?	?	—	—
San Salvador, El Salvador	?	—	?	?	√	√
Sao Bernardo do Campo, Brazil	?	√√	?	?	√	?

√ Positive √√ More positive √√√ Most positive
— Negative
? Not known
Source: Bleviss, 2004.

The study also found that assistance needs for a city or group of cities are often too large for one donor to handle. Hence, there is a need to work together among MDBs and donors to deliver assistance effectively. The ongoing assistance in Lima, Peru, on implementing a sustainable transportation system is a good, but rare, example of the type of cooperation needed. At present, assistance is being provided in Lima by the World Bank, the IDB, the United States Agency for International Development, and Swiss and Japanese bilateral aid agencies.

Finally, the study identified ways in which MDBs and donors that are working in developing country cities in other areas can include in their assistance components that will benefit transportation and climate change. By doing so, donors' strategies can yield bigger benefits. With regard to air pollution from transportation, for example, at present, most donor support has concentrated almost exclusively on vehicles themselves, rather than the overarching transportation system in which they operate. In particular, the activities have concentrated on the development of minimum vehicle emissions standards and associated vehicle inspection and maintenance, and the development of cleaner fuels. Major air pollution reductions that can accrue through transportation systemwide improvements, such as the widespread use of bus rapid transit systems, are not adequately assessed in these narrow programs.

Similarly, donors' support of government reform should include the development of local transportation planning and regulatory agencies. Traditionally, donors providing assistance on government reform have concentrated on finance and budget agencies, since they are critical to the long-term financial sustainability of governments. Expanding the donor effort to include transportation and planning agencies can also achieve important results, and it can open a door to strengthening other government functions. It is also critical that donors assist in identifying independent sources of financing for these agencies in order to lessen the political pressures on them.

Yet another focus for donor assistance in governance reform should be an assessment of administrative options to address the challenges of multiple governmental jurisdictions over urban transportation. In most urban areas, transportation systems extend beyond the jurisdiction of a single municipal government. While the United States dealt with this challenge with the creation of metropolitan planning organizations governed by each of the relevant municipal governments in an urban region, similar structures rarely exist today in developing countries.

In assistance on urban development reform and investment, donors can also catalyze a linkage between Urban Development Master Plans and Transportation Master Plans. Donors often focus their assistance on development of one of these plans, but linkages between the two are frequently missed. Unless land use and development is explicitly addressed in Transportation Master Plans, cities risk seeing their efforts to rehabilitate their public transportation systems fall short as the cities sprawl and the costs of the systems skyrocket in trying to serve these sprawled populations. Similarly, unless transportation needs are explicitly addressed in Urban Development Master Plans, city planners could well see goals fail because the needed transportation systems do not reach urban subcenters away from downtowns created by urban master plans.

Finally, donors can encourage cities to develop integrated Transportation Master Plans, which address all modes of transportation. At present, most donor assistance in transportation focuses only on infrastructure building—primarily roads and bridges—and, more recently, on reform of the public transportation systems. Without planning transportation systems as a whole, the interactive effects of all modes of transportation are frequently missed. All the benefits, for example, of integrating a bicycle path system or pedestrian system with public transportation can be lost by concentrating on only one mode of transportation. It is also important that donors provide assistance to analyze the environmental, social, and economic impacts of transportation master plans, including the impacts on greenhouse gas emissions.

References

Bleviss, D. *The Opportunities for Sustainable Urban Transportation in Medium-Sized Cities in Latin America and the Caribbean.* Washington, D.C., November 2004. www.iadb.org/sds/doc/Bleviss.pdf.

The Sustainable Mobility Project 2004. *Mobility 2030: Meeting the Challenges to Sustainability.* World Business Council for Sustainable Mobility, July 2004. www.wbcsd.org/DocRoot/nq0qcbBcHFhW4iBEYWTs/mobility-full.pdf.

U.S. Energy Information Administration (EIA). *International Energy Outlook, 2005.* Washington, D.C.: EIA, July 2005. www.eia.doe.gov/oiaf/ieo/pdf/0484(2005).pdf.

World Business Council for Sustainable Development (WBCSD). *Mobility 2030: Meeting the Challenges to Sustainability.* 2004. www.wbcsd.org/web/publications/mobility/overview.pdf.

CHAPTER 12

From Public Understanding to Public Policy: Public Views on Energy, Technology, and Climate Science in the United States

David M. Reiner

As in many political disputes, opponents in partisan battles over energy and environmental policy often invoke public opinion to justify their preferred position or policy choice. Being able to cite favorable public opinion polls or other indicators of public concern can provide an important source of legitimation for arguments in support of specific policies. The link between public opinion and policies can, however, be problematic. There are a host of reasons, including biases in the questions asked to assess public opinion, why indicators of public approval or disapproval might not offer a basis for action. This chapter explores some of the difficulties of translating public awareness and understanding first into public opinion and then into policy actions. In particular, the chapter presents evidence of public attitudes on the question of climate change because the associated policy debates are often viewed in a strongly partisan light in political debates in the United States.

An issue as technically complicated as climate change poses a series of demanding conditions in translating public support into public policy. The first hurdle is one of awareness. Many people simply do not pay attention to the issues involved in energy or environmental policy and so will be unaware of many of the issues of concern relevant to climate policy. Basic awareness does not imply understanding of the basic scientific facts or the underlying mechanisms that lead to climate change. In the absence of a firm grasp of the facts, there is a wide range of cognitive biases that plague individual assessment, and complicated technical problems are especially prone to such biases (Kahneman, Slovic, and Tversky, 1982).

In turn, basic understanding does not imply clear opinions on associated policies. A scientist or technologist steeped in technical details, for example, might well be indifferent with regard to which specific policies are enacted. By the same token, ignorance does not preclude strong opinions. More generally, positive or negative views do not reveal the strength of those views. Strong opinions do not necessarily translate into political support, nor does affiliation with a political party necessarily imply support for their party's position on every issue. Next, general views do not always translate into support for specific action, particularly those that will have a direct personal or local impact. Finally, opinion need not translate into individual actions or active political support that might move a policy forward. It is these missing links in the chain linking public understanding with public policy that will be explored by drawing upon public opinion results in the areas of science and technology, energy, and the environment.

Applied to the question of energy choices and their interaction with climate change policy, one might expect that some disconnects are severe. Public attention to both science and technology and to environmental issues is relatively low. Moreover, the underlying scientific evidence for climate change that would motivate action has been contested, at least in the United States, leading to further uncertainty. Possible solutions to the climate problem are often of a technical nature and themselves subject to considerable confusion. Translating those imperfect understandings into opinions can be affected both by political affiliation and personal interests. The environment has become a partisan issue in the United States that can then impact perceptions and attitudes. Even the term *environmentalist* has become laden with different meanings, with diverse implications for attitudes and policy choices. Many environmental issues also come with their own set of local and personal repercussions, such as siting of facilities or impacts on individual behavior. Finally, there may still be notable differences between the public's preferences and the choices made by their elected representatives.

Establishing the links between understanding and policy choice has to date been imperfect at best. Any inquiry, therefore, must rely on incomplete evidence. Probing these links is vital, however, to producing a richer appreciation of how the public can influence policy debates and, equally, to understanding how ongoing policy debates might influence the public.

Public Awareness

There is a relatively small informed audience for many policy questions. Researchers have concluded that less than one-fifth of U.S. residents meet

TABLE 12-1. Levels of Public Attention to Policy Issues (2002)

Issue	Attentive Public	Interested Public	Residual Public
Local schools	31	28	41
Foreign policy	5	23	72
New scientific discoveries	7	39	53
The use of new inventions	6	36	58
Science and technology	10	48	42
Space exploration	5	21	74
New medical discoveries	14	51	35
Environmental pollution	10	38	52
Economic issues	12	33	55
Agriculture	6	23	71
Military/defense	7	31	62

Source: NSB, 2002, Appendix Table 7-7.

a minimal standard of civic scientific literacy (Miller, Pardo, and Niwa, 1997). Looking across a range of policy issues, the National Science Foundation's Survey of Public Attitudes towards Science and Technology found that roughly 10 percent of a representative sample of the U.S. population could be categorized as "attentive" on both science and technology and on energy and environmental issues. New medical discoveries score slightly higher (14 percent), but it was local issues such as schools that registered by far the highest level of attention (31 percent).

In Table 12-1, the attentive public is defined as those who express a high level of interest in a particular issue, feel very well informed about the issue, read a newspaper on a daily basis, and regularly pursue news magazines or a magazine relevant to the issue. By contrast, the interested public consists of those who claim to have a high level of interest in a particular issue but who do not feel very well informed about it. The residual public consists of those who are neither interested in nor feel very well informed about a particular issue.

Science-attentive members of the public are most likely to be male, young, better educated, and affluent. They are also likely to vote, be politically active, be savvy about technology, and understand scientific information with minimal explanation (Borchelt, 2002). Even among those with graduate or professional degrees, however, less than 25 percent are considered attentive to science and technology issues. Similarly, attentiveness is low even among those with a "high" level of education in science and mathematics defined as having taken nine or more high school or university level science or math courses. Only 15 percent are considered attentive, although this percentage is still three times the level for those with "low" levels of science and math education (NSB, 2002, Appendix Table 7-8).

With regard to particular technologies, some register much greater awareness than others. Transport and renewable energy received above-average attention among a long list of energy-related technologies in studies at MIT (Curry et al., 2004; Curry, 2004). The highest level of awareness registered was for more efficient cars, which 70 percent of the U.S. public admitted to having heard or read of in the past year. Next were solar energy (65 percent public recognition), nuclear energy (55 percent), and wind energy, energy-efficient appliances, and hydrogen cars, all registering public recognition just below 50 percent. By contrast, technologies such as biomass or carbon capture and storage technologies were acknowledged with a high level of awareness by 10 percent or less of those people surveyed. Even these figures are likely to be inflated, since respondents are often reluctant to admit ignorance. Close to 20 percent of the public admitted not to have heard or read of any of the listed technologies.

Public Understanding

Given the low level of attentiveness to science or environmental issues among even highly educated members of the public, it is not surprising that many people have fundamental misunderstandings of basic scientific facts. For a number of years, the National Science Foundation (NSF) has tested public understanding of basic scientific terms and concepts (NSB, 2004). It has found that some concepts are understood by 80 to 90 percent of the public, including that the center of the Earth is very hot and that the oxygen breathed by humans is emitted from plants. Similarly, almost 80 percent of the public is aware that the Earth goes around the Sun. Other factual questions scored closer to 50 percent, indicating ignorance and, in some cases, that myths have taken root, displacing the scientifically correct answer. The idea that earliest humans lived at the same time as the dinosaurs, that antibiotics kill viruses, and that lasers focus sound waves are areas where misconception was common. Responses to identical surveys in the European Union are similar or, quite often, worse. Furthermore, polls in both the United States and Europe find that these correct or incorrect conceptions are virtually unchanged over time, indicating the level of public knowledge is not increasing in either place.

As part of its studies of environmental literacy, the National Environmental Education and Training Foundation (NEETF) has sought to develop a set of knowledge questions on basic environmental and energy facts (NEETF/Roper, 2002). Given a list of ten factual questions on energy, the average respondent was only able to identify slightly more than four correct answers on average, as shown in Table 12-2. Respondents were asked, for example, to describe the leading source of electricity in the United States and to identify examples of renewable resources. While almost two-thirds were able to recognize that heating and cooling used the most energy in the home, only one-third recognized that burning coal, oil, and wood is by far

TABLE 12-2. Public Assessment of Basic Energy Facts

Energy Knowledge Question	Percent Correct
Source of most energy usage in average home	66%
Percentage of oil imported from foreign sources	52%
Percentage of world's energy consumed by United States	50%
Disposal of nuclear waste in the United States	47%
Fastest and most cost-effective way to address energy needs	39%
U.S. industry increased energy demands the most in past 10 years	39%
Fuel used to generate most energy in United States	36%
How most electricity in the United States is generated	36%
Sector of U.S. economy consuming greatest percentage of petroleum	33%
Average miles per gallon used by vehicles in past 10 years	17%
Average number of correct answers:	4.1

Source: NEETF/Roper, 2002, pp. 4–5.

the largest source of electricity, accounting for roughly 60 percent of total generation. The top choice was hydroelectric power, which was selected by 39 percent of the respondents. A further 12 percent chose nuclear power as the leading power source (Coyle, 2004, p. 35). In spite of high levels of awareness of transport issues, some of the most persistent misperceptions were with respect to transport. For example, only 17 percent were aware that the average fuel efficiency of vehicles decreased over the course of the 1990s.

Only 1 percent of the respondents achieved an overall score of 9 out of 10 correct answers, 3 percent answered 8 out of 10 correctly, 8 percent had 7 correct answers, and 13 percent had 6 out of 10 answers correct. Men scored notably better, with 68 percent failing by scoring below 60 percent, compared to 84 percent of women. Perhaps unexpectedly, the youngest demographic group, aged 18 to 34, did not score best on environmental or energy knowledge. That distinction went to somewhat older respondents aged 35 to 64. The eldest age group performed worst on the knowledge question (Coyle, 2004, pp. 16–19).

The public demonstrated the same mix of understanding and confusion on specific questions linking energy technologies and their environmental impacts. Asked how different technologies contribute to carbon dioxide levels, the vast majority of respondents were able to identify that cars, coal power plants, and steel mills increased carbon dioxide levels, that trees reduced carbon dioxide levels, and that wind turbines did not contribute to an increase in carbon dioxide. The one area where respondents showed considerable confusion was with regard to nuclear power plants, where the majority of respondents either did not know or gave the wrong answer (Curry et al., 2004, pp. 4–5).

Similar results were reported in the 2003 Eurobarometer on Energy, which found that nuclear power was perceived as having a significant impact on global warming across most member states in the European Union, including France, with heavy reliance on nuclear power, and Germany, which had an acrimonious political debate over phasing out nuclear power (EC, 2003). Only in Sweden and Finland, which conducted intensive national dialogues over nuclear power over the course of many years, did the majority understand that nuclear power does not contribute to global warming.

The complicated, nonlinear, and contested nature of climate change means that there are relatively few that have accurate conceptions of either the problem or of possible solutions. There is an obvious confusion between weather and climate. Given the variability of both weather and climate, there is a tendency to confuse the two. Moreover, there is an inevitable attraction to ascribe causes even when it is difficult or impossible for even experts to discern a specific cause. Just prior to the United Nations climate conference in Kyoto in 1997, during an El Niño year, U.S. interviewees were asked, "What is the major source of the recent 'strange weather'?" (CBS News/NY Times, 1997). Thirty four percent cited natural variability, and 17 percent named El Niño—both plausible explanations. Many others offered a variety of other causes, including pollution or resource degradation (11 percent), space junk (10 percent), divine intervention (8 percent), and ozone depletion (8 percent). Global warming was suggested by only 5 percent of the respondents, the same percentage that answered, "Don't know."

Establishing the correct answer for such a question is virtually impossible, since many scientists themselves do not agree. At best it might be possible to say that the weather in 1997 was being affected by a variety of factors or that such unusual events might become more likely as a result of global warming. Many of the answers are clearly incorrect, but particularly notable is the reluctance to admit they simply "don't know" to complicated technical questions, which testifies to the difficulty in accepting uncertainty.

Finally, it is not only science or technology that can be the subject of misunderstanding. There is also considerable confusion over policy. Many studies, for example, find that the public favors U.S. participation in the Kyoto Protocol (Harris Interactive, 2002). When asked President Bush's position on the Kyoto Protocol, however, there was an even split in the general population. Roughly 60 percent of Republicans espoused the belief that President Bush supported the agreement, which he, in fact, emphatically rejected in the first days of his administration (PIPA/KN, 2005, p. 4). Willingness to support action on climate change also was related to beliefs over U.S. actions relative to those of other developed countries. While 44 percent believed that the U.S. effort was comparable to other developed countries, 24 percent believed that the United States was doing more than average to limit its greenhouse gases, which included 14 percent of Democrats and 38 percent of Republicans. Only 27 percent thought the United States was doing less than average, including 40 percent of Democrats and 16 percent of Republicans.

Impact of Information on Public Opinion

Understanding the impact of information is critical for complicated technical questions that are subject to so many misconceptions. New information can sometimes affect public opinion in important ways. For example, although confusion over facts remains, there has been a notable shift recently on the acceptance of the science of climate change in the United States. Public recognition that "there is a consensus among the great majority of scientists that global warming exists and could do significant damage" has grown from 28 percent in 1994 to 43 percent in 2004 and to 52 percent in 2005. At the same time, the view that scientists are divided on the existence of global warming and its impact fell from 58 percent in 1994 to 50 percent in 2004 and 39 percent in 2005. It appears that Republicans account for most of the recent shift. Republican believers in a scientific consensus on global climate change increased from 30 percent to 41 percent, while doubters fell from 63 to 46 percent (PIPA/KN, 2005, p. 5). Although the U.S. political parties remain divided, their partisan supporters are actually moving closer.

Table 12-3 summarizes the results of an investigation into the impact that increased recognition of the scientific consensus on global climate change could have on public policy preferences. The half of the sample that was asked to assume there was "a survey of scientists that found that an overwhelming majority have concluded that global warming is occurring and poses a significant threat" was notably more amenable to taking more aggressive action than those that were not asked to make that assumption.

TABLE 12-3. Effect of Asking Respondents to Presume a Consensus on Science of Global Warming

Preferred Policy	*Not Asked to Assume Consensus (half of sample)*	*When Asked to Assume Consensus (half of sample)*
Not take any steps to reduce greenhouse gases that would have economic costs	21%	6%
Take steps to reduce greenhouse gases, but only those that are low in cost	42%	35%
Take steps to reduce greenhouse gases, even if this involves significant costs	34%	56%

Source: PIPA/KN, 2005.

While the majority might now recognize the scientific consensus, there is still a small but respectable core of skeptics. In spite of three major international assessments of the science done by the Intergovernmental Panel on Climate Change involving many leading U.S. scientists over the last decade and a U.S. National Research Council report commissioned by President George W. Bush in 2001, public awareness of the consensus statements from the scientific community is still relatively low, undoubtedly affected by the partisan political disputes over the Kyoto Protocol.

In terms of solutions for addressing climate change, there is a strong and clear preference for new renewable energy technologies, such as wind and solar, and considerable optimism about the costs of these technologies. An MIT study on the future of nuclear power found that providing information had an impact on support for nuclear energy. The largest shift to nuclear power occurred when information on relative prices was provided (MIT, 2003). Similarly, if information on the technology including costs and emissions was provided, another MIT study found that support for both carbon capture and storage and nuclear energy increased substantially (Curry et al., 2004). Nevertheless, in both surveys, support for renewables remained quite strong even in the face of information showing much higher costs than many might have expected. Another notable finding of the NEETF survey cited earlier was that knowledge improved the espoused "pro-environment" response by 10 to 15 percent in many areas such as stated willingness to recycle, turn off lights and appliances when not in use, and lower the thermostat to conserve energy. Nonetheless, greater knowledge of energy and environment problems did not increase the willingness for other desirable behaviors, most notably using other forms of transport instead of driving or to accelerate slowly to conserve gasoline (NEETF/Roper, 2002, Figure 22).

Strength of Opinion

What does it mean to say that the public "favors renewables" or "supports the Kyoto Protocol"?

There are many economic or regional explanations that might be invoked to explain the current situation, but even a narrow focus on public opinion will give pause to the view that public support means that a policy will inevitably be enacted. Support or opposition does not indicate the depth of support or the reaction to moving forward with a policy. Simply lining up supporters against opponents may not offer much insight into the resulting political dynamic.

The issue of opening up the Arctic National Wildlife Refuge (ANWR) to oil exploration provides an example. Opposition to exploration changed slightly from 56 percent opposing exploration in 2002 to 53 percent in 2005, while support for exploration moved from 35 percent to 42 percent. More relevant for policymakers, however, was the fact that almost all that opposed exploration said they would be upset if drilling was allowed, whereas less than half of those that supported drilling said they would be upset if the refuge was not opened to drilling (Moore, 2005). Fully one-third do not care whether oil exploration proceeds or not, although almost all respondents were willing to voice an opinion.

There can also be significant differences between general beliefs and those at the local level. For example, 54 percent of U.S. residents favor the use of nuclear power to provide electricity for the country, including 17 percent who "strongly favor" nuclear power, against 43 percent who oppose the idea, including 22 percent who voice strong opposition. By contrast, 63 percent oppose building a nuclear power plant in their area, including 4 in 10 who oppose the idea strongly. By contrast, only 35 percent favor the construction of a plant in their area (Carlson, 2005).

In spite of generalized support, there may also be differences in approval of different policies to accomplish the same objective. A survey conducted by Yale University in May 2005 sought to identify preferences regarding policies to address U.S. dependence on imported oil. It found that mandating fuel-efficient vehicles was the leading option among survey respondents. Requiring the auto industry to make more fuel-efficient cars was viewed favorably by 93 percent of the national sample, including 85 percent of Republicans and 90 percent of sport utility vehicle owners. By comparison, 40 percent of the respondents favored a tax on cars with poor gas mileage, and only 15 percent supported increasing the tax on gasoline (Yale, 2005, p. 5). Thus, there may be tensions between the preferred policies of economists or policy analysts and the public. Another example involves the idea of emissions trading. When first proposed by policy analysts, there was considerable opposition. When informed that emissions trading would reduce the costs of compliance from $50 a month to $10 a month, support rose from 34 to 66 percent (PIPA/KN, 2004, pp. 11-12).

Changing Behavior and Perceptions of the Role of the Consumer

Most ambitious proposals to address climate change inevitably require some degree of personal sacrifice or inconvenience for consumers, whether that means higher prices or changing behavior patterns. The biggest problem in studying behavioral changes is that usually all that can be discerned is professed behavior. Thus, it is impossible to know if the NEETF study, which showed that more educated respondents were more likely to say they turned off the lights to save energy, actually showed that this group did, in fact, turn off lights to the same extent (NEETF/Roper, 2002). Perhaps their greater awareness of energy issues made them recognize that they should turn off lights more frequently. This could convince them to answer such a question in the affirmative, regardless of their personal behavior.

There is an inevitable tendency to underreport consuming behavior, particularly when conservation and efficiency are perceived as virtues. In certain cases, it is possible to compare the professed behavior against actual data, but this is usually quite difficult. Nevertheless, it can still be informative to examine how individuals claim to react to changes in their environment.

Issue framing can be particularly important in designing behavioral questions. For example, at the time of the surge in gasoline prices in the United States during the summer of 2005, 60 to 70 percent said that the price increases had led them to "cut down" on driving (CBS News, 2005). Yet, in that same period, only 32 percent said they drove less than the previous year when the price rise was not mentioned, while 50 percent said they drove the same amount and 14 percent said they drove more (ABC News, 2005). Final measurement of actual vehicle-miles traveled in 2005 might show some decline compared to 2004 when fuel prices were cheaper, but not nearly to the extent that the first polls might indicate.

Similarly, two-thirds of those polled in August and September 2005 agreed that gasoline price increases had led to "financial hardship" when the average U.S. retail price was just over $3.00 (CNN/USA Today/Gallup Poll 2005). However, almost 50 percent agreed with the same statement in March 2004 when the average price of gasoline was $1.70 and in May 2001, when the price was about $1.66 (EIA, 2005).

The other aspect of views regarding personal behavior that is relevant is the extent to which consumer demand is recognized as having an important role in contributing to either environmental change or to price hikes. Asked in an Associated Press/America Online poll, "Which ONE of the following would you say deserves the most blame for higher energy prices?," fewer than 10 percent were willing to hold the drivers of "gas-guzzling vehicles" responsible, compared to the 30 percent that blamed "oil companies that want to make too much profit," or the 20 percent of the respondents that attributed most blame to the "foreign countries that dominate oil reserves" or to "politicians" (AP/AOL, 2005). Similarly, when a Fox News/Opinion Dynamics survey asked who had "the most control over gas prices," 36 percent named domestic oil companies or producers as having the greatest role, compared to 13 percent that cited government, 12 percent that cited OPEC members or the Middle East, and 10 percent that cited the President or the Bush Administration. Only 5 percent cited consumers as having the most control (FOX News/Opinion Dynamics, 2005).

When posited in a more evenhanded manner, many respondents are more willing to admit that consumers do play some role. Asked who "should share the blame for the rise in gas and oil prices," 31 percent said "waste by consumers," and 20 percent said that this waste had at least some role. Only 7 percent said that consumers played no role (CBS News, 2005).

Identity Politics: Death of Environmentalism?

Another reason for exploring public attitudes in greater depth is the inclination of advocates on one side or another to invoke public opinion to support their position. One of the most influential, or at least one of the most discussed, challenges to the environmental movement came not from its traditional opponents on the right but from two avowed progressives, Ted Nordhaus and Michael Shellenberger. They argue that modern environmentalism, as embodied in the activities of the major national environmental groups, is "no longer capable of dealing with the world's most serious ecological crisis"—namely global warming (Shellenberger and Nordhaus, 2004, p. 6). Rather, after 15 years of campaigning and having spent hundreds of millions of dollars of funding from private donors and major foundations, environmental groups have "strikingly little to show for it." Their suggestion for moving forward relies on a heavily partisan assessment that calls for mobilizing the left rather than winning over moderates or conservatives.

The furor over their self-published paper, "Death of Environmentalism: Global Warming Politics in a Post-Environmental World," led to a rebuke from many environmental groups following its release in October 2004. Carl Pope, president of the Sierra Club, described the attack as "unclear, unfair and divisive." Pope took issue with many of the article's presumptions and conclusions and argued that the problem could be attributed to the left more generally, not just environmentalism. Pope did agree that environmental groups had "still not come up with an inspiring vision, much less a legislative proposal, that a majority of Americans could get excited about" (Mieszkowski, 2005).

Critical to the thesis of Shellenberger and Nordhaus are several assumptions about public opinion. To overcome the slow progress of action on climate change, they promoted a broader progressive agenda that sought to leverage coalitions with unions, minorities, and other stakeholder groups. On this point at least, Pope supported the idea of a coalition of progressive organizations as an important step forward.

It is useful to review the evidence that Nordhaus and Shellenberger cite to support their case. They argue that conservatives have been more successful in crafting their message that members of the environmental movement hold dissimilar values from the average U.S. citizen. As evidence of this rightward shift, they noted the following:

> *The number of Americans who agree with the statement, "To preserve people's jobs in this country, we must accept higher levels of pollution in the future," increased from 17 percent in 1996 to 26 percent in 2000. The number of Americans who agreed that, "Most of the people actively involved in environmental groups are extremists, not reasonable people," leapt from 32 percent in 1996 to 41 percent in 2000. (Shellenberger and Nordhaus, 2004, p. 11)*

One of the core problems in gauging support for environmental measures and the environmental movement in particular is that there are wide differences over what being an "environmentalist" means. A Gallup poll in 2000 found that while 83 percent said they agreed with goals of the environmental movement, only 16 percent described themselves as active participants in the environmental movement. Similarly, in a 2002 poll, Gallup found that 70 percent of Americans described themselves as either active in the environmental movement or sympathetic to it. About one-quarter were neutral, leaving only some 5 percent who were unsympathetic (Saad, 2003; Harris, 2005).

Even more problematic for the Nordhaus and Shellenberger thesis that success lies with a progressive coalition, support for environmentalism is virtually unrelated to political persuasion. Work by Gallup in 2003 found that of the 14 percent of the sample that said they were an "active participant in the environmental movement," 37 percent were self-described conservatives, 39 percent were moderates, and 20 percent liberals. These ideological breakdowns are not very different from those of the general population—17 percent liberal, 41 percent moderate, and 39 percent conservative (Crabtree, 2003). An additional reason to be skeptical of redefining environmentalism as a progressive issue is found in the earlier finding that Republicans accounted for the bulk of the shift towards greater credence in the scientific consensus "that global warming exists and could do significant damage."

Based on interviews with active members of some 20 environmental groups in the northeastern United States and control groups, Tesch and Kempton identify four quite distinct groups that fall under the catchall term of "environmentalist" (Tesch and Kempton, 2004). Based on their categorization, an environmentalist can be someone who does the following:

- Claims to be concerned about the environment, but takes no action
- Acts to preserve local habitat usually through private actions, also called a conservationist
- Participates in the political process by writing to public officials or attending hearings, also called an activist
- Participates in various forms of direct action, such as civil disobedience, also called a radical

These groups do not share common views, nor do they even all accept the label of environmentalist. With the exception of those belonging to radical or national environmental groups, 8 to 25 percent of environmental groups did not consider themselves environmentalists.

As one of Tesch and Kempton's interviewees described her dilemma over using the label, "You know you have to watch out for terms anyway because to term yourself or somebody else as an 'environmentalist,' 'religious fanatic,' or to put a label on somebody . . . [is] limiting and it's because you have an idea about what an environmentalist is and it might not be the same idea that I have, which may not be the same idea as somebody else has" (Tesch and Kempton, 2004, p. 77).

In that same vein, the issue is not only what people think, but what their leaders or elected representatives think they think. There is a notable disparity between the public's views and what leaders and political leaders expect those answers to be. Interestingly, not only did the vast majority of Democratic congressional staffers (94 percent) support the Kyoto Protocol, but so too did the majority of Bush administration officials (68 percent). Only Republican staffers, 21 percent of whom supported the Kyoto Protocol, were largely hostile (CCFR, 2004). Nevertheless, regardless of their personal views, most leaders presumed that the public was opposed to the agreement even though their overall level of support (71 percent) was almost identical to that expressed by the political leaders surveyed. Still, many more Democratic staffers (45 percent) and Bush administration officials (41 percent) correctly estimated at least the direction of support compared to Republican staffers (15 percent). The greater recognition of the public support of the Kyoto Protocol by Bush administration officials may help explain their personal backing for the treaty in spite of the administration's continued official opposition.

Shellenberger and Nordhaus argue that "the truth is that for the vast majority of Americans, the environment never makes it into their top ten list of things to worry about. Protecting the environment is indeed supported by a large majority—it's just not supported very strongly. Once you understand this, it's much easier to understand why it's been so easy for antienvironmental interests to gut 30 years of environmental protections" (Shellenberger and Nordhaus, 2004, p. 11). Using their logic, if environmental concerns "never make it into their top ten," one might ask why the environment should be a political priority and why politicians should respond to such a low priority with substantial resources and aggressive regulation. The reality is that there has been longstanding bipartisan support for environmental regulation and it is strong bipartisan support that led to the passage of strong environmental legislation such as the Clean Air Act and the Clean Water Act. To imagine that regulation of greenhouse gases can be accomplished without enlisting moderates and conservatives defies history and more importantly ignores current trends in public opinion.

Conclusion

Simply because opinions may be influenced by cognitive bias, misinformation, or ignorance does not mean they are not legitimate. Every election and referendum is contested with imperfect information. Many more decisions, made on a daily basis by elected leaders and appointed regulators, are taken without explicit recourse to public opinion.

Nevertheless, one might still ask to what extent policymakers should seek to correct misunderstandings or simply proceed as if the public were more fully informed. The "paradox of representation" has long held that while legislators are elected to represent the views of their constituents, they are also elected to govern and are expected to lead and thereby influence public opinion rather than simply voting in the same manner as the public would on every issue. Leaders, therefore, may find themselves unaware of what views are held by the electorate.

Public support is neither necessary nor sufficient for a technology to succeed or fail, but public opinion can influence votes over legislation, research and development funding levels, and regulatory decisions, especially when issues rise to public attention. The level of knowledge of both science and technology and energy and environmental issues affects behavior and ultimately erodes support for the difficult decisions that will be needed on climate change. Current levels of both awareness and understanding of basic scientific facts are low and have not changed significantly in a decade. Knowledge is also sporadic, as some facts are well understood while others are subject to persistent myths.

Education clearly has a role to play. One might contrast the confusion over nuclear power with the remarkably high levels of public understanding of the role that trees play in the carbon cycle. At the same time, few have heard of such technical terms as biomass or carbon sequestration. For whatever reasons, certain basic facts are successfully imparted to the vast majority of the public. Both the media and the education system influence public awareness and understanding, and more effort should be given to better understanding that process.

On the issue of climate change, there remains a stronger level of public support for taking action than realized by many politicians. More importantly, the public has increasingly begun to accept the scientific consensus about global climate change, and with that acceptance comes a greater willingness to support more aggressive action. Much like environmentalism, calls for action on climate change, at least among the public, remain deeply bipartisan.

References

ABC News. Poll conducted by TNS Intersearch. August 18–21, 2005.

AP/AOL. Associated Press(AP)/America Online (AOL) poll conducted by Ipsos-Public Affairs. August 9–11, 2005.

Borchelt, R. "Research Roadmap for Communicating Science and Technology in the 21st Century." In *Communicating the Future: Best Practices for Communication of Science and Technology to the Public*. Gaithersburg, MD: National Institute of Standards and Technology, 2002.

Carlson, Darren K. "Public Warm to Nuclear Power, Cool to Nearby Plants: Gender, Politics Affect Responses." Gallup News Service, May 3, 2005.

CBS News. Polls conducted on October 3–5 and August 29–31, 2005.

CBS News/New York (NY) Times. Poll conducted on November 23–24, 1997. Roper Center Question ID: USCBSNYT.97011B, Q06.

Chicago Council on Foreign Relations (CCFR). *Global Views 2004: American Public Opinion and Foreign Policy.* Chicago: CCFR, 2004. Available at www.ccfr.org/globalviews2004/sub/pdf/Global_Views_2004_US.pdf.

CNN/USA. Today/Gallup. Polls conducted on September 26–28, September 12–15, and August 28–30, 2005.

Coyle, K. J. *Understanding Environmental Literacy and Making It a Reality: What Ten Years of NEETF/Roper Research and Related Studies Tell Us about How to Achieve Environmental Literacy in America.* Washington, D.C.: National Environmental Education & Training Foundation, 2004.

Crabtree, S. "Surprising Stats on 'Active' Environmentalists," Gallup News Service, April 8, 2003.

Curry, T. E. "Public Awareness of Carbon Capture and Storage: A Survey of Attitudes toward Climate Change Mitigation." Master's Thesis. Cambridge, MA: Massachusetts Institute of Technology, June 2004.

Curry, T. E., D. M. Reiner, S. Ansolabehere, and H. J. Herzog. "How Aware Is the Public of Carbon Capture and Storage?" In E. S. Rubin, D. W. Keith, and C. F. Gilboy, eds., Proceedings of 7th International Conference on Greenhouse Gas Control Technologies. "Volume 1: Peer-Reviewed Papers and Plenary Presentations." Cheltenham, UK: IEA Greenhouse Gas Programme, 2004.

Energy Information Administration (EIA). "Retail Gasoline Historical Prices." 2005. Available at www.eia.doe.gov/oil_gas/petroleum/data_publications/wrgp/mogas_history.html.

European Commission Directorate General for Research (EC). *Eurobarometer—Energy: Issues, Options and Technologies, Science and Society.* Report EUR20624. Luxembourg: European Communities, 2003.

FOX News/Opinion Dynamics. Poll conducted on September 13–14, 2005.

Harris. Poll conducted on August 9–16, 2005.

Harris Interactive. "Majorities Continue to Believe in Global Warming and Support Kyoto Treaty." 2002. Available at www.harrisinteractive.com/harris_poll/index.asp?PID=335.

Kahneman, D., Paul S., and A. Tversky, eds. *Judgment under Uncertainty: Heuristics and Biases.* New York: Cambridge University Press, 1982.

Mieszkowski, K. "Dead Movement Walking?" Salon, January 14, 2005. Available at archive.salon.com/tech/feature/2005/01/14/death_of_environmentalism/print.html.

Miller, J. D., R. Pardo, and F. Niwa. *Public Perceptions of Science and Technology: A Comparative Study of the European Union, the United States, Japan, and Canada.* Chicago: Chicago Academy of Sciences, 1997.

MIT. *The Future of Nuclear Power: An Interdisciplinary MIT Study.* 2003. Available at web.mit.edu/nuclearpower.

Moore, D. W. "Public Leans Against Oil Exploration in Arctic Reserve: About a Third of Americans Would Not Be Upset if Oil Exploration Occurs or Not." Gallup News Service, March 18, 2005.

National Environmental Education and Training Foundation (NEETF)/Roper. "Americans' Low Energy IQ: A Risk to Our Energy Future." Washington, D.C.: NEETF, August 2002. Available at http://www.neetf.org/roper/Roper2002.pdf.

National Science Board (NSB). *National Science Foundation Science and Engineering Indicators 2002.*

———. *National Science Foundation Science and Engineering Indicators 2004.*

Program on International Policy Attitudes (PIPA)/Knowledge Networks (KN). *Americans on Climate Change.* Washington, D.C.: PIPA, June 2004. Available at www.pipa.org/OnlineReports/ ClimateChange04_Jun04/ClimateChange04_Jun04_rpt.pdf.

Program on International Policy Attitudes (PIPA)/Knowledge Networks (KN). *Americans on Climate Change: 2005.* July 2005. p. 4.

Saad, L. "On Linking the Environment with Energy Policy." Gallup News Service, March 4, 2003.

Shellenberger, M., and T. Nordhaus. "Death of Environmentalism: Global Warming Politics in a Post-Environmental World." 2004. Available at www.thebreakthrough.org/images/Death_of_Environmentalism.pdf.

Tesch, D. and W. Kempton. "Who Is an Environmentalist? The Polysemy of Environmentalist Terms and Correlated Environmental Actions," *Journal of Ecological Anthropology* 8: 67–83, 2004.

Yale University School of Forestry & Environmental Studies. *Survey on American Attitudes on the Environment—Key Findings.* May 2005. p. 5. Available at www.yale.edu/envirocenter/poll2key.prn.pdf.

CHAPTER 13

Narrative Self-Identity and Societal Goals: Automotive Fuel Economy and Global Warming Policy

Kenneth S. Kurani, Thomas S. Turrentine, and Reid R. Heffner

> "As simple as it may seem, in the face of prevalent discourses and dominant knowledges, simply listening to the story someone tells us constitutes a revolutionary act."
> —Freedman and Combs, 1996

The "rational actor" model from economics has served as nearly the sole model of human behavior in transportation energy analysis and policy in the United States. Research by Kurani and Turrentine (2004) indicated that the behavior of automotive consumers is unlikely to conform to this model in either their vehicle purchase or use, at least as regards fuel economy. This disconnect between theory and behavior causes those working in the "rational analytic" framework to struggle to explain, for example, the growing popularity of high-fuel economy hybrid electric vehicles (HEVs). Unable to explain such behavior, policy recommendations in this framework to limit emissions of greenhouse gases (GHG) and reduce petroleum consumption fail to account for the full value that consumer/citizens may place on automotive fuel economy and other strategies. Arguing over cost-effectiveness, policymakers and politicians miss the full variety of policies and the full extent of changes that citizen/consumers will support.

From the perspective of rational economic choice, people are assumed to move through life making a series of utility-maximizing or satisficing or bounded rational choices. As developed later in this chapter, these choices are assumed to be based on preferences. However, preferences are generally taken as given; their origins and changes over time are rarely discussed. Even developments such as random utility theory are more concerned with the distribution of preferences at a point in time than in processes over time (see, for example, McFadden, 1986). Therefore transportation and energy analysis and policy have been blind to the development of new values and tastes that occur during rapid technology, market, and policy change. The reliance on the rational economic perspective in transport appears to stem more from a desire to quantify behavior for mathematical models, and less from an interest in how people behave.

Solutions to these problems may be found by enriching the behavioral approaches applied to transportation energy analysis and policymaking. In this chapter, such an alternative approach is developed and an example of its analytical and policy implications is presented. The alternative is more cultural, psychological, and process oriented. It focuses on the transforming influence of markets and mobility on ways of life. In particular, personal construction and communication of an authentic identity narrative gains importance in market-oriented, highly mobile societies. Moreover, identity increasingly is constructed and expressed through buying, owning, and using products. Thus, consumer products, especially automobiles and homes, acquire symbolic meanings of considerable importance.

Even seemingly functional and financial attributes of products, such as automotive fuel economy, may be evaluated symbolically for what they mean rather than algorithmically for what they cost or contribute to "utility." For example, a household that buys a high fuel economy HEV may apply the meaning "lower resource consumption/living lighter" to the HEV and never calculate the fuel savings due to its higher fuel economy. Lowering their resource consumption becomes an important subplot in a narrative self identity, first symbolized then made real by the availability, purchase, and use of a high fuel economy HEV.

The symbolic value of automobiles within identity narratives is a crucial perspective for policymaking regarding societal problems such as global warming. Social policies may be enhanced if consumers are offered products and policies that facilitate more interesting, compelling, and meaningful stories about themselves, rather than policy, legislative, and industry elites arguing over cost effectiveness. Rationality is less of a primary analytical framework for research and more a normative discourse for households and individuals. If some households strive to achieve this norm, others act on example, faith, impulse, intuition, or the advice of a stranger. Both groups share a desire to construct and represent meaningful lives.

In support of an approach based on narrative identities, the results of two series of household interviews conducted by researchers at ITS-Davis are presented. One focused on the role of fuel economy in household vehicle purchase and use (Kurani and Turrentine, 2004; Turrentine and Kurani, 2005). This research was based on in-home interviews with 57 households in northern California in 2003 and 2004. All interviewees had purchased a new or used automobile within the preceding year; eight had purchased HEVs. Prior to their interviews, households constructed complete vehicle ownership histories. These histories deepened the context for examining the recent vehicle purchase and provided additional vehicle purchases to discuss. The other series was conducted during the fall of 2004 and winter of 2005 and focused on vehicle purchase and use by buyers of the early "economy-tuned" HEVs: the Honda Insight and Civic Hybrid and the Toyota Prius (Heffner, Kurani, and Turrentine, 2005). Twenty-five additional HEV-owning households in northern California were interviewed. This study focused on carefully eliciting from households a range of functional and financial attributes and symbolic meanings and their roles in vehicle purchase, use, and postpurchase evaluation of their HEVs.

What Is a "Rational" Consumer and Does This Idea Dominate Transportation Energy Analysis?

The term *rational* is used here to refer to the model of consumer behavior from neoclassical microeconomics. In the words of J. P. Quirk, "Each individual is motivated by self interest and acts in response to it." Next, "decision makers' choices are consistent with their evaluations of their self-interest." And finally, for purposes of this discussion, "these choices could be predicted simply from a knowledge of their preferences and the relevant features of their alternatives" (Quirk, 1982).

Policymaking and analysis regarding consumers, automobiles, energy, and GHG emissions are waged largely within a "rational analytic" framework based on the rational consumer model just presented. For example, California's Health and Safety Code Section 43018.5 requires that "maximum feasible and cost-effective" measures be adopted to limit GHG emissions from passenger vehicles, light-duty trucks, and other vehicles the state determines to be used primarily for noncommercial, personal use. The explicit requirement that reductions be "cost effective" means that any citizen/consumer values not captured by the price of automobiles and fuels—for example, more meaningful and fulfilling identity narratives—won't count toward how much GHG emissions must be reduced.

Examples of analyses conducted within a rational analytic framework include studies of the effects of fuel economy standards on consumers (see, for example, Sennauer, Kinsey, and Roe, 1984; Greene and Liu, 1988; Goldberg, 1998; and Yun, 2002). Such analyses often identify particular "problems" for, or of, consumers. Discount rates for investments in energy conserving technologies are widely inferred and a rebound effect on travel from higher fuel economy is believed to be revealed (see, for example,

Greene, 1983; Train, 1985; Goldberg, 1998; Verboven, 1999; Greening, Greene, and Difiglio, 2000; Espy and Nair, 2005). The National Research Council's (2002) report on Corporate Average Fuel Economy (CAFE) standards explicitly assumes that consumers compare an upfront investment in higher fuel economy against a discounted future stream of fuel savings and used vehicle prices.

Further analytical examples from the record of this conference series include Miles-McLean, Haltmaier, and Shelby (1993), who address both consumer discount rates and rebound effects. Cameron (1995) employed a consumer surplus approach to estimate societal benefits of travel demand management policies—in particular a fee on vehicle miles traveled. In their analysis of the potential for pricing to reduce gasoline consumption over the near term and implicitly over the long term, DeCicco and Gordon (1995) invoke the rational actor model and point out that "if consumers were rational, cost minimizing, utility maximizers, the cost advantage of choosing higher fuel economy is relatively small in the context of the total cost of owning and operating a car."

DeCicco and Gordon's qualifier, "if consumers were rational," is crucial. Analyses such as those just cited are seldom conducted in settings that allow observation of whether or not consumers are behaving according to the rational actor model. Rather, payback periods, attribute tradeoffs, interest rates, and elasticities are inferred based on the assumption consumers are behaving rationally. Efforts to manage systemic, societal effects such as global warming depend on pulling back this "as if" veil of assumed economic rationality to reach an understanding of how people act.

Despite the near universality of the rational actor model in transportation energy policy analysis and policymaking, research by the authors (Kurani and Turrentine, 2004; Turrentine and Kurani, 2005) finds that few, if any, consumers actually treat their transportation energy expenditures, or any other vehicle expenditures, in accordance with this model. Based on detailed examinations of 120 household vehicle purchases by the 57 households, the authors concluded that households in the sample were not "rational." They found no instances in which fuel or travel expenditures were incorporated into payback period or net present value calculations or cost minimization or utility maximization algorithms. This does not mean households did not respond to changes in prices or other factors. Their responses, however, were not typically evaluated for their actual impact on household budgets, nor compared to other aspects of household spending. Rather, the households treated gasoline prices, the cost of a tank of gasoline, fuel economy ratings, or a particular trip as symbols of a good or bad decision or action. The results of the second set of interviews with HEV buyers also suggested the purchases of HEVs were better explained by something other than the rational actor model.

Economic rationality is apparently too scarce to serve as the sole behavioral model. The sample of 57 households in the fuel economy study is small, but the authors contend its design overrepresents subpopulations who have the knowledge and skills to implement rational analyses. More than one-third of the households contained at least one member who was a financial services professional, had some collegiate coursework on the topics of payback periods and net present value calculations, or ran a small business or farm. Also, at the time of the interviews gasoline prices on average were higher in California than nationally.

The research design and context were biased towards finding rational consumers, yet none were found. Even those respondents who do possess economic skills have not applied them to an automobile purchase. One of the financial service sector respondents replied to the question about how long to wait to be paid back by fuel cost savings by saying, "Oh, the payback period. I never thought of it that way." A few respondents track fuel use and expenses, but did so for maintenance purposes, and none could recall details or summaries of the costs or fuel amounts.

Hypothetically, sustained increases in gasoline prices could encourage more rational calculations by more automotive consumers as over time people could learn and apply rational analytic tools and methods. The interviews suggest however, that such change is unlikely. Further, such a hypothesis would still overlook the more compelling behavioral approach presented next.

An Alternative Behavioral Approach

The alternative approach presented in this chapter draws on several theoretical perspectives. First, social constructionist approaches "argue that people perceive the world the way they do because they participate in socially shared practices and interact with the world in terms of meaning systems which are simultaneously transmitted, reproduced and transformed in direct and symbolic social interchanges" (Dittmar, 1992). Further, a social constructionist perspective "views material possessions as socially shared symbols for identity" (Dittmar, 1992). When consumer goods serve as symbols, function and meaning coexist. For example, sport utility vehicles (SUVs) may signify "independence" because high ground clearance and four-wheel drive give them the capability to drive off-road. Heffner, Turrentine, and Kurani (2006) provide a more extensive review of social constructionist approaches and markets for automobiles.

Second, Nobel Prize–winning economist Gary Becker (1992) proposed that a household should be treated as a collective of individuals that acts to produce consumption from inputs such as income, time, goods, services, skills, and knowledge. Combined with social constructionist ideas, households can be seen to be creating novel value and new elements of their narratives through the purchase and use of goods and services.

Third, another Nobel Prize winner in economics, Daniel Kahneman, argues that intuition based on accessible information is the more common

TABLE 13-1. Two Behavioral Approaches

Rational Analytical	"Social Constructionist"
1. Decisions	Actions
2. Individual	Social creation and exchange of symbols
3. Self-interest	Self-identity
4. (Preexisting) preferences	Investment in consumption outputs
5. Relevant attributes of alternatives	Symbols and accessible attributes of perceived alternatives

basis for judgments than is reason, which includes such things as rational calculations (Kahneman, 2002; Kahneman and Frederick, 2002).

Last, the approach presented here draws on Anthony Giddens's structuration approach (1984, 1991), which defines and elaborates how the forces driving modernization link with social constructionist approaches to personal identity. Giddens identifies the conditions under which a person's identity has been transformed from a socially defined role into a personal project. Specifically, he defines "the reflexive project of the self" as "the process whereby self-identity is constituted by the reflexive ordering of self-narratives" (Giddens, 1991). Only under such conditions does it make sense to discuss the role of consumer products in self-constructed narratives of identity. Table 13-1 provides a contrast between the major points of the rational actor model just presented and the alternative approach to be developed in the remainder of this section.

Decisions and Actions

The first point of departure from the rational analytic framework is that not all behaviors are preceded by a decision. Therefore, a theory of decision making will omit from analysis some relevant actions. "Action" refers to a broader set of behavioral outcomes than "decisions," such that actions are roughly categorized into the subset of actions taken because of decisions to act and the subset of actions taken in the absence of decisions to act. Olshavsky and Granbois (1979) distilled "decision processes" into four components:

1. Two or more alternative actions exist, and, therefore, choice must occur.
2. Evaluative criteria facilitate the forecasting of each alternative's consequences for the consumer's goals or objectives.
3. The chosen alternative is determined by a decision rule or evaluative procedure.
4. Information sought from external sources and/or retrieved from memory is processed in the application of the decision rule or evaluation procedure.

They concluded that "a synthesis of research on consumers' prepurchase behavior suggests that a substantial proportion of purchases do not involve decision making, not even on the first purchase." One can fairly ask, do people buy automobiles without deciding to do so?

When symbol attribution and exchange substitute for rational algorithm, vehicle transactions may be made without decision making or with sharply attenuated decision making. What could be more linked to a rational decision about an automotive purchase than prices? Yet, changing vehicle and gasoline prices create emotional states—anticipation, excitement, anxiety, or anger, for example—that reveal symbolic meaning attached to products through their prices.

Such symbols can shift not only which alternatives are being considered, but can change, attenuate, or substitute for a decision-making process. Zero percent financing can be a symbol of a "good deal" but a misleading one. Take the case from the fuel economy interviews of a young household looking to replace a Honda Civic their family was outgrowing. They knew about the Civic Hybrid, but it was not bigger than the Civic they already owned. Sitting in a restaurant one morning, they noticed the Toyota dealer across the street. They walked over after breakfast, with no intention of buying. Toyota was offering zero percent financing on the 4Runner SUV. The couple became excited. When they got home, the husband—an accountant—made some quick calculations. The financed price of the 4Runner seemed so reasonable compared to the midsize sedans they had been considering. They bought the 4Runner.

The couple regretted their action within weeks. In his excitement, the husband had only considered the purchase price. Compared to their Civic, the increased cost per tank of gasoline and the increased frequency of refueling now serve as constant reminders that they acted impulsively under the guise of a calculated decision.

In other cases, the nonfinancial symbolism of prices moved respondents into more fuel economical vehicles. One of the HEV-owning households interviewed had initially been shopping for a new SUV to displace an old one. They had been involved in what Kahneman (2002) would call "reasoned judgment" and what Olshavsky and Granbois (1979) would recognize as decision making—a deliberative, slow, serial process involving evaluation of alternatives and exploration of future uses for the new vehicle. Together the male and female heads of the household developed evaluative criteria and rules: Their old SUV was too small, and its fuel economy was low and getting lower; they wanted more room and fuel economy of 20 miles per gallon, and they still wanted four-wheel drive. They gathered information, visited dealerships, and initially decided on a new Toyota Sequoia.

Before they made a down payment, however, the price of gasoline rose past $2.00 per gallon. The male head of household "flipped out," in his words. He came home and made a unilateral announcement that they would be buying a Prius. He said, "This is going to be my car. It is the car I have to drive every day. I'm buying a hybrid."

The prepurchase behavior switched from calculated decision making to noncalculated, emotion-driven action. He did not deliberately award a new higher value for a fuel economy evaluative criterion. He did not seek more information. He did not investigate if there were variants of the Sequoia or other SUVs with higher fuel economy. He did not seek any new information about the Prius. He did not calculate fuel cost savings. His action was about his anger at oil companies and oil-producing regimes in the Middle East. His Prius purchase was based on the symbolic meaning of readily accessible information—rising gasoline prices posted at every gasoline station—and associative emotion. His judgment was made quickly, even if he had to wait months for the vehicle. It was intellectually easy, if emotionally charged.

These examples show that emotion can give symbols great power. Bagozzi et al. (1999) define emotion as "a mental state of readiness that arises from cognitive appraisals of events or thoughts . . . [that] may result in specific actions to affirm or cope with the emotion." It is this "readiness to act" we heard preceding many HEV purchases. The HEV symbolizes a desired goal. The goal has an emotional component creating readiness to act. The readiness to act helps to prompt the act of buying the HEV in order to affirm that emotion, perhaps hope, or to cope with it, in the case of anger.

In the rational analytic framework, a properly constructed evaluation by a household considering the purchase of an HEV would compare a "with HEV" and a "without HEV" cost and benefit stream. Many households related that their HEV prompted its own transaction—that is, without the HEV, no transaction for a vehicle would have occurred. In such cases, rational analysts might think they should choose a preexisting vehicle in the household fleet for the "without HEV" case, on the argument the HEV displaced an existing vehicle—leaving aside for the moment those households in which the HEV both prompted a transaction and was an addition to the household fleet. If this preexisting vehicle is already fully paid for, and if the only positive differential value of the HEV is private gasoline cost savings, the HEV looks very expensive.

However, households in which the HEV prompted its own transaction tell a different story. First, no households talked about their HEVs only in terms of fuel cost savings. Few said that fuel cost savings were most important. Even the ones who did mention fuel savings tended to speak of low fuel cost or high fuel economy as symbols of smart consumerism. These households did not calculate fuel cost savings and factor these into a car buying decision. To them, "hybrid" meant "low fuel use achieved through high technology." There were two things about HEVs that allowed for this symbolism—a distinct technology and a nonincremental increase in fuel economy compared to the vehicles they were already driving. With this meaning, an HEV was seen as something a smart consumer would buy in a period of high and rising gasoline prices. In addition to "smart consumerism," the early fuel-economical, low-polluting HEVs symbolized "living lighter," "lower resource consumption," "clean air," "concern for others," "civic mindedness," and "high-technology." Of some concern for global warming policy, few households said HEVs symbolized "reduced risk of climate change."

Even when the HEV displaced a preexisting vehicle, keeping the former vehicle is not considered by its owners to be an alternative to the HEV. The HEV becomes a new symbol in the narrative identity of at least one member of the household and thus the narrative of the whole household. Interpreted this way, these households did not acquire an HEV as the result of a comparison to an existing household vehicle, but because the HEV uniquely symbolized a new narrative, a new story about whom the owner wanted to be. Since few vehicles prior to HEVs allowed for the incorporation of an automobile into a narrative of, for example, a high-tech nerd who wants to save the planet, few if any previous vehicles are regarded as alternatives.

A subset of prompted transactions are those households whose "choice set" consisted of only one vehicle. These households did not even compare across HEVs. Efforts by salespeople to steer these consumers to other vehicles were sometimes resented. One young woman went to a dealership to shop for a Honda Civic Hybrid, and, according to her, the salesman's efforts to sell her a nonhybrid variant—because it was a better buy—nearly cost him a sale. She was not at the dealership just to get a good deal on a car but to enrich and extend her narrative of her identity. The salesperson made the mistake of thinking she needed more and better choices, and perhaps of thinking he knew what choices were better for her.

Individuals in a Social Context

In the rational analytic framework, social exchanges may be used to gather information to support an individual's algorithmic decision. In the alternative approach developed here, the role of symbol creation and social exchange of symbols is more central to creating and expressing narrative identities. Such narratives develop in a specific cultural context. In this case, it is our consumer society.

Exchange is a fundamental social activity and has been the focus of cultural anthropology. Whether between kin, nations, kings and subjects, humans and supernatural beings, or marriage partners, social exchange initiates, sustains, and signifies relationships, which are real or imagined, economic, political, or spiritual. Social exchange—fundamentally symbolic—takes the form of gifts, songs, labor, meals, and even revenge. Even the self is sustained by exchange: a worker puts in eight hours of hard work and rewards himself with a beer.

Modern market exchange is often regarded as "nonsocial" or even "antisocial" in that exchange is immediate, so the relationship ends at the moment of exchange. However, many companies and buyers develop a relationship. Further, identity becomes entangled in not just the buying of goods but in their use over time. The focus here will be on personal exchange—that is, personal communication, the networks in which it occurs, and its role in narrative identity.

One example of the creation and exchange of symbols at the personal level is imitation. This imitation is not mere mimicry. The action being imitated may symbolize values and goals important to the group, thereby sustaining or initiating membership in that group. Several households that bought HEVs either imitated or were imitated by others in their social networks. The imitator often imagined that the people they imitated made a "good decision." Imagined decision making by someone else substitutes for some or all of the imitator's prepurchase process. Imitators are quite clear about this, and they talk unselfconsciously about following the lead of someone they know and trust. Comments included "She can buy anything she wants, and she bought this car" and "They are the sorts of people who would have investigated this carefully. Her husband is an engineer."

The clearest cases of imitation of HEV purchases were observed in pre-existing groups in which members knew each other, bonds were strong, and social exchange was frequent. And while imitative acts and their opposite—a desire to not emulate some group of "others"—occurred most often among groups known to each other, it also occurred among impersonal groups. One buyer of a Honda Civic Hybrid claimed he could visually distinguish his vehicle from the nonhybrid Civics owned by a group of young men for whom he had low respect.

Imitation is far from the sole form of personal social exchange. For one HEV buyer in our survey—who was unhappy with her experience shopping for an HEV at one dealership—a conversation at a cocktail party provided her information about a neighbor's son selling hybrids of a different make at a different dealership. The importance of this example is not the specific information but (1) that the information was conveyed at a neighborhood party from the owner of an HEV to the respondent, who (2) had been thinking about HEVs since the introduction of the Honda Insight. Regarding the first point, HEVs, fuel economy, international oil politics, and environmental impacts of automobility have risen to the level of casual conversation. Regarding the second, the respondent had been enacting a narrative of "a person who would like to own an HEV" for a few years. The information from her neighbor allowed her to change her narrative to one of "a person who owns an HEV." More than simply informing a utility-maximizing decision, the information she received facilitated a narrative that she had been looking to enact.

Self-Interest and Self-Identity

Dittmar (1992) explains the social constructionist "view that, among the wide-ranging symbolic significances of material possessions, it is their relation to our sense of identity, and to the identity of others, that is particularly important." In several instances recorded in the two interview-based studies, the stories people told about themselves and their automobiles were clearly about their use of automobiles to create and project part of their identity. While rational people act in ways that are consistent with their self-interest, self-interest is better defined as a process of creating and expressing identity through a narrative rather than maximizing utility.

This is easily seen by contrasting two households interviewed in our survey. First, take the case of a household and personal narrative told during one of the fuel economy interviews. Their narratives are conservative in the general sense of being resistant to change, and certainly in the sense of not seeking change for its own sake. The interviewees were a farming couple living in the same area in which they were born and raised. For several years, they have bought the same type of vehicles, although different models for each of them, from the same dealership.

The couple was interviewed because they had recently purchased her another Ford Mustang, her fourth in 12 years. She had wrecked two of the previous three in single-vehicle accidents on country roads. At each purchase, the only "decision" made was whether they could afford a new Mustang with whatever high-power engine was offered that model year. Throughout the interview, she spoke about her fast driving with nonchalance and disdain about the consequences for her and others. "People in this neighborhood hear me coming and they know to get out of the way," she said. The repeated purchase of a type of vehicle she had admired when she was young, the absence of alternatives during each purchase, the repetition of dangerous driving practices, and the lack of reflection on consequences all indicate a person who has settled into a specific idea of who she is and a specific representation of that person.

In contrast, one of the HEV buyers had throughout his life experimented with many types of automobiles. In a vehicle purchase history spanning 50 years and more than 20 vehicles, he rarely made consecutive purchases of similar types. He switched from big cars built in the United States to small, economical, Japanese cars in the 1970s. A stint in the U.S. Navy as a young man allowed him to see something of the world. Until recently, a pilot's license and a small plane allowed him to see the world from a different perspective. His life story also includes a religious conversion.

His purchase of an HEV fit his lifelong pattern of experimentation and exploration. He was intrigued by hybrid electric technology and admired the technical accomplishment of increased efficiency. Furthermore, the space around the driver reminded him of the cockpit of his last small airplane—a pleasant reminder of flying. The HEV also renewed his desire to explore the world around him. He said he drives more now, traveling the region to discover what it holds. His increased driving is not a product of an economic "rebound" effect because of higher fuel economy. Rather, he feels he can drive his HEV without polluting.

Several other interviews provide examples of what are often taken in rational analyses to be functional attributes of vehicles serving instead as elements of personal or household narratives. SUVs commonly provide symbolic rather than functional benefit. Several households interviewed had bought four-wheel drive SUVs, believing these would be good vehicles to drive to winter skiing. Unfortunately for many of these would-be skiers, children were born into the family, shifting expenditure and activity patterns for years—such that these households rarely or never ski. Even so, some continue to buy SUVs, using them mostly to commute to work in distinctly nonsnowy suburban and urban settings, all the while maintaining the image of themselves as people who ski. Towing capability is another vehicle attribute that appears to serve a purely symbolic role in some individual and household narratives. In these cases, the narrative is that "we are people who tow, and someday, we'll own something to tow." Years later, many interviewees who bought expensive towing packages for a truck or SUV still own nothing to tow.

These examples all show people buying particular automobiles that supported their narrative identities. The construction of a narrative identity can also explain why people won't buy a particular vehicle. This is relevant to the discussion of the types of vehicles that rational analysts typically use for purposes of comparison when arguing that HEVs don't make economic sense at current or historical gasoline prices. Such analyses typically assume that consumers should compare the closest functionally equivalent vehicle to the HEV—for example, the Civic Hybrid to the Civic, the generation one Prius to the Corolla, or the generation two Prius to the Camry (see, for example, Edmunds.com, 2005).

Despite similar performance, handling, size, and passenger and payload capabilities in these vehicle pairings, few households interviewed made such comparisons. Despite their generally favorable past experiences with Corollas, for example, few households who bought a first generation Toyota Prius compared it to a Toyota Corolla. When asked why, most dismissed the comparison as irrelevant.

The purchase of a small, economical, inexpensive vehicle often symbolized a time in the household's history when they were struggling financially—a chapter in their story they don't wish to repeat. As one buyer of a first generation Prius said, "After my divorce, I was so broke I had to beg two different car dealers to sell me a used Corolla. I vowed I'd never be that poor again." She recalled the Corolla had worked well, and recently she helped her daughter buy an "economy car," but rejected such a car for herself. Her HEV represents her years of political activism, an homage to her daughter's ongoing environmental activism, and a better world for her new grandson; a Corolla represents a time she'd rather forget.

Preexisting Preferences and Investment in Consumption Outputs

Kurani, Turrentine, and Sperling (1996) discuss preference formation prior to purchase of a novel vehicle and during vehicle use. Kurani (1992) and Turrentine (1994) discussed distinctions between search attributes that can be evaluated during a prepurchase process and experience attributes that can only be evaluated with experience of the goods or services. Here, the discussion of preference formation is extended using Becker's (1992) argument that households be treated as producers of consumption and reinterpreted within Giddens' (1991) discussion of the increasingly "precise time-space zoning of personal life." Specifically, these zones are important as settings for narratives and as elements of the narratives themselves.

One interviewed HEV buyer had purchased a first generation Prius to replace a conventional Honda Accord. This is a household of two retired persons. The female head of household drives the Prius. She plays an active role in the day-to-day care of her grandchildren, picking them up from school most days. She started her story by saying, "You know the thing I like best about my car? That it shuts off completely at a stop and only uses the electric motor to creep along." The engine shutoff at first seemed strange when she test drove the vehicle. Now, it is her favorite feature, particularly when she waits in the line of cars and trucks at her grandchildren's school. She and her husband have taken their resources and roles to create one consumption output, her Prius. She has taken her Prius and its strange new capability, combined them with her role as caregiver for her grandchildren and a specific time and place in which she fulfills that role, and has created a new consumption output—lower pollution at her grandchildren's school. Now, she sits in that line of cars and trucks and wonders why everyone else doesn't buy an HEV.

Other examples of new consumption outputs include surprises and contradictions to predictions from rational analytic approaches. As discussed earlier, inferences from within a rational analytic approach include estimates of a "rebound effect" from increased fuel economy, or more accurately, reduced cost of travel. In several households interviewed, the symbolic value of "reduced resource consumption" attached to their HEVs led to reinforcing rather than rebounding behaviors. For example, since acquiring her HEV, one respondent is now looking for other ways she can "live lighter," including driving her HEV less than the midsize sedan it displaced. She is walking to more nearby destinations and investigating transit for her commute between home and work. Other HEV owners reported that the comprehensive energy use instrumentation in their HEVs allowed them to learn to drive more economically. They now apply these lessons to driving

230 *Driving Climate Change*

TABLE 13-2. Selected Vehicle Transitions in Household Buying Hybrid Vehicles

With Hybrid	Without Hybrid
'01 Prius displaces Jaguar XJ6	No transaction. Keep Jaguar XJ6.
'01 Prius displaces Isuzu Trooper	Chrysler PT Cruiser to displace Isuzu Trooper
'01 Prius displaces Toyota Camry	Toyota RAV4 to displace Toyota Camry
'02 Prius (used) displaces Toyota Camry	Toyota Camry (used) to replace Toyota Camry
'03 Prius displaces Honda Accord	No transaction. Keep Honda Accord.
'03 Civic Hybrid displaces Ford Ranger	VW Jetta (used) to displace Ford Ranger
'04 Prius displaces Toyota Avalon and Chevrolet Astrovan	Keep Toyota Avalon, sell Chevrolet Astrovan; or Toyota Sienna displaces both Avalon and Astrovan.
'04 Prius displaces Honda Accord	No transaction. Keep Honda Accord.
'04 Prius added to household fleet	Toyota Sequoia to displace Toyota 4Runner
'04 Civic Hybrid displaces Civic	No transaction. Keep Honda Civic.

other vehicles as well. In both these situations, the new "consumption output" these households are producing with their HEV is further reduced energy use for their daily travel.

Relevant Attributes of Alternatives and Symbols, and Accessible Attributes of Perceived Attributes

One strength of the methodological approach taken in both sets of interviews is it reveals otherwise unobserved decisions and actions that people take on their way to observed vehicle transactions. For example, it can uncover the types of vehicles people did not buy in addition to the one they did. In doing so, it uncovers additional reasons why common assumptions by rational analysts regarding buyers of HEVs are not satisfied in many households. As discussed above, numerous analysts compare the estimated cost-benefit streams of buying HEVs to the estimated streams of some "equivalent" vehicles that buyers should have considered when buying their HEVs. Table 13-2 lists some examples from interviews with HEV buyers of observed "with hybrid" and stated or conjectured "without hybrid" transitions.

The discovery of HEVs by a household can affect their consideration of vehicle alternatives in many ways that lie outside typical rational analyses. In some households, HEVs prompted vehicle purchases that otherwise would not have occurred. In others, HEVs were purchased with little or no comparison to other vehicles. In still other cases, HEVs displaced vehicles entirely dissimilar to themselves, including multiple vehicles in a household that reached retirement age and reduced the number of vehicles they own from two to one.

It might seem that the rational analytic approach could simply be enriched by a subtler accounting of the types of actions households take when they buy vehicles. For example, a population-level study could be conducted based on the comparison of HEV models to a broader sample of vehicle types.

However, the problems with the rational analytic approach are more fundamental than such an adjustment suggests. First, a comparison of an HEV to another vehicle is one indication that a decision process is being enacted, but, as just discussed, decision processes do not precede all vehicle transactions.

Second, as also just discussed, several households had no "without hybrid" vehicle. Here, the HEV prompted a transaction that would not otherwise have occurred. Picking a prior vehicle in the household misses what many of these households were really doing. Once some people became aware of HEVs, keeping their prior vehicles was not an option because those vehicles were no longer part of narratives those people wanted to tell and enact.

Third, the HEV may be incorporated into an ongoing narrative but in a novel way that could not be predicted from an analytical perspective that does not allow for the substitution of one attribute for another in a narrative structure. That is, the presumptive comparisons of HEVs to "similar" vehicles obscures, rather than reveals, why some people buy HEVs. The first example in Table 13-2 is a household that displaced a Jaguar XJ6 with a first generation Prius. When driving the Jaguar, the male head of household imagined both he and the car projected power and mastery over the road and other drivers and vehicles. Unable to compete for road position as aggressively in the Prius, he now glowers at drivers of large SUVs and pickup trucks, still imagining himself the victor—only now in a battle for the planet's resources. He maintains a narrative of road warrior, but through his Prius he has recast himself as a high-tech, über-conserver.

Such substitution is possible because the HEV is rich in symbols, including the hybrid drivetrain system, a particularly low-polluting emission rating, and high fuel economy. This is a new combination of symbols. Kurani and Turrentine (2004) discuss that prior to the introduction of HEVs, "fuel economy" was a phrase largely reserved by the public for small, inexpensive, low-powered cars. HEVs, on the other hand, were seen by both buyers and nonbuyers of HEVs to be "high-efficiency" vehicles. The term "high-efficiency" carried connotations of advanced engineering and technology and high-quality manufacturing. Thus, the high fuel economy of HEVs is associated with a new set of meanings not previously available in the marketplace—even if previously there have been cars with high miles per gallon.

232 Driving Climate Change

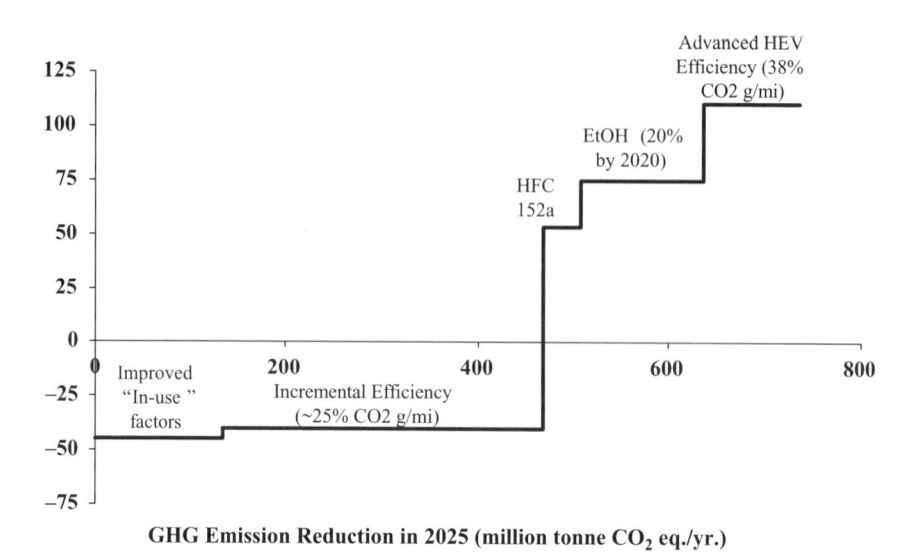

FIGURE 13-1. Cost-effectiveness of CO_2 emission reductions in light-duty vehicles. Source: Lutsey, 2005.

An Application of the Alternative Approach

Some implications for analysis and policymaking of the alternative approach developed here can be illustrated by Figure 13-1. The figure shows Lutsey's (2005) midlevel estimates of the cost effectiveness of deploying a variety of technologies across the light-duty vehicle fleet in a manner to reduce GHG emissions. "Improved 'in-use' factors" are primarily vehicle and tire maintenance programs. "Incremental efficiency" gains are based on technologies currently available in at least some vehicles. These include variable valve timing, variable valve lift, six-speed automatic transmissions, and others. However, it does not include hybridization beyond integrated starter-generators. Next are a change of air conditioning refrigerant to HFC152a, increased use of cellulosic ethanol (EtOH) for fuel, and finally deployment of advanced hybrids at least conceptually similar to the economy-tuned HEVs offered to date.

The negative cost effectiveness for improved in-use factors and incremental efficiency gains mean these would pay society back in monetary terms, in addition to reducing GHG emissions. Because his analysis is cumulative up to the year 2025, Lutsey's midlevel results indicate that improving in-use factors and making incremental improvements in fuel economy could reduce GHG emissions by 9 billion metric tons of carbon dioxide equivalent and yield almost $400 billion in net benefits.

The existence of such large potential GHG reductions and monetary payoffs demonstrates the extent to which markets and rational analytic policy are failing. Despite three decades attempting to shape markets through analysis and policy based on a rational analytic framework, several maintenance practices and known technologies that would increase efficiency, reduce GHG emissions, and pay society net benefits remain underdeployed. Why?

There is no explicit time dimension in Figure 13-1, but the rational analytic framework implies time runs from left to right. To the left are actions that provide GHG reductions and monetary benefits—these should be done first. Moving to the right, actions to reduce GHG emissions are more costly and should be done only after the measures to the left. In a rational analytic world, the failure to make the incremental changes that actually pay financial benefits means there is little hope of making big changes, such as deployment of advanced HEVs, that cost money to implement. Thus, the rational analytic framework also fails to explain why some consumers are buying these advanced HEVs now.

A solution arises if a "social constructionist" time dimension is assumed to run in the opposite direction, from right to left in Figure 13-1. In a world where fuel economy and nonincremental increases in fuel economy are valued for far more than their effects on private fuel costs, at least some consumers will pay more now for advanced HEVs, even though rational analysis indicates this is the last thing they and society should consider. If people are assumed to tell stories of and about themselves and are looking for symbols to incorporate into those stories, then large costly changes are possible because such changes symbolize broader goals and better stories. It may be true now that few consumers will pay hundreds of dollars extra for an automobile to save a few dollars a week on gasoline. However, in a world where actions are based on symbols, emotions, and narratives of identity, some people will pay disproportionately more for larger increases in fuel economy.

If time runs from right to left in Figure 13-1, then large changes capable of symbolizing transformative life stories may make the world safe for making the incremental changes. By introducing peace, climate change, clean air, ecosystem protection, and reduced resource consumption into a public conversation about fuel economy and automobile purchases, people who act to acquire expensive, nonincremental increases in fuel economy may empower other people to act, too. These other people may not all be able or willing to go so far as to buy the expensive advanced HEVs, but because they start to value fuel economy more and for a broader set of reasons, they may choose to act to acquire other means and symbols to achieve and represent those values.

Conclusions

Transportation energy analysis and policymaking are overly reliant on the "rational actor" model of human behavior, a model that fails to fully explain household vehicle purchase and use, in particular those aspects that are socially and culturally transforming. This chapter presents an alternative behavioral approach in which individuals and households acquire and apply symbolic meanings of automobiles, as well as other consumer goods and services, to construct narrative identities. While people's budgets and competing wants affect their decisions, the symbolic approach opens research to other important determinants of behavior in the automotive market and particularly an automotive market in a period of rapid change.

The rising and volatile price of gasoline is not just an economic issue but a social one as well; that is, people are talking about it. Gasoline prices, petroleum geopolitics, and automotive technologies like hybrid electric drivetrains are in the news most days, books about oil politics are on bestseller lists, and editorials regularly focus on petroleum and gasoline supplies and prices. As this conversation continues, the type of automobile each person drives increasingly will come to symbolize a position on such issues.

The symbolic meaning of fuel economy in vehicle purchase decision making demonstrates that nonincremental alternatives allow and may foster nonincremental thinking and behavior. In the case of HEVs, fuel economy is valued, in 2005 at least, as more than fuel cost savings. Early HEV buyers didn't buy just lower private fuel cost when they bought their cars. They bought, instead, a piece of a much broader future, including a less-consumptive lifestyle, smart consumer choices, clean air, lower oil consumption, and less terrorism.

Individual lifestyle choices intersect with world politics, forming a "reflexive" zone of choice that Giddens (1991) has called "life politics." Characterizing consumers solely as economically rational actors will not only be technically incorrect but will also ignore these life politics. When experts measure fuel economy only as miles per gallon and value it only as cents per mile, they confound what they can measure for the totality of what matters. A social constructionist perspective and the technical variants discussed here offer rich research opportunities to discuss life politics and lifestyle choices with automobile buyers, to explore alternative policy and technology futures that appear increasingly essential and important.

Life politics involves exchange and mutual effect between societal and institutional goals and the day-to-day lived behaviors of individuals' narratives. The narratives of individuals and households are connected to societal policy through discourses. Hare-Mustin (1994) defines discourses as systems "of statements, practices, and institutional structures that share common values." Social discourses are subject to change through the telling of new narratives that challenge prevailing discourses. Weingarten (1991) states, "Changes of discourse occur when the collective conversations people have about their lives transform culturally available dominant narratives about people's lives."

To date, buyers of the fuel efficient HEVs are telling stories about their lives in which they have connected their choices about automobiles to local, regional, and global concerns. Telling those stories more broadly, to more people, gives strength to a discourse about consumer choices being made to address these concerns. These stories are instigated and facilitated by a new technological choice that has become a symbol capable of incorporation into their narratives.

It is not only from the telling of individual and household narratives that a prevailing discourse can be changed or dislodged. A complementary strategy would be for leaders to tell better stories. In the United States, federal energy and climate change policies are currently conflicted, at best. Leaders at intermediate levels of governance, such as mayors and governors, are beginning to tell civic narratives linking their choices to swap full-size SUVs for HEVs to civic and fiscal responsibility, reduced resource consumption, and lower GHG emissions. Telling such stories can be vital to initiating and sustaining a national conversation, a new discourse about energy, energy efficiency, carbon-free energy, global warming, future transportation, and the possibilities for positive societal changes.

Few narratives and symbols are shared by "the American people" simply because Americans are so diverse. Still, there are discourses—for example, protecting the family or land, progress, and prosperity—that are broadly appealing. Discourses and symbols can, and should, be directed to different groups. For example, all automobile drivers may need to value fuel economy more highly if increased fuel economy is to be a successful policy to address global warming. But automobile drivers are diverse people who may need to be addressed by different discourses through different media. Glossing over differences in Americans' narratives is precisely how opponents of fuel economy standards diffuse arguments for higher standards. They charge, for example, that higher fuel economy standards will force "Americans" to buy vehicles they don't want.

Altering existing symbolic meanings is likely to create some turmoil. Different interest groups can be expected to contest the meaning of symbols, especially if they have strong interest in an existing symbol. This will most certainly be the case if proposed new symbols and meanings conflict with the interests of industry, which is a strong source of symbols and images through advertising. Attempting to link negative ideas to certain vehicles may spur countermeasures from the automakers that attempt to reinforce links between those vehicles and favorable ideas, even the ideas that other interest groups are using to devalue the vehicles. For example, what is the best way to "protect your family": ensconce them in the perceived safety of a full-size SUV or reduce petroleum consumption and emissions of GHGs by driving something with higher fuel economy?

On the other hand, the initial experience with high fuel economy HEVs shows how corporations can participate profitably to advance societal goals. Manufacturers' participation depends on their belief in the transformative power of new narratives. Policy initiatives built around symbols, narratives, and discourse will need to be supported by some measure of protection for the link between societal goals and specific symbols. Until now, "hybrid" has stood for a high-technology approach to fuel economy, a lighter way to live, a cleaner car. That meaning is likely to shift as performance-tuned HEVs and hybrid electric SUVs enter the marketplace.

The point is not that symbols can replace standards but that by helping citizen/consumers link broad social goals and values to their day-to-day lives through new societal discourses and personal narratives, people as citizens will support policies that shape markets to provide products that they as consumers will buy. Unfortunately, the past symbols around automobiles for many Americans are about power and prestige. This history is not an optimal source of symbols to support climate change policy. However, there are other symbols from past and present U.S. culture to draw on to steer toward the future, including innovativeness, thrift, fairness, and hard work.

Increasing automotive fuel economy can effect immediate reductions in GHG emissions. The size and pace of changes in fuel economy have been debated solely in terms of private financial returns to rational consumers. The alternative behavioral approach and supporting data presented in this chapter show that it may be easier to sell higher fuel economy to a larger number of people by addressing the symbolic meanings of large improvements in fuel economy than by arguing over the cost effectiveness of small ones. More generally, any policy intended to address global warming is more likely to succeed if it addresses peoples' desires to tell better stories about themselves.

Acknowledgment

The U.S. Department of Energy and the Energy Foundation supported research into the role of fuel economy in household vehicle purchase and use. Our thanks to David Greene at Oak Ridge National Laboratory for his intellectual impetus. Research into the vehicle purchase and use behavior of early buyers of HEVs was supported by Toyota Motor Sales, USA Inc. We would like to thank the 82 households from both studies for opening their homes to us for an evening. The statements and conclusions offered here are solely those of the authors; nothing here should be construed as a statement of policy or opinion by any supporting agency or organization or by the University of California.

References

Bagozzi, R. P., M. Gopinath, and P. U. Nyer. "The Role of Emotions in Marketing." *Journal of the Academy of Marketing Science*, v. 27, n. 2, 1999.

Becker, G. S. The Economic Way of Looking at Life. 1992. Available at http://nobelprize.org/economics/laureates/1992/becker-lecture.pdf.

Cameron, M. "A Consumer Surplus Analysis of Market-Based Demand Management Policies for Southern California." In D. Sperling and S. A. Shaheen, eds., *Transportation and Energy: Strategies for a Sustainable Transportation System.* Washington, D.C.: American Council for an Energy Efficient Economy, 1995.

DeCicco, J. M., and D. Gordon. "Steering with Prices: Fuel and Vehicle Taxation as Market Incentives for Higher Fuel Economy." In D. Sperling and S. A. Shaheen, eds., *Transportation and Energy: Strategies for a Sustainable Transportation System.* Washington, D.C.: American Council for an Energy Efficient Economy, 1995.

Dittmar, H. *The Social Psychology of Material Possessions: To Have Is to Be.* New York: St. Martin's Press, 1992.

Edmunds.com. Most Hybrid Vehicles Not as Cost-Effective as They Seem, Reports Edmunds.com. 2005. Available at www.edmunds.com/help/about/press/105827/article.html?tid=edmunds.h..pressrelease..1.*

Espy, M., and S. Nair. "Automobile Fuel Economy: What Is It Worth?" *Contemporary Economic Policy* 23, 2005. pp. 317–323.

Freedman, J., and G. Combs. *Narrative Therapy: The Social Construction of Preferred Realities.* New York: W.W. Norton & Co., 1996.

Giddens, A. *The Constitution of Society.* Berkeley, CA: University of California Press, 1984.

———. *Modernity and Self-Identity.* Stanford, CA: Stanford University Press, 1991.

Goldberg, P. K. "The Effects of the Corporate Average Fuel Efficiency Standards in the US." *The Journal of Industrial Economics* 46:1, 1998. pp. 1–33.

Greene, D. L. "A Note on Implicit Consumer Discounting of Automobile Fuel Economy: Reviewing the Available Evidence." *Transportation Research* 17B:6, 1983. pp. 491–499.

Greene, D. L., and J. Liu. "Automotive Fuel Economy Improvements and Consumer Surplus." *Transportation Research A* 22A:3, 1988. pp. 203–218.

Greening, L. A., D. L. Greene, and C. Difiglio. "Energy Efficiency and Consumption—The Rebound Effect—A Survey." *Energy Policy* 28, 2000. pp. 389–401.

Hare-Mustin, R. "Discourses in the Mirrored Room: A Postmodern Analysis of Therapy." *Family Process* 33, 1994. pp. 19–35.

Heffner, R. R., K. S. Kurani, and T. S. Turrentine. "Vehicle Image in Hybrid Electric Vehicles." Proceedings of the 21st Electric and Hybrid Vehicle Symposium. Monaco, 2005.

Heffner, R. R., T. S. Turrentine, and K. S. Kurani. *A Primer on Automobile Semiotics.* UCD-ITS-RR-06-01. Davis, CA: Institute of Transportation Studies, University of California, 2006.

Kahneman, D. Maps of Bounded Rationality: A Perspective on Intuitive Judgement and Choice. 2002. Available at http://nobelprize.org/economics/laureates/2002/kahnemann-lecture.pdf.

Kahneman, D., and S. Frederick. "Representativeness Revisited: Attribute Substitution in Intuitive Judgment." In Gilovich, T., D. Griffin, and D. Kahneman, eds., *Heuristics and Biases* (pp. 49–81). New York: Cambridge University Press, 2002.

Kurani, K. S. "Application of a Behavioral Market Segmentation Theory to New Transportation Fuels in New Zealand." Ph.D. Dissertation. ITS-RR-92-05. Davis, CA: University of California, Institute of Transportation Studies, 1992.

Kurani, K. S., and T. S. Turrentine. *Automobile Buyer Decisions about Fuel Economy and Fuel Efficiency.* Final Report to United States Department of Energy and Energy Foundation. Research Report ITS-RR-04-31. Davis, CA: University of California, Institute of Transportation Studies, 2004.

Kurani, K. S., T. Turrentine, and D. Sperling. "Testing Electric Vehicle Demand in 'Hybrid Households' Using a Reflexive Survey." *Transportation Research D.* 1: 2, 1996.

Lutsey, N. Personal communication, Institute of Transportation Studies, University of California, Davis. 2005.

McFadden, D. "The Choice Theory Approach to Market Research." *Marketing Science* 5:4, 1986. pp. 275–297.

Miles-McLean, R., S. M. Haltmaier, and M. G. Shelby. "Designing Incentive-Based Approaches to Limit Carbon Dioxide Emissions from the Light-Duty Vehicle Fleet." In D. L. Greene and D. J. Santini, eds., *Transportation and Global Climate Change*. Washington, D.C.: American Council on an Energy Efficient Economy, 1993.

National Research Council. Effectiveness and Impact of Corporate Average Fuel Economy (CAFE) Standards. Washington, D.C.: National Academy Press, 2002.

Olshavsky, R. W., and D. H. Granbois. "Consumer Decision Making—Fact or Fiction?" *Journal of Consumer Research* 6:2, 1979. pp. 93–100.

Quirk, J. P. *Intermediate Microeconomics*, 2nd ed. Chicago, IL: Science Research Associates, 1982.

Senauer, B., J. Kinsey, and T. Roe. "The Cost of Inaccurate Consumer Information: The Case of the EPA Mileage Figures." *The Journal of Consumer Affairs* 18:2, 1984. pp. 193–212.

Train, K. "Discount Rates in Consumers' Energy-related Decisions: A Review of the Literature." *Energy* 10:12, 1985. pp. 1243–1253.

Turrentine, T. S. "Lifestyle and Life Politics: Towards a Green Car Market." Ph.D. dissertation, UCD-ITS-RR-94-30. Davis, CA: Institute of Transportation Studies, University of California, 1994.

Turrentine, T. S., and K. S. Kurani. "Car Buyers and Fuel Economy." *Energy Policy*, 2005.

Verboven, F. "Implicit Interest Rates in Consumer Durables Purchasing Decisions: Evidence from Automobiles." Antwerp, Belgium: Center for Economic Research, University of Antwerp, 1999.

Weingarten, K. "The Discourses of Intimacy: Adding a social constructionist and feminist view." *Family Process* 31, 1991. pp. 45–59.

Yun, J. M. "Offsetting Behavior Effect of the Corporate Average Fuel Economy Standards." *Economic Inquiry* 40:2, 2002. pp. 260–270.

CHAPTER 14

Lost in Option Space: Risk Partitioning to Guide Climate and Energy Policy

David L. Bodde

Every action taken or not taken, every investment made or not made, every capability gained or lost brings consequences that reach far into the future and remain unforeseen and unforeseeable. Yet, policy choices must be made, and even inaction becomes, in reality, another form of strategic choice. Climate and energy policy, strongly linked through the combustion of carbonaceous fuels, requires planners to persuade a properly skeptical public and their elected officials that the policy "bets" they must place now will perform well far into an unknowable future. To accomplish this fully, they must consider the full spectrum of likelihoods and outcomes, and employ analytical tools better suited to the task.

Two modes of thought tend to underlie the choices made or implied by energy and climate policies: a focus on the likelihood of some future event or a focus on the possible outcomes of future events. "Likelihood" thinking provides the foundation for much of the technology-based regulation predominant in the developed countries—transportation safety and nuclear reactor regulation are two examples. Likelihood thinking tends to the analytical and implicitly discounts future events whose probabilities cannot be quantified. In contrast, "outcome" thinking concerns itself chiefly with possibilities—the consequences of an airplane crash, for example, or a terrorist attack on a nuclear power plant. Outcome thinking need not always be negative, but it tends to lightly regard means and likelihood. Witness, for example, the highly positive State of the Union speech by President Bush in January 2001, which focused on the advantages of the hydrogen economy but provided little technical or policy detail on how to bring this new economy into existence.

Policy planners must integrate their thinking about both likelihood and outcomes if they are to design options that are robust against an uncertain climatic future and if they are to help elected officials and private citizens understand the choices before them. The planners' concern should not be to predict the nature and consequences of future societal risks. Instead, energy planners must focus on options that are wise and that can endure well into an unknowable future. Whether inclined toward likelihood or outcomes thinking, those charged with energy/climate policy must include in their plans the full spectrum of likelihoods and outcomes—the future implications of present decisions.

Uncertainty, Ambiguity, and Ignorance: The Monsters under the Bed

Risk is a time-dependent concept. Whether the hurricane will strike tomorrow and with what consequence can be forecast today, though not precisely. By the end of the week, however, all that will be known—only to be replaced by other unknowns about the future. Similarly, the passage of enough time will eventually illuminate each present-day energy/climate uncertainty, ultimately revealing what could have been done if perfect foresight had been granted the decision maker.

In the case of energy/climate issues, however, waiting for the passage of time to provide clarity raises the prospect of severe and irreversible damage—yet, the nature and extent of this damage cannot be demonstrated convincingly in advance. Thus, the general policy problem persists: Appropriate actions must be taken before it can be shown that they really are appropriate (Bodde et al., 2005).

In the Domain of Risk and Beyond

Where the consequences of prospective hazards can be identified and where their probabilities of occurrence can reasonably be estimated, policy decisions fall into the domain of "risk" as it is properly understood. For decisions that incur this kind of risk, historical experience serves well as the basis for a priori probability estimates and thus provides an invaluable guide to the future. Financial markets, for example, commonly employ the variance of return around the historical mean as a measure of risk for portfolios of financial assets. Thus, an investment in energy conserving technology can be made with a reasonably clear view of its risks and range of payoffs.

Many important policy issues, however, do not fall within the domain of risk because there is a monster hiding under the policy bed: the prospect that events beyond the range of historical experience and unknowable at the time that a decision must be made can emerge to influence its outcome. This prospect cannot properly be called "risk" because neither the probabilities nor the outcomes of these events can be understood adequately in advance. Indeed, they might not be understandable, since historical experience provides no insight into events that never happened or that occurred when nobody was taking the data. Under such circumstances, internally self-referencing tools for managing risk become blind guides to decision making.

Policy planners must think outside the risk box, searching out and planning for strategic surprise, which dwells outside the domain of risk, in order to prepare an option portfolio for issues that are poorly understood, both in terms of the likelihood of the event in question and of its consequences if it should occur. They must consider the monsters under the bed:

- Uncertainty, where reliable estimates cannot be made for the likelihood of the outcomes identified
- Ambiguity, where the outcomes cannot be closely characterized because they cannot be imagined or because such characterization depends upon the perspective of the observer
- Ignorance, where neither likelihood estimates nor well-characterized outcomes enjoy sufficient credibility to guide policy or to motivate action

Risk Partitioning in the Energy/Climate Dilemma

These components of what is commonly called "risk" can be organized into a full-spectrum risk space (Awerbuch et al., 2006), as shown in Figure 14-1. In particular, policymakers would gain insight if the larger energy/climate problem were partitioned into the categories shown in Figure 14-1. The following four examples—energy efficiency, oil peaking, terrorist attack, and climate change—demonstrate how risk partitioning can be done and suggest how two policy tools—scenario analysis and real options analysis—can illuminate the full implications of unknowable futures.

Investing in Energy Efficiency: A Case of Risk

Public policies can apply two basic levers to increase the efficiency with which energy is converted into products and services—influencing the cost of energy and regulating its use. With regard to cost, a complex web of subsidies and taxes, many in conflict with one another, influence the cost of energy at the point of use. Nevertheless, specialists in the field understand these complexities well and are quite capable of estimating the value of the energy saved at any cost level. Even when new regulations, like the Corporate Average Fuel Economy (CAFE) standards imposed by the Energy Policy and Conservation Act of 1975, required changes in energy-using products, the competitors quickly adjusted to the new rules of competition. Once the new rules become apparent, analysts can estimate their impact reasonably well.

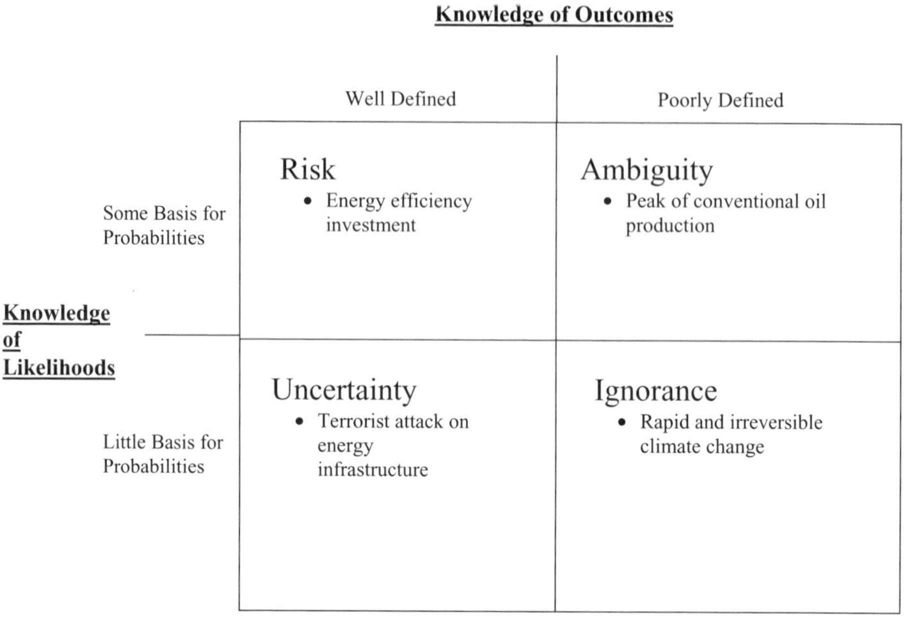

242 *Driving Climate Change*

Knowledge of Outcomes

	Well Defined	Poorly Defined
Some Basis for Probabilities	**Risk** • Energy efficiency investment	**Ambiguity** • Peak of conventional oil production
Little Basis for Probabilities	**Uncertainty** • Terrorist attack on energy infrastructure	**Ignorance** • Rapid and irreversible climate change

Knowledge of Likelihoods

FIGURE 14-1. Knowledge about likelihoods and outcomes. *Source:* Awerbuch et al., 2006.

Thus, investors in energy efficiency, either public or private, can reasonably understand the consequences of their investments and the likelihood that they will achieve the desired returns, and decision makers can find reasonable estimates of the effects of alternative policies on the use of energy (CBO, 2004). For these reasons, this component of the larger energy/climate problem generally falls within the "Risk" quadrant of Figure 14-1. Here, the unknowables do not inhibit action, and policies ensure the clarity that allows the private economy to accommodate the public's goals most efficiently.

Other energy/climate issues are not so conveniently arranged. Either the consequences of alternative actions are poorly characterized or the likelihood of game-changing events cannot be estimated, or both. These issues fall into the less certain quadrants of the full-spectrum risk space defined in Figure 14-1—ambiguity, uncertainty, and ignorance.

Oil Peaking: A Case of Ambiguity

Ambiguity characterizes the northeast quadrant of the risk space shown in Figure 14-1. Issues falling into this quadrant generally have sufficient evidence for most observers to form opinions regarding the likelihood of events, but views diverge wildly regarding their consequences. Thus, a primary objective of policies under ambiguity is to better characterize the economic and environmental consequences of events that can be reasonably foreseen.

Consider, for example, the peaking of conventional world oil production, now predicted by most analysts. About 72 million barrels per day (MMbpd) of conventional oil were pumped out of the ground in 2004, according to the United States (U.S.) Energy Information Administration (EIA). World oil supply, which adds to conventional production, liquids produced from solid hydrocarbons, natural gas plant liquids, other hydrogen and hydrocarbons for refinery feedstocks, and refinery processing gain, was estimated at 84 MMbpd (EIA, 2005). If the current limit to conventional production capacity really is only 10 percent above that, then a peaking of conventional world oil would seem quite near. Indeed, the durability with which price exceeds marginal production cost, apparent in late 2005 prices above $60 per barrel, suggests that financial markets include production limitations, together with disruption risk, in the price calculus.

Other analysts, however, foresee world production capacity in the range of 100 to 120 MMbpd, achievable with investments in new production technologies likely to come online in the next few years (*The Economist*, 2005). Thus, much disagreement remains concerning the timing of the transition from increasing to declining conventional production. But most geologists seem to have reached sufficient consensus that a peaking point exists and that reasonably available signposts—discovery rates for new fields, or projections of petroleum demand, for example—can guide the astute observer in estimating its timing, even if approximately. Thus, the likelihood dimension of the risk space can be estimated.

In contrast, views of the possible consequences of a downturn in conventional production vary sharply with the perspective of the observer. On the one hand, geologists and those holding a science-based perspective tend to view the coming peak as catastrophic (see the Hirsch chapter in this book). They warn that the downturn will be steeper than many realize and that unconventional production of liquid fuels could arrive too late and in insufficient quantity to make a difference. Consequently, this school of thought foresees sharply curtailed economic activity, especially in demand-inelastic sectors like transportation. Some analysts of more apocalyptic persuasion imagine worldwide economic collapse.

On the other hand, many economists imagine a smoother transition as fuel price increases motivate unconventional sources of hydrocarbons—chiefly coal, shale, and tar sands—to replace the conventional. Indeed, the excess of price over the production cost of oil when prices are above $60 per barrel suggests an ample incentive to produce liquid fuels from tar sands, estimated to cost between $30 and $40 per barrel. At worst, this would mitigate the fall-off of conventional oil production, and at best it might provide for continued growth in liquid fuels consumption. However, this cheerful view requires two assumptions:

- An acceptable way can be found to sequester the massive amounts of carbon dioxide that would be released from unconventional hydrocarbon feedstocks or that carbon dioxide release ultimately proves to be less of a concern than economic downturn.
- Sufficient and timely investment in unconventional hydrocarbon sources will be forthcoming in response to the price signals.

Thus, the estimated consequences of the inevitable peak in production of conventional oil depends closely on the point of view of the observer. This has significant implications in the energy/climate policy debate. For energy/climate issues that fall within the Ambiguity quadrant of Figure 14-1, energy policy should assume primary responsibility for relieving the unknowns that inhibit action. In the case of oil peaking, several elements of a transition policy emerge as essential to move this issue into the Risk quadrant, where private investment can more confidently provide public service:

- Cost and Feasibility of Carbon Capture and Sequestration (CCS): Boundaries around the cost to capture and permanently sequester the carbon emissions from unconventional fuel production would enable private investors to place more intelligent "bets" on these resources. This implies an acceleration of CCS demonstrations with the technologies currently in the mainstream—storing carbon dioxide in underground formations.
- Licensing and Regulating Carbon Repositories: Cost is only one dimension of the uncertainties surrounding CCS. The other is regulatory uncertainty. Any carbon repository will surely require some kind of license and receive regulatory scrutiny, thus adding the politics of regulation to the list of hazards facing potential investors. Early policy attention to the conditions for public acceptability and oversight might forestall the protracted legal warfare that made the introduction of nuclear power so painful for its proponents.
- Precompetitive Research and Development: Informed by attention to public requirements like CCS, precompetitive research can lower the risk of investment in unconventional fuel production once its results become widely available.
- Risk Sharing for Early Production: The intent of public policies should not be the removal of all risk—indeed, risk taking is the societal function of private capital. Rather, risk sharing would seek to remove the prospect that catastrophic events outside the scope of private markets will emerge to upset the business case for investment. The Price-Anderson Act, for example, accomplished this, although imperfectly, by placing a cap on liability for accidents at nuclear power plants. Similarly, policy might help move the unconventional production of motor fuels to the Risk quadrant, where private investment can marshal the resources and the skills.

Oil Disruption: A Case of Uncertainty

The world's largest oil processing facility, Saudi Arabia's Abqaiq complex, sits about 24 miles north of the Gulf of Bahrain. The flow of petroleum through Abqaiq is comparable to the entire U.S. production in 2004 of around eight million barrels per day. The entire petroleum output from the southern oil fields in Saudi Arabia flows through this facility to the loading terminals at Ju'aymah and Ras Tanura. The consequences of a successful terrorist attack on any of these facilities can be understood with grim certitude. The likelihood of such an event, however, remains obscure, so the issue of oil disruption must reside in the "Uncertainty" quadrant of Figure 14-1.

On the one hand, many analysts argue that our current understanding of the terrorist threat makes the likelihood of successful disruption remote. This thinking emphasizes the strength of the Saudi security forces and notes that they are composed exclusively of ethnic Saudis, all of the Sunni persuasion. A gloomier outlook notes that Saudi Sunnis also perpetrated the September 11 terrorist attacks on New York and Washington, D.C. Other observers accuse the House of Saud of buying temporary political stability and security by providing financial support to terrorist groups, an arrangement unlikely to prove either stable or secure over the longer term (Baer, 2003).

Better intelligence would, of course, help matters, but past surprises, from the 1941 attack on Pearl Harbor to the 2005 destruction caused by Hurricane Katrina, suggest that plenty of information is often available before the disaster—the difficulty lies in its interpretation and acceptance. To the extent that this remains true, preparations for rapid response are the strongest policies to prevent future energy supply disruptions. Some specific energy policies would emphasize the following:

- Building a much stronger strategic petroleum reserve, perhaps as much as two billion barrels
- Diversifying and dispersing the sources of petroleum, including the expansion of unconventional production
- Eliminating petroleum use from its most demand-inelastic sector—transportation—over the long term

Rapid and Irreversible Climate Change: A Case of Ignorance

Climate change could occur too rapidly for effective adjustment. Such an event would create winners as well as losers, but the latter would probably outnumber the former and would include the poorest people around the globe, always the most vulnerable to environmental catastrophe.

This gloomy prospect, however, has not yet motivated effective policies. Some critics quite correctly note that the scientific case for causation has not been made and that much ignorance surrounds global climate change phenomena (Schlesinger, 2005). They contend that the prospective benefits of protecting against a vaguely specified disaster far in the future must be weighed against the certain costs that would be borne today.

The Ignorance quadrant of the larger risk space tends to dominate much of the landscape in energy/climate policy. The most obvious response would be to accelerate research on climate change in order to expand understanding of the underlying physical phenomena. But until sufficient learning can be accumulated, decisions will still be made—or not made, which amounts to the same thing. The quality of decision making for issues in which ignorance dominates the risk space can be improved through the use of two planning tools commonly employed by industry but neglected or misapplied by the federal government: scenario planning and real options analysis, discussed in the sections that follow.

Scenario Planning

Some federal agencies, notably the U.S. Department of Energy (DOE), use scenarios for planning, but they do not use them as well as they might. The difficulty springs from an excessive emphasis on price, especially the price of oil. Typically, scenarios are built around a set of production and consumption assumptions that yield a base case, with high-price cases and low-price cases on either side (EIA, 2005). Quite properly, these are not offered as forecasts but rather as plausible market outcomes in the future. Nevertheless, the focus remains on the price trajectories and not on the primary forces that drive these outcomes. Those driving forces that are not closely connected to price thereby suffer neglect. An alternative approach to scenario planning, commonly practiced in industry, could raise the quality of the policy debate by building the analysis around those primary forces (Bodde et al., 2005).

The Practice of Planning

Scenarios are stories about the future, a way to understand the impact of conditions that, while perfectly plausible, might never come to pass. Most importantly for policies aimed at an unknowable future, they are not stories about the policies themselves but rather about the context in which those measures dwell. Thus, scenarios, properly done, focus on the external world and the implications of alternative futures for the policies being considered. They offer decision makers and policy analysts a systematic way to ask, "What would we do if . . . ?" and hence provide unique value in managing the Ignorance quadrant of risk space. In effect, scenarios create an intellectual wind tunnel in which new policies and program concepts can be inexpensively tested. For example, Shell Oil, one of the pioneers of scenario analysis, used the method to avoid strategic surprise when prices on the world oil market collapsed to the $10 per barrel range in the 1990s, an event previously held to be unthinkable.

A large literature has grown up around constructing scenarios and using them effectively, though the practice remains more art than science (Schwartz, 1996; van der Heijden, 2002). Here, it suffices to note the common principles that underlie scenario applications, especially in cases where mission and necessity force energy planners to operate in the Ignorance sector of risk space. These principles include the following:

- Operational relevance: Good answers only follow good questions. Thus, scenarios must be built around a key policy issue for which some decision is imminent. A single focusing question should address this decision directly and ask the implications of strategic actions or portfolios of actions that might be taken.
- Causal forces: The scenario planning team must identify the forces operating in the external environment that will most strongly influence the outcome of policy decisions made now. The range of plausible effects of these causal forces then defines the scope of the scenarios. Here the analysis must capture the unpredictable, those forces firmly within the domain of uncertainty, ambiguity, and especially ignorance.
- A learning platform: The chief value of scenarios derives from making them, not from having them. Properly done, a scenario analysis becomes a platform through which organizations and their policy analysts learn about their external environment and the connections between that environment and the prospects for the success of the proposed actions. The very act of constructing the scenarios serves to communicate the policy process throughout the organization that builds them. Thus, the process itself can help build an operational consensus and clarify areas of policy disagreement, where those exist. This means that scenario planning must become a team sport played with intramural talent for its learning value to be fully realized.
- Continuity: Scenario planning only offers value as an ongoing way of mapping how actions that can be taken now ramify into an unknowable future. It does not work well as a one-off activity. It might be more accurate to rename the process "scenario thinking." The scenarios must be updated and revised as the passage of time reveals more about once-obscure events.
- Organizing observations: Scenarios provide a framework for organizing perceptions of unforeseeable events. The future itself will unfold eventually, and scenario analysis can provide signposts to sharpen our ability to discern the emerging patterns of events before they are fully developed. Current knowledge must be scanned for clues to the future and the scenarios frequently updated to account for real observations.

Scenario Principles in Action: A Brief Example

The scenario matrix shown in Figure 14-2 offers an example to illustrate the scenario method and the principles just sketched. It focuses on this question: What policies could be implemented within the next ten years to accelerate the transition of the automobile away from petroleum dependence? Even at this high, strategic level, the focal question still requires a key assumption: that constraints on the supply of conventional petroleum make this transition something that should be accelerated. More tightly focused questions could also be posed, but these would also require more assumptions. The constraints arising from climate change appear as one of the variables in the analysis.

Thus, scenario analysis requires considerable judgment to pose a question sufficiently focused that it illuminates meaningful distinctions among policy choices, yet broad enough to encompass the key issues. There is no formula for striking the proper balance, but if analysts and decision makers take the time to thoughtfully debate alternative framings of the issues, then those efforts will probably meet success. Two primary forces in the external environment will influence the answers to this question within the ten-year event horizon:

- Limits on carbon emissions imposed by climate change concerns. These could vary from strong constraints, the upper half of the policy framework of Figure 14-2, to essentially no constraints, in the lower quadrants. Thus, carbon constraints, or their absence, form the vertical axis.
- Constraints on the supply of conventional petroleum. These range from disruptive interventions, perhaps by terrorists, on the right side of

Fuel Use Strongly Constrained by Climate Change

Malthus's Revenge
- Nuclear/electric
- Renewable energy in all forms
- Carbon capture & sequestration

The Very Visible Hand
- Mass transit
- Telecom substitutes for travel
- Renewable energy
- Strategic reserves
- Carbon capture & sequestration

Fuel Availability Constrained by Resource Base

Fuel Availability Constrained by Disruption

The Invisible Hand
- Unconventional hydrocarbons
- Market-paced hybrid electric vehicles
- Fuel cell and H_2 storage

Law and Order
- Strategic reserves
- Diversity of petroleum sources
- War on terror

Fuel Use Unconstrained by Climate Change

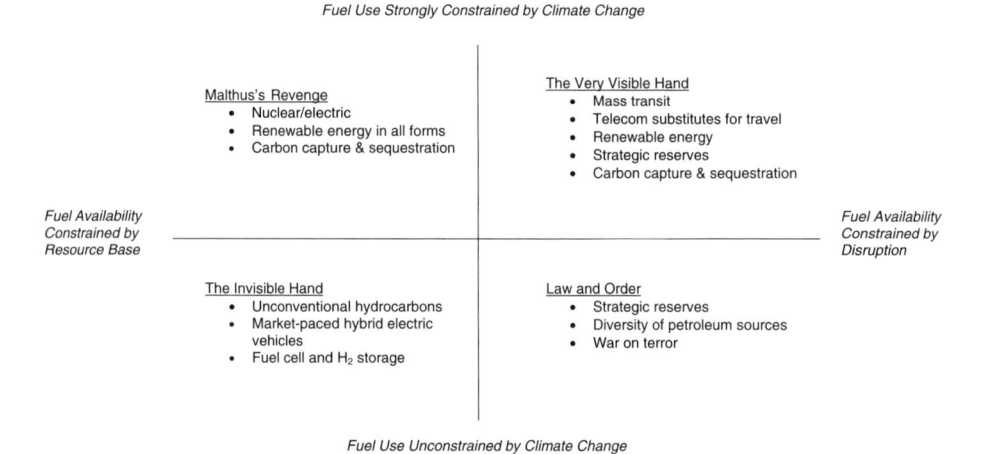

FIGURE 14-2. A policy matrix for the auto transition.

Figure 14-2 to simple resource inadequacy on the left. This builds the horizontal axis.

Taken as the axes of the matrix in Figure 14-2, these forces define a set of four distinct event patterns and capture much of the ambiguity, uncertainty, and ignorance of the risk space of Figure 14-1. Of course, more could be imagined, and "wild card" scenarios are frequently used to capture the impact of occurrences that might be unlikely but would have a severe impact if they did occur.

The essential characteristics of the four scenarios would then be set out as stories about the future, each one labeled with a characteristic name. These stories must be plausible and hold a reasonable prospect of occurring, even though many will not be congenial to the personal wishes of the analyst. In practice, the scenario stories often run several pages in length, but here a simple summary will suffice.

The scenario in the upper left corner of Figure 14-2, "Malthus's Revenge," is an unhappy world. Severe concerns with climate change mean that the atmosphere can no longer be used as a carbon sink. At the same time, resource constraints on conventional petroleum raise the cost of motor fuels and industrial petroleum to levels that cause a global recession. The policies with greatest leverage here would include the following:

- Carbon capture and sequestration (CCS). If this can be done satisfactorily, then the entire hydrocarbon resource base would be open to relieve the resource constraints on conventional fuels.
- Expanded renewable energy use for producing hydrogen or electricity. This important set of technologies provides a partial hedge against the failure of CCS.
- Increased reliance on nuclear energy for producing hydrogen or electricity—also a hedge.

Diagonally across the matrix, the lower right quadrant frames a scenario called "Law and Order." Here, chronic supply disruption, rather than resource depletion, motivates the transition in the auto sector. Environmental considerations do not inhibit petroleum use, either because of offsetting climatic events or simply because concern for the economy has trumped concern for the environment. In this world, CCS offers little value—a striking contrast to the high value of the technology in "Malthus's Revenge" scenario. Renewable energy must compete in the marketplace, but nuclear power remains inhibited by concerns with terrorism and rogue states. The policy options offering traction in such a world include a large strategic petroleum reserve, perhaps on the order of 2 billion barrels, and the capacity to use it as an effective price and supply shock absorber. Other options are diversification of conventional oil supply into politically stable regions and antiterrorist campaigns.

The lower left quadrant shows the "Invisible Hand" scenario. In this world, as in "Law and Order," concern with climate change does not drive policy. Therefore, carbon release does not constrain the search for unconventional hydrocarbon feedstocks, and the use of these hydrocarbon fuels enables a smooth transition away from conventional petroleum.

Early evidence of the feasibility of CCS offers much less value in the "Invisible Hand" scenario because carbon release is not an issue. Renewable and nuclear energy enter the market, but only as their cost competitiveness allows. Hybrid electric vehicles enter the market in proportion to the services they offer—onboard electronic capabilities, improved torque at each wheel, and so forth.

In the "Invisible Hand" scenario, research to improve the competitive status of fuel cells and onboard hydrogen storage might find a higher payoff than in, say, "Malthus's Revenge." This is because the desperate circumstances of the "Malthus's Revenge" scenario would encourage storing hydrogen onboard vehicles in pressurized tanks and burning it in internal combustion engines—both bringing enormous efficiency losses. By contrast, the hydrogen vehicles under the "Invisible Hand" scenario must compete in the marketplace with hybrid electric vehicles on the basis of consumer services. Publicly funded research would be the only way to accelerate that.

Finally, the upper right quadrant shows the "Very Visible Hand" scenario. Public needs drive this scenario, in contrast with the market orientation of the "Invisible Hand." Though conventional petroleum resources remain available, concerns with global climate change and terrorism sharply inhibit their use. As in the "Malthus's Revenge" scenario, early resolution of the questions surrounding CCS offer extraordinary policy value. Renewable energy would be encouraged by policy fiat, though nuclear would remain constrained by terrorist fears, thus removing an important hedge against the failure of CCS. Alternative hedges, such as mass transit and reduced vehicle travel, would rise in importance, and policies to encourage them would find value.

In sum, scenarios provide a systematic way to test how policy alternatives would work under sharply varied, but equally plausible, circumstances. In general, two kinds of policy options emerge: those that are robust across two or more scenarios, like CCS in the preceding example, and those that provide an essential hedge against disaster in one scenario, like nuclear energy or diversification of conventional petroleum supply. A well-balanced policy portfolio would include both kinds of options.

Real Options Analysis

In contrast with scenarios, real options analysis finds its intellectual roots in trading financial options, and hence the method has retained a reputation as a quantitative tool. Like scenario planning, its chief value lies in the thought process, not the numbers. Applied to policy decisions, this concept of options analysis seeks to estimate the value of creating a new "real option": the technological or political capacity to do something that cannot now be done, like permanently and safely disposing of the carbon wastes generated in the manufacture of hydrogen. The knowledge created by the research becomes the real equivalent to a financial call option—the ability, but not the requirement, to invest. Negative results matter, too, and much value can be derived from learning that a desired goal cannot be achieved within reasonable boundaries of cost, time, and public acceptability.

The option value of the knowledge gained from research will vary with future circumstances. In a world characterized by abundant renewable or nuclear energy, for example, an option to dispose of the carbon from fossil-based hydrogen production would add little value. On the other hand, learning sooner rather than later that such disposal is not achievable would have great value, as it would lend urgency to the development of nuclear and renewable alternatives. Thus, real options analysis can help decision makers balance their bets placed throughout the energy portfolio.

In addition to balancing the strategic portfolio, real options analysis offers insights to the managers of individual programs. Take, for example, FutureGen, a $1 billion project, initiated by the DOE with the support of industrial partners, to build and operate a zero carbon emission, coal-fueled power plant. The project bundles two important, but separable, objectives. First, FutureGen seeks to demonstrate that useful energy products—such as electric energy, hydrogen, and process heat—can be produced reliably and economically from the gasification of coal. Second, it would achieve zero carbon emissions by capturing the carbon dioxide effluent and permanently sequestering it from the biosphere in some geologic formation, such as depleted oil and gas reservoirs, unmineable coal seams, deep saline aquifers, or basalt formations. These reservoirs would require monitoring for an extended period to verify that the carbon dioxide did indeed remain in place.

Importantly for policy, these two chief benefits of the FutureGen project are quite separable. The DOE can demonstrate the option of geological sequestration without producing the energy products simply by purchasing the carbon dioxide from conventional power plants or even in industrial gas markets and then pumping it underground. Similarly, the department can demonstrate the chemical refinery approach to processing coal without attempting to dispose of its byproducts. Thus, each of these project components carries a distinct option value that does not depend on the other component—either the value of having the capability or the value of knowing that it cannot be achieved and that something else must be substituted.

As the FutureGen demonstration is currently managed, reliable knowledge about carbon storage must await both the completion and operation of the power plant and the protracted period of monitoring. The high option value of this knowledge would imply that the DOE should consider a separate and early effort in carbon storage. A real option analysis, once accomplished, might lead the DOE to hold that same opinion.

Taking Thought

Though this uncertain world might not be arranged as conveniently as humans might like, neither is it arranged as perversely as we might fear. Putting on a clean shirt does not always attract the soup-of-the-day, even though it might seem to. By taking thought about a future characterized by uncertainty, ambiguity, and ignorance, those charged with policy leadership can place more intelligent bets about the future and, in doing so, better persuade those charged with political leadership of the actions required.

References

Awerbuch, S., A. C. Stirling, J. Jansen, and L. Beurskens. "Portfolio and Diversity Analysis of Energy Technologies Using Full-Spectrum Risk Measures." In D. Bodde, K. Leggio, and M. Taylor (eds.), *Managing Enterprise Risk: What the Electric Industry Experience Implies for Contemporary Business*, Burlington, MA: Elsevier, 2006.
Baer, R. *Sleeping with the Devil*. New York: Crown Books, 2003.
Bodde, D. L., K. Leggio, and M. Taylor. "Speaking Uncertainty to Power." *Oil and Gas Energy Quarterly*, vol. 53, no. 4, June 2005. pp. 911–921.
Hirsch, R., R. Bezdek, and R. Wendling. "Peaking of World Oil Production and its Mitigation.
Schlesinger, J. "The Theology of Global Warming," *The Wall Street Journal*, August 8, 2005. page A10.
Schwartz, P. *The Art of the Long View*. New York: Currency-Doubleday, 1996.
The Economist. "Oil in Troubled Waters." April 28, 2005.
U.S. Congressional Budget Office. Fuel Economy Standards Versus a Gasoline Tax. U.S. Congress, March 2004. Available at www.cbo.gov/publications.
U.S. Energy Information Administration (EIA). *Annual Energy Outlook, 2005*. Washington, D.C.: EIA, 2005.
Van der Heijden, K. *The Sixth Sense*. Hoboken, New Jersey: Wiley, 2002.

CHAPTER 15

Toward a Transportation Policy Agenda for Climate Change

David Burwell and Daniel Sperling

How might we move forward in reducing greenhouse gases? Participants at the tenth Biennial Asilomar Conference on Climate Change Policy did not come to many definitive conclusions. But they did agree that climate change is an issue of pressing public concern that calls for innovative solutions. The conference outlined many potential strategies to address this problem from a wide variety of perspectives, including regulatory and voluntary approaches; technology-based approaches for both vehicles and fuels; and market and policy approaches to increase energy supply and reduce consumption. As the public becomes better informed on both the potential impacts of climate change and the contributing role of transportation in generating greenhouse gas (GHG) emissions, many of these strategies, alone or in combination, will help define strategic remediation plans to protect the global climate.

Opportunities abound. Most transport-related GHG strategies are synergistic with existing policy initiatives; solutions to traffic congestion and air pollution, and measures to improve transportation efficiency are each generally consistent with the goal of reducing transportation GHG emissions. International and local initiatives are expanding and will eventually force a coherent national policy to emerge within the United States and other nations. The public is demanding corporate responsibility in this area, and both energy and transportation companies are responding with their own roadmaps and narratives.

This chapter is intended as a synthesis of the various conference presentations and of the issues raised during conference discussions. Data and other quantitative measures are drawn from previous chapters and from conference presentations listed at the end of this chapter. These presentations addressed a wide range of strategies for reducing transportation-related GHG emissions, some in combination with each other and some self-standing. A summary of these strategies includes the following:

- Improved fuel economy, accomplished through improved fuel systems and vehicle technologies (Sloane, 2005)
- Alternative fuel technologies including electric and plug-in hybrids, biofuels, biodiesel, and hydrogen (Jackson, 2005)
- Tailpipe emissions regulations coupled with consumer education campaigns and incentives to promote and reward more fuel-efficient purchasing and driving behavior (Dumas, 2005; Reilly-Roe, 2005)
- Regulatory programs including possible application of stationary "cap-and-trade" programs to mobile sources of GHG emissions (German, 2005)
- Marketing campaigns to associate fuel-efficient purchasing behavior with self-identity, values, and peer group association (Kurani, 2005)
- System integration and management including improved road connectivity, inter-modal connectivity, bicycle and pedestrian access, and a variety of transit, paratransit and bus investments (Toth, 2005)
- Transportation Demand Management (TDM) programs to improve transportation and land use planning at the project and regional level, access management, zoning, redevelopment planning, transit-oriented development, and regional growth management (Ewing, 2005; Garry, 2005)
- Integration of transportation and urban development planning, improved system and financial management, consolidation and coordination of competing private transit systems, and investment in a variety of transit improvements including express bus, bus rapid transit (BRT), nonmotorized transport (NMT), and TDM (Bleviss, 2005; Hook, 2005; Mehndiratta, 2005; Schipper, 2005)

Crisis and Opportunity: Numbers, Needs, and the Not Particularly Rational Transportation Consumer

Transportation-related energy use and resulting GHG emissions pose looming threats to economic growth, the global environment, public health, and overall quality of life. The consequences for national economies vary by region. Transportation energy use will increase most dramatically in the developing economies, especially in Asia and Latin America.

The Numbers: Growth in Demand

Vehicle usage around the world continues to increase—rapidly in some regions. Increased vehicle travel is swamping vehicle efficiency improvements, with the result that GHG emissions continue to increase. Unless low-carbon fuels begin to replace petroleum and substantial improvements are made in fuel efficiency, this trend will continue.

Total carbon emissions from passenger transportation would actually decline if, for instance, average fleet efficiency in North America improved from 20.5 miles per gallon (mpg) in 2003 to 29 mpg by 2030, new car efficiency improved from 21 mpg to 38 mpg, and advanced gasoline and diesel burning internal combustion engines (ICEs) grew from 1 to 42 percent of total sales (Johnston, 2005). These ambitious targets for the United States would bring the country to efficiency levels that already exist in Europe. In 2003 European passenger fleet efficiency was 31.5 mpg, new car efficiency was 35 mpg, and advanced gasoline and diesel ICE comprised 39 percent of the market.

To reduce GHG emissions from the entire transport sector, though, broader changes are needed. Light-duty passenger vehicles consume only about half the energy used for transportation. GHG emissions from freight trucks, bus and rail, off-road vehicles, aviation, and marine transportation would remain as additional challenges.

Another factor to consider is upstream emissions. So far, only GHG emissions from combustion of fuel in the vehicle—the "pump-to-wheel" portion of the fuel cycle—have been considered. A considerable quantity of GHG emissions are also generated upstream in the transportation fuel supply chain, during the extraction, refining, and transporting of fuel to the pump, known as the "well-to-wheel" fuel cycle. And still more GHG emissions result from fossil fuels consumed in the materials and construction of vehicles and infrastructure for roads, rail, aviation, and marine transport. While these tend to be secondary sources, they need to be considered in crafting effective action plans.

Many at Asilomar argued that more emphasis needs to be given to vehicle usage. Reid Ewing, relying on U.S. Energy Information Administration data (AEO, 2004), argued that increased travel will swamp expected technology improvements (Ewing, 2005). Indeed, that will be the case if technology and fuel improvements do not accelerate.

The most rapid increases in transportation-related CO_2 emissions are in the densely populated Asia Pacific region, which includes India, China, and Japan. The world population is forecasted to grow from 6.3 billion people in 2003 to 8.0 billion in 2030, with the Asia Pacific region increasing from 3.5 billion people to 4.4 billion. Asia will likely account for well over 50 percent of total world population in 2030. Car ownership in the region is expected to soar from 15 per thousand population to 100 per thousand. The result would be 420 million vehicles by 2030. Optimistically assuming significant fuel efficiency improvements and increased use of diesel engines in the Asia Pacific region, emissions will still increase roughly fourfold.

Even with this increasing rate of car ownership in the Asia Pacific region, vehicle ownership and emissions per capita would fall far short of those in the United States and Europe. Johnston estimated that vehicle ownership would be less than 12 percent of the North American per capita car ownership and less than 22 percent of European car ownership (Johnston, 2005).

Transportation growth is not limited to the Asia Pacific region. It is a significant component of growth in energy demand in virtually all rapidly developing economies. Meeting the expected growth in travel—and, therefore, energy—requires accelerated gains in vehicle efficiency, provision of timely and adequate alternative transportation services, increasing the diversity of the transportation energy mix, and integrating transport services to meet mobility and access needs through a seamless and efficient intermodal transportation network.

If supply or demand challenges are not met, economic growth is compromised. The transportation sector must accept its proportionate responsibility in addressing this threat by pursuing at least the following strategies (Johnston, 2005):

- Developing and adopting a portfolio of transportation management and innovation approaches that have proven effective in reducing transportation GHG
- Promoting research and development and improvements in governance that encourage deployment of successful innovations
- Acknowledging and addressing the critical role of consumer choice, at the individual and collective level, in defining successful policy approaches

Needs: Policy, Connectivity, Technologies, Fuels

To pursue those strategies, a broader and deeper portfolio of scientific and engineering initiatives is needed to inform the climate change debate. Clearly, the more we know about climate change science, the better. Initiatives in this area by the Bush administration are welcome (Mahoney, 2005). However, such inquiries must be accompanied by policy action.

An effective national research and policy agenda will need to reflect the accelerating global connectivity of Science, Technology, and Engineering (ST&E) resources and capabilities. Significant and essential intellectual capacity exists in all regions of the world. Increased connectivity creates synergies between technical communities in academia, industry, and government. Policymakers can help leverage this capacity by working to remove legal and trade barriers to joint development projects and by supporting long-term protection of intellectual property. Government funding of research related to energy and climate change needs to encourage international collaboration. The issue is *global* climate change, and programs to address it need global resources. This approach acknowledges the existence of an international marketplace for ideas as well as for technologies. In a global economy, capital will follow intellectual property and market opportunity, and capital is essential to sustained innovation.

Other actions are needed to understand and influence the factors that affect the rate of market penetration of new transportation technologies, the growth in transportation demand, and how transportation choices are made. We need to understand purchasing behavior, mode choice, driving behavior, and choice of fuel technologies. An informed analysis of the political and economic risks of various courses of action is needed, with the results being transparent and widely distributed to interested audiences, including policymakers and the general public.

The global reliance on oil as the almost exclusive transportation fuel must be addressed in any program to limit global climate change. In the United States, for example, oil accounts for 97 percent of all transportation fuel. Success in reducing vehicle-related emissions will require a larger diversity in the transportation energy mix. Even if, optimistically, 50 percent of vehicles in 2025 had 50 percent better fuel economy than in 2002, the United States would still experience a 35 percent increase in total transportation fuel use, given projected growth in vehicles (Sloane, 2005).

Transportation is the fastest growing source of GHG emissions in the world. It already accounts for more than 20 percent of global GHG emissions and more than 30 percent of U.S. GHG emissions.

To stabilize carbon dioxide concentrations in the atmosphere at even twice pre-industrial levels would require a sharp reduction in emissions across all economic sectors, on the order of 50 percent by 2050. Incremental improvement in conventional vehicle technology is not enough. New fuel and vehicle technologies will be required, along with other strategies, if the transportation sector is to contribute its proportionate share to this reduction. Based on the present level of vehicle manufacturer investment in hydrogen fuels, market introduction of hydrogen-based fuels should begin between 2010 and 2015. In the meantime, a focus on other fuel technologies such as biofuels can help lower the rate of growth of transportation-related GHG emissions (Sloane, 2005).

The Human Factor

While good data and science, and deployment of innovative technologies can go far in improving the energy efficiency of transportation and in assessing the risk of climate change from the transportation sector, consumer behavior also plays a central role. Unfortunately, consumer response to new technologies and shifting public priorities is not well understood. As Kurani argued in his presentation, and earlier in this book, vehicles have symbolic meaning to consumers beyond their utility in providing access to goods and services. Convincing consumers to be more energy sensitive in their vehicle purchases and use is difficult when energy is just one of many factors they consider.

More generally, economic self-interest is not always the dominant factor in consumer behavior. Symbolic meanings—transportation as a statement of self-identity, values, and peer-group association—are also important. In addition, the cost savings gained from purchasing fuel-efficient vehicles is less of a factor in consumer behavior than feelings about being a "smart consumer," or "buying a piece of the future" for your children. Stories or narratives that appeal to consumer self-identity or interest in creating a better world may achieve better results in improving the efficiency of consumer transportation behavior than appeals that focus on self-interested arguments such as improved fuel efficiency. These stories and narratives may also work at the community level and affect the collective behavior of communities, governments, and corporations. Research, including modeling, can help these entities articulate their preferred future narratives or community "scenarios," and even lead to changes in behavior (Johnston, 2005; Kurani, 2005).

The Regulatory Landscape for Transportation, Energy, and Climate Change

Policy places an important role in shaping the behavior of individuals and companies. DeCicco highlighted the scale of the energy and GHG challenge by noting that oil consumption from motor vehicles increased 25 percent between 1990 and 2003 to 8.6 million barrels per day (bpd). This total oil consumption figure is approximately equal to the average annual oil production of Saudi Arabia over the last 15 years. An increasing share is consumed by light-duty trucks, which due to their lower fuel economy emit, on average, 39 percent more CO_2 per mile than passenger cars. Light trucks now represent 59 percent of total vehicle fleet CO_2 emissions. DeCicco argued that these numbers support action to reduce emissions from transportation vehicles (DeCicco, 2005).

There is increasing support nationally for a regulatory approach to transportation-related CO_2, according to Grundler (2005). Leadership is emerging within Congress and the Bush administration for action. However, any regulatory scheme must be based on a solid understanding of externalities, and the costs and savings must be transparent to the consumer.

Technology policy can play a particularly significant role in developing innovative GHG-reduction strategies (Rubin, 2005). It can help smooth out the innovation process through such interventions as research and development support to promote invention, patent protection to foster innovation, tax credits or procurement support to favor adoption, and education to encourage diffusion. Historically, public policy has contributed to technology innovation. For example, patent filings for clean air technologies increased from less than 10 to more than 100 annually in the ten years after enactment of the Clean Air Act. Both regulatory and nonregulatory policies will be needed to stabilize transportation-related GHG emissions. However, little is known about the relative efficacy of such policies, their most effective sequencing, or the potential benefits and risks of various combinations of them. Ongoing research is needed but does not eliminate the need for action (Rubin, 2005).

Vehicle regulation is the most prominent and widely used tool to improve vehicle fuel consumption and reduce carbon emissions. Regulations have been adopted in nine countries and regions, as presented by An (2005). Efficiency goals vary widely by region and implementation is nonuniform, however. Efficiency is regulated by a variety of standards including average fleet efficiency, vehicle category, total weight, and engine weight. The European Union (EU) and Japan have the most stringent vehicle standards, calling for between 16 and 19 percent reduction over 2002 vehicle fleet efficiency, respectively, by 2008 and 2010, respectively. The 16 percent target would reduce vehicle CO_2 emissions to 140 grams per kilometer of driving. Japan is on track to meet the standards, but the EU might fall somewhat short.

According to An (2005), the EU, China, Canada, and California would improve fuel efficiency by at least 20 percent by 2016 if all presently enacted standards for future years were met. He reported that the United States is projected to have both the lowest fuel economy rating and the lowest gains, just a 3 percent gain in fuel efficiency (An, 2005).

Energy companies also have a key role to play in both balancing transportation energy supply and demand and in addressing energy security issues. With respect to transportation energy, a short-term priority is to develop clean fuel technologies with lower carbon emissions. In the longer term, energy companies can assist in the gradual transition in engine technology from the conventional ICE drivetrain to hybrid electric drives, and ultimately to fuel cell systems. Whether this transition will be driven by the regulatory environment or by voluntary industry innovation is as yet uncertain (Eggar, 2005).

Other regulatory options, such as carbon trading and so-called "feebate" programs, may also play a role in reducing GHG emissions. Both options offer creative approaches, but also come with implementation and political uncertainties (Dumas, 2005).

Integration of transportation into a national carbon trading program, including a cap-and-trade program, is possible but problematic (German, 2005). According to German, downstream systems that focus on consumer purchasing and driving behavior face large political and administrative barriers. Upstream trading systems focusing on fuel suppliers are possible but work only by limiting fuel availability. Sector strategies that focus on vehicle manufacturers have a variety of barriers—including double counting, allocating responsibility between manufacturers and oil producers, and uncertainty about future emissions based on user behavior—that make this strategy very difficult. A system where manufacturers buy or sell credits to the government based on relative fuel intensity may be the most promising strategy.

Feebate programs, where fees are imposed on purchasers of vehicles that fail to meet a set fuel economy or emissions standard and cash rewards or rebates are granted for the purchase of vehicles that exceed the standard, are under active consideration in Canada but are not presently imposed anywhere in the world (Dumas, 2005). Factors influencing the design of a feebate program include price elasticities, the cost of technologies, feebate structure, selection of the standard, and the quality and availability of fuel economy data. Feebates can result in significant reductions in fuel consumption but only with a large transfer of payments between consumers or between customers and governments. There are many variations to such a program, including imposing the charges only on the purchase of vehicles with very high or low fuel efficiency. The efficacy and acceptability of such a program varies depending on whether it is specific to a state or province or encompasses an entire country or group of countries.

Despite the plethora of potential policy approaches, such as those highlighted here, policy makers in the United States have, to date, largely ignored the transportation sector in developing a national climate change strategy. National policy on transportation CO_2 is characterized by resistance to increases on fuel economy standards, gas taxes, energy taxes, and carbon-reduction requirements. The only progress nationally in the United States at this time is a minor increase in fuel economy standards for light-duty trucks. Miller argued that policy in this area is highly influenced by party politics, leading to a probable stalemate in the near future (Miller, 2005). While the U.S. Congressional Budget Office prefers gasoline taxes over corporate average fuel economy (CAFE) standards, neither is receiving much political support.

The Promise of Integrated Transportation Solutions

A recurring theme of the Asilomar Conference was the need for more and stronger collaborations and partnerships—between vehicle and fuel suppliers, emerging and economically advanced nations, and the many public and private entities investing in and managing transport services.

Vehicle and System Efficiency: A New Partnership

The most promising strategies to reduce transportation-related GHG emissions in the highly motorized societies of North America and Europe are those based on technology improvement and new fuels. This is true not just for passenger vehicles but also for large trucks. These advances increasingly require partnerships between fuel and vehicle suppliers, with investments in new and improved engines linked to new and improved fuels. Examples include diesel engines with low-sulfur diesel fuel, biofuels with flexible fuel engines, natural gas vehicles with natural gas stations, electric vehicles with recharging locations, and fuel cell vehicles with hydrogen supply.

Other strategies targeted at vehicle use are also promising. They may or may not be as effective at reducing GHG emissions, but they undoubtedly could make a substantial contribution. These other strategies include managing vehicle use while enhancing access. Importantly, these other demand-based strategies can contribute to other important metropolitan goals. Key players in the United States are state departments of transportation (DOTs), which are rapidly switching from a focus on system expansion to system management (Toth, 2005). The impetus for this transition comes from several sources, including completion of the interstate highway system; rising land costs in metropolitan areas that often render road expansion prohibitively expensive; and an emerging realization that beyond a certain level of development system expansion is no longer a cost-effective means of reducing congestion and improving access.

Improved system management as a focus of state DOT efforts to address congestion is also advanced by the recent realization that system expansion has done little to reduce congestion. According to the Texas Transportation Institute (TTI), the total delay experienced by a peak hour urban traveler rose from 16 hours in 1982 to 62 hours in 2000. In addition, the period of delay during rush hours expanded from 4.5 hours to 7.0 hours and the extent of the system experiencing congestion increased from 34 to 58 percent over the same time period (Lomax and Schrank, 2005). These results have encouraged state DOTs to give increased attention to strategies that manage overall travel demand, rather than increase total system capacity. This new approach has collateral energy conservation and climate benefits.

Toth described New Jersey DOT's policy of integrating transportation and land use planning. Their goal is to encourage the use of alternative modes, especially walking and bicycling; improve road connectivity to diffuse trips across the network; and coordinate transportation and land use planning and invest in low-cost incremental improvements that support efficient land uses (Toth, 2005). The New Jersey DOT is promoting this new policy through the New Jersey Future in Transportation (NJFIT) campaign, which is designed to solicit local planning partners to help the agency advance these objectives. It has also initiated a cooperative training program in support of this effort.

Better data and modeling at the regional level can support both technological innovation and state DOT "smart transportation" investment initiatives. Regional planning agencies are now using integrated transportation and land use planning models to conduct "scenario planning" of alternative regional growth plans. These models provide projections of increases in VMT and GHG emissions resulting from alternative land use and transportation investment scenarios with sound planning principles. These exercises suggest the potential for reduced land consumption, VMT growth, and GHG emissions (Garry, 2005).

In the Sacramento, California, region, for example, a two-year "regional blueprint" scenario planning exercise combined widespread public participation with three different types of models that evaluated alternative spatial distribution of economic activity, trips, and land use parcels, based on density, design, diversity of uses, and destinations—the four Ds. Results indicate that the consumer-preferred growth scenario reduced the share of trips by personal vehicles from 93.7 percent in 2050, under base case conditions, to 83.9 percent in the preferred scenario. VMT dropped by 25 percent and GHG emissions fell by 15 percent. These reductions assumed no efficiencies resulting from CO_2 tailpipe emission reductions, improvements in fuel economy, or change in regional growth rate or economic structure (Garry, 2005).

National surveys of regional and local planning initiatives show even more significant VMT and GHG emissions reductions from specific development projects. In these surveys comparisons were made between total VMT generated by projects under an unconstrained land use regulatory system, where development occurred primarily at the urban fringe on undeveloped land, called greenfield development, versus similar project location in an urban, transit-oriented location on a redeveloped site, called brownfield development. These comparisons showed VMT and GHG emissions reductions of about 50 percent for the brownfield developments when compared to greenfield development (Ewing, 2005). In the case of a specific mixed use redevelopment of a brownfield site in downtown Atlanta, a 33 percent reduction in VMT was projected due to improved regional accessibility of the site, and another 5 percent reduction was projected from the adoption of favorable density, design, and diversity of use criteria. While such results are site dependent, the travel demand management and GHG emissions reduction potential of redevelopment planning appears significant (Ewing, 2005).

Careful studies are needed that integrate analyses of transportation energy and GHG reduction benefits resulting from both technology and fuel improvements and from land use and growth management initiatives. The two strategies have some countervailing tendencies, since improved road system efficiency tends to increase total travel demand and encourage dispersed settlement patterns. Also, since vehicle manufacturers, energy companies, state DOTs, and regional land use planners operate within entirely separate regulatory structures, come from different professional disciplines, and manage for almost separate outcomes, coordination between these groups is extremely difficult.

Lessons from Abroad: Thinking Like a System

While more advanced economies can significantly reduce GHG emissions from mobile sources through technological innovation, the situation is different in developing regions. Emerging economies where motorization is just getting underway have limited ability to reduce GHG emissions from fuel- and vehicle-based strategies alone. In these societies the most promising strategy for improving transportation efficiency appears to be in system development and management, also known as sustainable transportation systems or simply sustainable transport.

Except for a few island nations, climate protection is not a policy priority in the developing world. Development of reliable transportation systems is the overriding priority. Yet, sustainable transport is the "horse" that can pull the climate "cart" (Schipper, 2005). The three pillars of sustainable transport are environmental protection, with a focus on safety, public health, and air pollution; social equity, to ensure reliable access for the poor, nondrivers, and all races and genders; and economic sustainability, creating a level playing field for all modes and producing financially sustainable public and private operators.

Many of the strategies to reduce GHG emissions are the same as those that lead to economically, socially, and environmentally sustainable transport services. The challenge is to identify these strategies and to create the effective governance structure with clear laws to implement and enforce them.

Establishing sustainable transportation systems is not easy. Developing economies, by their very nature, have relatively undeveloped and uncoordinated transportation infrastructure. Management of the infrastructure that does exist is often chaotic. In these countries, avoiding the GHG emissions that have not yet occurred is the best GHG reduction strategy (Mehndiratta, 2005; Schipper, 2005). This can be advanced through improved system planning, design, development, finance, and management (Bleviss, 2005). The key is to blend climate protection into locally relevant issues such as public transport, congestion, local air quality, and urban livability.

The demographic shifts of developing economies support transportation policies that have climate benefits. As these economies grow people migrate to urban areas, causing stress on underdeveloped public infrastructure. In Asia alone, at least seven cities—Jakarta, Shanghai, Hyderabad, Bejing, Tokyo, Seoul, and Bangkok—will have populations exceeding 10 million by 2015 (Mehndiratta, 2005). The population densities in these cities are so high that meeting transportation needs through automobiles is not feasible. Sustainable transportation systems are needed to improve air quality and reduce traffic congestion and fuel costs. Air pollution from dirty, two-stroke motorbikes and old vehicles is an acute problem. Congestion cripples economic development and fuel costs are a burden on fragile economies. The public benefits associated with sustainable transportation systems are therefore more tangible in developing as compared to developed economies. Climate change prevention is simply a beneficiary of these efforts.

Examination of the Latin American and Caribbean region offers an opportunity to study a system development and management approach for transportation and climate planning. According to Bleviss (2005), to be successful, such an approach must include decentralization of transportation system management from the federal to the regional and local levels; regional transportation planning capacity; land use and urban development planning that is effective at, for instance, eliminating the barriers to mixed use developments at transit nodes; mixed transportation structures that include bus rapid transit, nonmotorized transport, and consolidation of redundant private taxi and minibus fleets into efficient publicly regulated and scheduled routes; and transportation demand management measures at the project level.

The focus of multilateral aid, according to Bleviss (2005), should be on medium-sized cities that reflect these transportation characteristics and have the financial ability to build and manage their transportation systems. Cities in Latin America that have achieved success with this model include Bogota, Colombia; Curitiba, Brazil; and Cuenca, Ecuador.

Mediating organizations exist to facilitate collaboration between developed and emerging economies in reducing transportation energy use and GHG emissions. The Global Environmental Facility (GEF) provides public funding for environmentally sustainable transportation projects funded by the World Bank, regional development banks, the UN Development Program, and the UN Environmental Program. However, an early GEF focus on hydrogen-fueled buses has failed to yield any net GHG emissions reductions (Hook, 2005). Bleviss suggests that similar one-shot funding of NMT and BRT projects have yielded few positive GHG emissions results, and in some cases the results have been negative.

Successful projects require integrated, coordinated action. In Bogota, Colombia, a coordinated program to integrate NMT and BRT projects, coupled with a campaign to dampen growth in motor vehicle use, has proven effective.

The clean development mechanism (CDM) provides a method for governments that are signatories to the Kyoto Protocol and private CO_2 emitters to secure credits toward their GHG reduction targets. This is done by funding clean transportation projects in the developing world. However, the effectiveness of the CDM to reduce transportation-related GHG emissions is hampered by difficulties in establishing a baseline from which additional, or surplus, GHG emissions are calculated for credit. Determining the price of carbon credits is another challenge. No transportation project has yet qualified for CDM credits, and only three transportation projects are in the CDM pipeline, including the Bogota TransMilenio project (Winkelman, 2005).

The project-level focus of the CDM cripples its utility as a transportation-related GHG reduction strategy. Winkelman (2005) argues the CDM mechanism would be much more effective in reducing transportation emissions in developing countries and regions if CDM credits could be given for sectorwide transportation policies such as travel demand reduction, smart growth, renewable fuel standards, and fuel economy regulations. He suggests that this change be made if and when the Kyoto Treaty is renegotiated.

Overall, the key lesson learned from presentations on developing economies is that these areas face unique development challenges that require unique transportation solutions. While rapid mobilization will occur, population densities are already so high that a systems approach is the only feasible strategy for meeting transportation needs. This provides significant opportunities to control GHG emissions through a focus on preventing the emissions that don't yet exist. While climate change is of marginal concern in most developing economies, other considerations such as air quality, congestion, fuel costs, and social equity can drive transportation investments that support energy efficient and climate friendly outcomes.

The final session of the conference revisited barriers to, and opportunities for, implementing transportation energy and climate strategies in combination. Such combined strategies are beginning to appear in Europe under the label of integrated transportation strategies and are primarily promoted as a scheme for addressing congestion where road expansion is not feasible. The most visible example of integrated transportation strategies in operation is the London congestion-pricing scheme where cordon fees, parking restrictions, and increased transit service levels are applied in combination to achieve the desired reduction in congestion. This scheme has reduced car trips by 20 percent with resulting energy and climate benefits. While not expressly labeled as such, the Bogota, Colombia, TransMilenio project that includes bus rapid transit and nonmotorized transport improvements to provide better access to transit, and demand management in the form of auto-free zones, parking reforms prohibiting parking cars on sidewalks, and car-free days is another example of a combined strategy that has reduced congestion and increased access for the transit-dependent, while also yielding transportation energy and climate benefits (Hook, 2005).

Conclusion: Toward a Policy Agenda for Climate Change

Participants at the Asilomar Conference sought to explore the outlines of a policy agenda that would allow the transportation sector to reduce transportation-related GHG emissions. With climate science models suggesting that carbon dioxide and other greenhouse gas emissions must be reduced 50 percent or more from baseline projections by 2050 to stabilize atmospheric carbon dioxide levels at twice preindustrial levels, what is the role and responsibility of the transport sector? Can this 50 percent goal be achieved? The unique characteristic of this book and the Asilomar Conference was the careful examination of a wide variety of possible strategies by a wide range of experts and leaders. Many examples were discussed, including efforts by individual companies such as FedEx and Kinkos to reduce GHG emissions from their transportation activities, the novel memorandum of understanding between the government of Canada and the auto manufacturers to voluntarily reduce GHG emissions from their vehicle product lines, and introduction of low-carbon fuels by energy companies. Likewise, the commitment of infrastructure suppliers such as the New Jersey DOT to build and manage more energy-efficient transportation systems, and a renewed commitment by regional agencies to growth management, provide another set of initiatives.

At the intersection of these three public and private groups—transportation and energy providers, infrastructure builders and managers, and land use planners and decision makers—lies the real responsible party, the consumer of transportation services. Ultimately it is personal behavior—the way we access transportation services and the way we settle upon the land—that dictates the actions of energy and transportation providers. Since virtually every citizen is a transportation planner and decision maker in meeting his or her transportation needs, the challenge of climate change can only be addressed by broad public participation in changing energy use and travel behavior.

Acknowledgments

The author wishes to thank John E. Johnston, Planning Executive, Exxon-Mobil Research and Engineering Company and Jamie Knapp, Principal, J Knapp Communications, for their help in the development of this chapter. Any error in representing the views of the various presenters is the author's sole responsibility.

References

All presentations that were made at the 10th Biennial Asilomar Conference, 2005, can be accessed at http://www.its.ucdavis.edu/events/outreachevents/asilomar2005/.
An, F. Presentation at tenth Biennial Asilomar Conference, 2005.
Bleviss, D. Presentation at tenth Biennial Asilomar Conference, 2005.
DeCicco, J. Presentation at tenth Biennial Asilomar Conference, 2005.
Dumas, A. Presentation at tenth Biennial Asilomar Conference, 2005.
Eggar, D. Presentation at tenth Biennial Asilomar Conference, 2005.
Ewing, R. Presentation at tenth Biennial Asilomar Conference, 2005.
Garry, G. Presentation at tenth Biennial Asilomar Conference, 2005.
German, J. Presentation at tenth Biennial Asilomar Conference, 2005.
Grundler, C. Presentation at tenth Biennial Asilomar Conference, 2005.
Hirsch, R. Presentation at tenth Biennial Asilomar Conference, 2005.
Hook, W. Presentation at tenth Biennial Asilomar Conference, 2005.
Jackson, M. Presentation at tenth Biennial Asilomar Conference, 2005.
Johnston, J. Presentation at tenth Biennial Asilomar Conference, 2005.
Kesghi, H. Presentation at tenth Biennial Asilomar Conference, 2005.
Kurani, K. Presentation at tenth Biennial Asilomar Conference, 2005.
Lomax, T. J., and D. Schrank. "The 2005 Urban Mobility Report." Texas Transportation Institute, May 2005.
Mahoney, J. Presentation at tenth Biennial Asilomar Conference, 2005.

Mehndiratta, S. Presentation at tenth Biennial Asilomar Conference, 2005.
Miller, C. Presentation at tenth Biennial Asilomar Conference, 2005.
Reilly-Roe, P. Presentation at tenth Biennial Asilomar Conference, 2005.
Rubin, E. Presentation at tenth Biennial Asilomar Conference, 2005.
Schipper, L. Presentation at tenth Biennial Asilomar Conference, 2005.
Sloane, C. Presentation at tenth Biennial Asilomar Conference, 2005.
Toth, G. Presentation at tenth Biennial Asilomar Conference, 2005.
U.S. Energy Information Administration (EIA). *Annual Energy Outlook: 2004.* Washington, D.C.: EIA, 2005.
Winkelman, S. Presentation at tenth Biennial Asilomar Conference, 2005.
World Resources Institute. Climate Analysis Indicators Tool (CAIT), version 3.0. Washington, D.C.: World Resources Institute, 2006. Available at www.cait.wri.org.

APPENDIX A

About the Editors and Authors

About the Editors

Daniel Sperling is a professor of civil engineering and environmental science and policy and founding director of the Institute of Transportation Studies (ITS-Davis) at the University of California, Davis. Dr. Sperling is recognized as a leading international expert on transportation technology assessment, energy and environmental aspects of transportation, and transportation policy. In the past 20 years, he has authored or coauthored over 160 technical papers and 8 books. Dr. Sperling is associate editor of *Transportation Research D* (Environment) and a current or recent editorial board member of five other scholarly journals. He is a recent member of U.S. National Academy of Sciences committees on Highway Finance (2003–2004), Hydrogen Production and Use (2002–2003), Personal Transport in China (2000–2002), Transportation Environmental Cooperative Research Program Advisory Board (1999–2001), Biomass Fuels R&D (1999), Enabling Transportation Technology R&D (1998), Transportation and a Sustainable Environment (1995–1997), Transportation Options for Megacities (1994), and Liquid Fuel Options (1989–1990). He is a founding chair and emeritus member of the Alternative Transportation Fuels Committee of the U.S. Transportation Research Board. Dr. Sperling consults for international automotive and energy companies, major environmental groups, and several national governments. He has testified numerous times to the U.S. Congress and various government agencies and provided keynote presentations and invited talks in recent years at international conferences in Asia, Europe, and North America.

James S. Cannon is an internationally recognized researcher specializing in energy development, environmental protection, and related public policy issues. He is president of Energy Futures, Inc., which he founded in 1979.

Among its activities, Energy Futures publishes the quarterly international journal *The Clean Fuels and Electric Vehicles Report* and the bimonthly newsletter *Hybrid Vehicles*. Mr. Cannon has written several books on alternative transportation fuels, including *The Drive for Clean Air* (1989), *Paving the Way to Natural Gas Vehicles* (1993), and *Harnessing Hydrogen: The Key to Sustainable Transportation* (1995). He previously collaborated with Dr. Sperling to edit another volume of presentations from the Asilomar Conference held in 2003, *The Hydrogen Energy Transition: Moving toward the Post Petroleum Age in Transportation*. Over the past decade, his research into alternative transportation fuels has taken him to over 20 countries on 5 continents. Mr. Cannon previously had a seven-year professional association with the U.S. Office of Technology Assessment, and for eight years was an energy policy analyst for the Energy, Minerals, and Natural Resources Department of the State of New Mexico. Mr. Cannon holds an A.B. degree in chemistry from Princeton University and an M.S. degree in biochemistry from the University of Pennsylvania. He lives with his family in Boulder, Colorado.

About the Authors

Feng An is an international transportation consultant and founder of Energy and Transportation Technologies LLC. In recent years, he has been actively involved in several key automotive fuel economy and GHG emissions studies around the world, especially for the United States, China, Mexico, and Brazil. He is a director of Auto Project on Energy and Climate Change (APECC), a nonprofit organization based in Beijing, China. He currently also serves as transportation consultant to Hewlett Foundation's Latin America program, Energy Foundation's China sustainable transportation program, DOE's Argonne National Laboratory and Environmental Defense. From 2000 to 2003, he served on a joint U.S.-China National Academy Committee and coauthored a book, *Personal Cars and China*. Dr. An received his Ph.D. from the University of Michigan in 1992 and M.S. from Tsinghua University in Beijing in 1986, both in Applied Physics. Dr. An has authored numerous publications, including 17 SAE papers, in the area of automotive technologies and their impacts on energy and environment. He is also a board director of Professional Association for China's Environment (PACE) and editor-in-chief of *Sinosphere Journal*, an online publication of PACE.

Anup P. Bandivadekar is a doctoral candidate at the Engineering Systems Division at the Massachusetts Institute of Technology (MIT). His research interests are focused around developing frameworks and methods to foster innovative solutions toward achieving a sustainable energy and transportation system. Currently, he is working with the MIT Laboratory for Energy and the Environment on evaluating fuels and vehicle technologies as well as policy alternatives that could significantly reduce greenhouse gas emissions of the U.S. light-duty vehicle fleet in the next 30 years. He holds a B.A. from University of Mumbai and an M.S. from Michigan Technological University in the field of mechanical engineering, and an M.S. in Technology and Policy from MIT.

Roger Bezdek has 30 years of experience in research and management in the energy, utility, environmental, and regulatory areas, serving in private industry, academia, and the federal government, and is the founder and president of Management Information Services, Inc., a Washington, D.C.–based economic and energy research firm. He has served as corporate director, corporate president and CEO, university professor, research director in ERDA/DOE, special advisor on energy in the Office of the Secretary of the Treasury, and U.S. energy delegate to the European Community and to NATO. He has also served as a consultant to the White House, federal and state government agencies, and various corporations and research organizations. Dr. Bezdek received his Ph.D. in economics from the University of Illinois (Urbana), is an internationally recognized expert in energy market analysis, R&D assessment, and energy forecasting. He is the author of 4 books and 200 articles in scientific and technical journals and serves as an editorial board member and peer-reviewer for various professional publications. He is the recipient of numerous honors and awards, has served as a U.S. representative to international organizations on energy and environmental issues, and lectures frequently on economic, energy, and environmental topics.

Deborah Lynn Bleviss has worked in the energy and environmental field for more than 20 years. Since late 2003, she has been a partner in a new consulting group, the BBG Group, which addresses sustainable urban transportation options, both domestically and internationally. Previously, Ms. Bleviss worked first as an advisor to (1996–1998) and then as program director of (1998–2001) the Inter-American Development Bank's "Sustainable Markets for Sustainable Energy" (SMSE) program. Prior to her work at the IDB, she worked at the U.S. Department of Energy of the United States (1995–1996) as an advisor to the assistant secretary of energy for Energy Efficiency and Renewable Energy, developing international and domestic clean transportation and energy financing initiatives. Ms. Bleviss was a lead author for the transportation mitigation chapter of the *Second Assessment Report of the Intergovernmental Panel on Climate Change* (published in 1995) and the author of *The New Oil Crisis and Fuel Economy Technologies: Preparing the Light Transportation Industry for the 1990s* (published in 1988). Ms. Bleviss is also an adjunct professor at the Johns Hopkins School for Advanced International Studies. Trained as a physicist, she received her education from UCLA and Princeton University.

David L. Bodde is a professor of engineering and business at Clemson University and a senior fellow at Clemson's Spiro Center for Entrepreneurial Leadership. Dr. Bodde serves on the Board of Directors of a variety of companies: Great Plains Energy (a diversified energy company and electric utility), the Commerce Funds (a mutual fund), and several privately held ventures. His past experience includes vice president of the Midwest Research Institute, assistant director of the U. S. Congressional Budget Office, and deputy assistant secretary in the Department of Energy. He recently served as chairman of the Environmental Management Board, advising the Department of Energy on the cleanup of the U.S. nuclear weapons complex and is a member of the National Research Council's Board on Energy and Environmental Systems. Dr. Bodde holds a D.B.A. from Harvard University (1976); M.S. degrees in nuclear engineering (1972) and management (1973), both from the Massachusetts Institute of Technology (MIT); and a B.S. from the United States Military Academy (1965). He is a veteran of the U.S. Army and served in Vietnam.

André Bourbeau is an energy economist and is currently Manager of Economic Analysis in the Environmental Affairs Directorate at the Federal Ministry of Transportation, Transport Canada. In this position, André oversees Transport Canada's Sustainable Transportation Research Initiative, which is intended to fill analytical gaps and contribute to a national perspective on climate change transportation solutions. Specific objectives of this initiative include facilitating the creation and improvement of sustainable transportation data and indicators, developing tools to improve decision making, and facilitating the assessment of climate change mitigation options and impacts, including cobenefits such as congestion reduction benefits. Prior to this, André worked in different ministries in the federal government, including Environment Canada and Natural Resources Canada. André holds an M.S. in economics from Laval University (Quebec City).

David Burwell is a partner in the BBG Group, an independent consulting firm providing services in sustainable transportation in North and South America. He also serves as the senior vice president for Transportation Programs at the Project for Public Spaces in New York City. He was formerly president and CEO of Rails to Trails Conservancy (1985–2001) and the Surface Transportation Policy Project (2001–2003). He has served on the TRB Executive Committee (1992–1998) and on the TRB Study Committee on Transportation and a Sustainable Environment (1997–1998). He is past chair of the TRB Sustainable Transportation Task Force (2003–2005). He was appointed a National Associate of the National Academies in 2003. He lives in Bethesda, Maryland.

John M. DeCicco, Ph.D., is a senior fellow who specializes in automotive strategies at Environmental Defense. His work entails technology assessment and policy analysis of ways to improve efficiency and reduce emissions of cars and light trucks. He has published extensively on the subject, with recent studies addressing options for improving the fuel economy of gasoline-powered vehicles, including conventional and hybrid powertrains and mass reduction, prospects for fuel cell vehicles, and market characterizations of automotive sector oil demand and carbon dioxide emissions. He has an interest in developing regulatory, market-oriented, and consumer educational strategies to foster progressive change in the auto market. Dr. DeCicco is active in the Society of Automotive Engineers and the National Research Council's Transportation Research Board, for which he chaired the Energy Committee from 1996 to 2000. He received his doctorate in mechanical engineering from Princeton University in 1988.

Alexandre Dumas is an energy economist and currently is acting senior economist in the Environmental Affairs Directorate at the Canadian federal Department of Transportation (Transport Canada). In this position, Alexandre is involved in Transport Canada's efforts to fill analytical gaps and contribute to a national perspective on climate change transportation solutions. His work focuses on the development of tools to improve decision making and the assessment of climate change mitigation options and impacts. Prior to this, Alexandre worked for different federal government departments including Natural Resources Canada and Statistics Canada. Alexandre holds a B.A. in economics from the University of Sherbrooke.

Duncan Eggar is a chartered civil engineer. He graduated from the University of Nottingham and gained his initial experience on various heavy engineering projects, including the Thames Barrier. He joined BP in 1981, and for ten years was involved with the design, construction, installation, and maintenance of offshore oil and gas production facilities in the North Sea and on a posting to New Zealand. He also participated in and led R&D on related topics. In 1990 he joined the global aviation fuels marketing unit of BP, where he held various roles in engineering management, Health Safety and Environment, project management, and operations management; this included a posting to South Africa. In April 2004 he returned to BP as the senior business advisor on Sustainable Mobility to the Refining and Marketing Segment of the BP group.

Freda Fung is an automotive analyst at Environmental Defense, working on issues related to carbon emissions and fuel consumption of cars and light trucks. She analyzes data and develops models to estimate carbon dioxide emissions and oil use from the U.S. auto sector and conducts research on policies and alternative strategies to reduce impacts of automotive use on climate change. Prior to joining Environmental Defense, Ms. Fung worked with Tellus Institute, a nonprofit research and consulting group in Boston, where she engaged in research on market-based supply chain management practices to reduce chemical use and waste in the automotive and aerospace sectors. She also worked as a consultant in a Hong Kong–based environmental consulting firm. Ms. Fung received an M.S. in environmental management and policy from Lund University, Sweden, in 1999 and an M.S. in economics from the Chinese University of Hong Kong in 1995.

John German is manager of environmental and energy analyses for American Honda Motor Company. His responsibilities include anything connected with environmental and energy matters, with an emphasis on being a liaison between Honda's R&D people in Japan and regulatory affairs. Mr. German has been involved with advanced technology and fuel economy since joining Chrysler in 1977, where he spent eight years in Powertrain Engineering working on fuel economy issues. Prior to joining Honda 7 years ago, he also spent 13 years doing research and writing regulations for EPA's Office of Mobile Sources' laboratory in Ann Arbor, Michigan. Mr. German is the author of a variety of technical papers and a recent book on hybrid gasoline-electric vehicles published by SAE. He was the first recipient of the recently established Barry D. McNutt award, presented annually by SAE for Excellence in Automotive Policy Analysis. He has a B.A. in physics from the University of Michigan and got halfway through an M.B.A. before he came to his senses.

David L. Greene is a corporate fellow of Oak Ridge National Laboratory (ORNL). He has spent 25 years researching transportation energy and policy issues. Dr. Greene received a B.A. from Columbia University in 1971, an M.A. from the University of Oregon in 1973, and a Ph.D. in geography and environmental engineering from Johns Hopkins University in 1978. He has published more than 175 professional publications, including three books (*Transportation and Energy, Transportation and Global Climate Change, and The Full Costs and Benefits of Transportation*), and is the recipient of four "best paper" awards from scientific societies. Dr. Greene has been active in the Transportation Research Board and National Research Council for over 25 years, serving on numerous standing and ad hoc committees dealing with energy and environmental research. In recognition of his service to the National Academy of Science and National Research Council, Dr. Greene has been designated a lifetime National Associate of the National Academies.

John B. Heywood is the director of the Sloan Automotive Laboratory and Sun Jae Professor of Mechanical Engineering at Massachusetts Institute of technology (MIT). Professor Heywood did his undergraduate work in Mechanical Engineering at Cambridge University and his graduate work at

MIT. His current research is focused on the operating, combustion, and emissions characteristics of internal combustion engines and their fuel requirements. He has also worked on issues relating to engine design in MIT's Leaders for Manufacturing Program. He was engineering codirector of the program from 1991 to 1993. He has published some 180 papers in the technical literature and has won several awards for his research publications. He holds an Sc.D. from Cambridge University for his published research contributions. He is an author of a major text and professional reference *Internal Combustion Engine Fundamentals* and coauthor with Professor Sher of *The Two-Stroke Cycle Engine: Its Development, Operation, and Design*. He is now directing MIT's Mechanical Engineering Department's Center for 21st Century Energy, which is developing a broader set of energy research initiatives. In January 2003, Professor Heywood was appointed codirector of the Ford-MIT Alliance.

Reid R. Heffner is a researcher and Ph.D. candidate at the University of California, Davis, Institute of Transportation Studies. His research focuses on the consumer response to advanced technology vehicles, including hybrid electric vehicles and hydrogen fuel cell vehicles. Currently, Mr. Heffner is investigating how advanced vehicles serve as symbols to consumers and the impact this symbolism has on consumers' decisions to purchase these vehicles. Before joining UCD-ITS in 2003, he spent six years designing, building, and marketing new technology products. Most recently, Mr. Heffner was a senior product manager at supply-chain software maker Manugistics, where he managed collaborative development projects with transportation clients that include Mitsubishi Motors and Harley-Davidson. Mr. Heffner has a B.A. from Colgate University (1993) and an M.B.A. from Georgetown University (1997).

Robert L. Hirsch is a senior energy program advisor at SAIC and a consultant in energy, technology, and management. Previously, he was a senior staff member at RAND (energy policy analysis), executive advisor at Advanced Power Technologies, Inc. (environmental and defense R & D), vice president of the Electric Power Research Institute, vice president and manager of Research and Technical Services for Atlantic Richfield Co. (oil and gas exploration and production), founder and CEO of APTI (commercial and Defense Department technologies), manager of Exxon's synthetic fuels research laboratory, manager of Petroleum Exploratory Research at Exxon (refining R & D), assistant administrator of the U.S. Energy Research and Development Administration responsible for renewables, fusion, and geothermal and basic research (presidential appointment), and director of fusion research at the U.S. Atomic Energy Commission and ERDA. He has served on advisory committees for Department of Energy programs and national laboratories, the General Accounting Office, the Office of Technology Assessment, the Gas Research Institute, and NASA. He holds 14 patents and has over 50 publications. He is immediate past chairman of the board on Energy and Environmental Systems of the National Research Council, the operating arm of the National Academies, has served on a number of National Research Council committees, and is a national associate of the National Academies.

Walter Hook has been the executive director of the Institute for Transportation and Development Policy (ITDP) since 1993. ITDP provides technical assistance to municipalities in developing countries working on bus rapid transit, nonmotorized transport, travel demand management, and brownfield revitalization. In the past, ITDP also played a key advocacy role in the development of the Global Environmental Facility's Operational Program 11 on Transport and earlier on the transport sector lending policies of the World Bank. He earned a Ph.D. in Urban Planning from Columbia University in 1996.

Kenneth S. Kurani is a member of the professional research staff at the Institute of Transportation Studies at the University of California, Davis. Working as part of a multidisciplinary team, he develops and applies theories and methods to evaluate user responses to new transportation and information technologies. While this research is largely conducted within a household activity-based approach to travel demand analysis and interactive stated preference and reflexive survey methodologies, one of the primary goals is to expand and enrich the behavioral models applied in transportation research. His research explores how citizen/consumers can use new technologies to shape both their own lives and efforts to market transportation and communication networks according to their collective benefits such as energy efficiency, air quality, safety, and social equity. For the past 20 years, he has primarily worked within the area of consumer markets for alternative fuel and electric-drive vehicles. His ongoing research includes household response to electric, hybrid, and fuel cell vehicles and consumer/citizen valuation of automotive fuel efficiency. Dr. Kurani holds a Ph.D. in civil and environmental engineering from UC Davis.

David M. Reiner is a lecturer in technology policy at the Judge Business School, University of Cambridge, the United Kingdom. He is also a research associate of the Centre for International Business and Management and the Electricity Policy Research Group at Cambridge, of the Centre for the Study of the United States at the French Institute for International Relations in Paris, and of MIT's Joint Program on the Science and Policy of Global Change and Carbon Sequestration Initiative. Dr. Reiner received his Ph.D. in political science at the Massachusetts Institute of Technology and has taught at Tufts University in Medford, Massachusetts, and the Graduate School of International Studies in Geneva, Switzerland. His areas of research include energy and environmental policy, technology policy, competition policy, and public perceptions. Current projects include a study on public and opinion leader perceptions of carbon capture and storage technologies (with Chalmers University of Technology in Sweden, the University of Tokyo, and MIT and funding from the Alliance for Global Sustainability and the electric power industry on three continents); institutional design for climate change policy; the potential role for the environmental aid in developing countries; and the politics and geopolitics of energy security in the United Kingdom.

Gary Toth has 32 years of experience within the New Jersey Department of Transportation (NJDOT) and is currently director of Project Planning and Development. He is a graduate of Stevens Institute of Technology in New Jersey in 1973 with a B.E. (major in civil engineering). He also is a graduate of the Environmental Management Institute at the University of Southern California in 1980. He is one of the originators of the NJDOT Task Force on Context Sensitive Design (CSD) and has participated in workshops or peer reviews on CSD or Community Impact Assessment (CIA) in Maryland, Connecticut, Washington D.C., Indiana, and Oregon. He has participated in panels on integrating transportation and land use at various locations around the country, including the New Partners for Smart Growth, Surface Transportation Policy Project, North Atlantic Transportation Planning Officials, National Community Impact Assessment Committee, Northeast Association of State Transportation Officials, and with the upcoming Executive Seminar on Transportation and Land Use, hosted by the National Cooperative Highway Research Project. Mr. Toth is the father of three children. He enjoys reading, coaching soccer, cooking, and wine.

Thomas S. Turrentine began his anthropology career studying processes of cultural change in the Peruvian Andes. For the past two decades he has been working with the Institute of Transportation Studies at the University of California at Davis integrating anthropological methods and theory with transportation research to create unique approaches to understanding the future of transportation systems, especially consumer and citizen response to new technologies and policies. Dr. Turrentine has conducted field research in the United States, Peru, Bolivia, Chile, Canada, and New Zealand, and has been an invited speaker to international conferences on the subject of the alternative fueled vehicles market and transportation research methods in Sweden, France, Canada, and England. He has served as a consultant to automobile and other corporations on the future of cars and personal transport. Recent work is focused on the market for hybrid electric and fuel cell vehicles.

Robert M. Wendling has 30 years of experience in consulting and management in the energy, environmental, statistical/econometric modeling, and regulatory areas, serving in private industry and the federal government. He has served in industry as corporate CEO and president and as corporate vice president and in senior economic positions in the U.S. Department of Commerce and the Department of Energy. Mr. Wendling was director of Commerce's STAT-USA office, cocreator of the NRIES Regional Econometric Impact Model, and also served as the lead U.S. representative for Asian Pacific Economic Cooperative (APEC) tariff and trade negotiations. He is the author of 75 reports and professional publications on energy and environmental topics and lectures frequently on various energy, environmental, regional economic analysis, and economic modeling and forecasting topics. He received his M.A. in economics from George Washington University.

APPENDIX B

Asilomar Attendee List: 2005

First & Last Name	Affiliation
Feng An	Argonne National Laboratory
Shosaku Ando	Nissan Technical Center North America
Michael A. Ball	Transport Canada
Anup Bandivadekar	Massachusetts Institute of Technology
Shannon F. Baxter Clemmons	California Environmental Protection Agency
Louise Bedsworth	Union of Concerned Scientists
William Black	Indiana University
Deborah Bleviss	BBG Group
Kate Blumberg	International Council on Clean Transportation
Carl Blumstein	California Institute for Energy & Environment
Dave Bodde	Clemson University
John Boesel	WestStart-CALSTART
André Bourbeau	Transport Canada
James Boyd	California Energy Commission
Aaron Brady	Cambridge Energy Research Association
Harold Brazil	Metropolitan Transportation Commission
Christie Joy Brodrick	James Madison University
Joe Browder	Dunlap & Browder
Susan Brown	California Energy Commission
Andy Burke	ITS-Davis
David Burwell	BBG Group
John Cabaniss	Association of International Automobile Manufacturers
Robert Campbell	Hydrogenics
James Cannon	Energy Futures, Inc.
Alex Charpentier	University of Toronto
Eileen Claussen	Pew Center on Global Climate Change
Michael Coates	Mightycomm/Bosch
James Corbett	University of Delaware
Paul Craig	UC Davis

Continued

Asilomar Attendee List: 2005—cont'd

First & Last Name	Affiliation
Joshua Cunningham	ITS-Davis
John DeCicco	Environmental Defense
Mark Delucchi	ITS-Davis
Richard Doctor	Argonne National Laboratory
Bill Drumheller	Oregon Department of Energy
Alexandre Dumas	Transport Canada
Sarah Dunham	U.S. Environmental Protection Agency
Louise Dunlap	Dunlap & Browder
Catherine Dunwoody	California Fuel Cell Partnership
Harry Dwyer	ITS-Davis
George Eads	CRA International
Duncan Eggar	BP
Anthony Eggert	ITS-Davis
Paul Erickson	ITS-Davis
Reid Ewing	National Center for Smart Growth
Alex Farrell	UC Berkeley
Lasse Fridstrom	Institute of Transport Economics
Danielle Fugere	Bluewater Network
Ichiro Fujimoto	Takushoku University
Tom Fulks	Mightycomm/Bosch
Mark Gaber	Center for Climate Change and Environmental Forecasting-U.S. DOT
Cynthia Gage	U.S. Environmental Protection Agency
Gordon Garry	Sacramento Area Council of Governments
John German	American Honda Motor Co.
Henry Gong, Jr. M.D.	California Air Resources Board
Konstadinos Goulias	UC Santa Barbara
S. William Gouse	United States Council for Automotive Research
Robert L. Graham	Electric Power Research Institute
Robert Greco	American Petroleum Institute
David Greene	Oak Ridge National Laboratory
Charles Griffith	Ecology Center
Alison Grigg	BC Hydrogen Highway, Canadian Transportation Fuel Cell Alliance
Tom Gross	Consultant
Christopher Grundler	U.S. Environmental Protection Agency
Anna Halpern-Lande	Massachusetts Institute of Technology
Ayame Hashinokuchi	Nissan Technical Center North America
Eric Haxthausen	Environmental Defense
Robert Hayden	DRH/NHA
Toshio Hirota	Nissan Motor Co.
Robert Hirsch	Science Applications International Corporation
Masahiro Hisatomi	Nissan Motor Co.
Walter Hook	Institute for Transportation and Development Policy
Robert Horn	Stanford University

Asilomar Attendee List: 2005—cont'd

First & Last Name	Affiliation
Shang Hsiung	U.S. DOT—Federal Transit Administration
Nan Humphrey	Transportation Research Board
John Hutchison	Ontario Ministry of Environment
Roland Hwang	Natural Resources Defense Council
Mitch Jackson	FedEx
Mike Jackson	TIAX
Jeffrey Jacobs	Chevron
Norman Johnson	Robert Bosch
Jack Johnston	ExxonMobil
Alan Jones	Nissan Technical Center North America
Andrew Jones	UC Berkeley
Emmanuel Kasseris	Massachusetts Institute of Technology
Dean Kato	Toyota Motor Sales
Paul Khanna	Natural Resources Canada
Haroon Kheshgi	ExxonMobil
Jamie Knapp	Knapp Communications
Benjamin Knight	Honda R&D Americas
Fumiaki Kobayashi	Toyota Motor Sales
Wilfrid Kohl	Johns Hopkins University—IEEP
Joe Krovoza	ITS-Davis
Ken Kurani	ITS-Davis
Therese Langer	American Council for an Energy-Efficient Economy
Bob Larson	U.S. Environmental Protection Agency
Michael Lawrence	Jack Faucett Associates
Arthur Lee	Chevron
Martin Lee-Gosselin	Université Laval
Paul Leiby	Oak Ridge National Laboratory
Lewison L. Lem	California State Automobile Association
Timothy Lipman	UC Berkeley & UC Davis
Alan Lloyd	California Environmental Protection Agency
Michael Love	Toyota Motor Sales
Jeffra Lyczko	DaimlerChrysler
Reza Mahdavi	California Air Resources Board
James R. Mahoney	National Oceanic & Atmospheric Administration
Eiji Makino	Nissan Technical Center North America
Brad Markell	United Auto Workers
Paul Marx	U.S. DOT—Federal Transit Administration
Shomik Mehndiratta	World Bank
Alan Meier	International Energy Agency
Marc Melaina	ITS-Davis
Martine Micozzi	Transportation Research Board
Christopher Miller	Senate Committee on Environment and Public Works
Marianne Mintz	Argonne National Laboratory
Reg Modlin	DaimlerChrysler

Continued

Asilomar Attendee List: 2005—cont'd

First & Last Name	Affiliation
Michael Murray	Sempra Energy
Paul Nieuwenhuis	Cardiff University
Joan Ogden	ITS-Davis
Margo Oge	U.S. Environmental Protection Agency
Patrick Oliva	Michelin
Gary Oshnock	DaimlerChrysler
Ben Ovshinsky	Energy Conversion Devices—Ovonics
Renee Pearl	ITS-Davis
Mark Pedersen	Air Products & Chemicals
D. Louis Peoples	Nyack Management Company
Joanne Pereira-Ekstrom	Natural Resources Canada
Joseph Perkowski	Idaho National Laboratory
Richard Plevin	UC Berkeley
Steve Plotkin	Argonne National Laboratory
Joanne Potter	Cambridge Systematics
Jim Ragland	Aramco Services Company
Chella Rajan	Tellus Institute
David Raney	American Honda Motor Co.
Sharima Rasanayagam	British Consulate
Pana Ratana	Shell Hydrogen
Peter Reilly-Roe	Natural Resources Canada
David Reiner	University of Cambridge
Birgitta Resvik	Confederation of Swedish Enterprise
Conor Reynolds	University of British Columbia
Edward Rubin	Carnegie Mellon University
Ichiro Sakai	American Honda Motor Co.
Lee Schipper	EMBARQ—World Resources Institute
George Schuette	ConocoPhillips
Yoshiki Sekiya	Nissan North America
Susan Shaheen	UC Berkeley
Jananne Sharpless	California Institute for Energy Efficiency
David Shearer	California Environmental Associations
Charles Shulock	California Air Resources Board
Clare Sierawski	U.S. Department of Transportation
Harrison Sigworth	Chevron
Fred Silver	WestStart-CALSTART
Christine Sloane	General Motors
Gail Slocum	Pacific Gas & Electric
George Smith	Caltrans
Sabrina Spatari	University of Toronto
Dan Sperling	ITS-Davis
Brian T. Stokes	Pacific Gas & Electric
George Sverdrup	National Renewable Energy Laboratory
Tabitha Takeda	Natural Resources Canada
Fujio Takimoto	Subaru Technical Research Center

Asilomar Attendee List: 2005—cont'd

First & Last Name	Affiliation
Dean Taylor	Southern California Edison
Margaret R. Taylor	UC Berkeley
Miles Tight	Leeds University—Institute for Transport Studies
Kazuo Tomita	Toyota Motor Sales
Luke Tonachel	Natural Resources Defense Council
Per-André Torper	Ministry of Transport and Communications Norway
Gary Toth	New Jersey Department of Transportation
Ahsha Tribble	National Oceanic & Atmospheric Administration
Thomas Turrentine	ITS-Davis
Eileen Tutt	California Environmental Protection Agency
Stefan Unnasch	TIAX
Henry Wedaa	Valley Environmental Associates
Peter Wells	Cardiff University
Bill West	Southern California Edison
Cato Wille	Statoil
Steve Winkelman	Center for Clean Air Policy
Emily Winston	ITS-Davis
John Woody	Pew Center on Global Climate Change
Fuminori Yamanashi	Nissan Technical Center North America
Christopher Yang	ITS-Davis
Akimasa Yasuoka	American Honda Motor Co.
Rick Zalesky	Chevron
Jimin Zhao	University of Michigan
Amy K. Zimpfer	U.S. Environmental Protection Agency

Index

Page numbers with "t" denote tables; those with "f" denote figures

A

ACEA agreement, 154–155
Africa, 184
Air pollution, 199, 263
Allocations, 96–97, 103–104
Alternative Motor Fuel Act of 1988, 83
Ambiguity
　description of, 241
　oil peaking as, 242–244
Arctic National Wildlife Refuge, 208–209
Asia, 129
Asian Development Bank, 173
Asilomar Conference, 253–265
Asilomar Declaration, 5–7, 60–61
Attentive public, 203
Australia, 145t, 147
Automobile(s). *See* Vehicle(s)
Automobile exhaust, 35
Automobile manufacturers. *See also specific manufacturer*
　allocations, 96–97, 103–104
　Canadian, 107
　carbon burdens of
　　customer interest in reducing, 86
　　description of, 76–81
　　public policy involvement in reducing, 86
　　reductions in, 85–86
　carbon dioxide emissions, 73–75
　consumer influence on, 58
　Corporate Average Fuel Economy standards. *See* Corporate Average Fuel Economy standards

Automobile manufacturers *(Continued)*
　emission reduction credits for, 152–153
　feebate effects on, 118
　flexible-fuel vehicle credits, 77
　tax incentives for, 54t, 57t

B

Behavioral approach, alternative
　application of, 232–233
　attributes, 230–231
　decisions and actions, 222–225
　individuals in a social context, 225–226
　overview of, 221–222
　summary of, 234
Bicycling, 174–178, 186
Biennial Conference on Transportation Energy and Environmental Policy, 2–3
Biodiesel, 19, 53t, 56t
Biofuels
　disadvantages of, 37–38
　land capacity constraints and, 38
　transportation fuel use reduced by using, 69
Biomass
　ethanol from, 16–17, 19
　mandate for, 62
BMW, 80–81
Bogota, 193
Brazil
　ethanol use in, 37
　per capita energy demand in, 33t

Bus rapid transit programs, 178–182
Bus transportation, 168, 169t

C

CAFE. *See* Corporate Average Fuel Economy
California
  emission reduction credits for manufacturers, 152–153
  fuel economy standards in, 145t, 151–153
  greenhouse gas emissions in, 146, 153
California Air Resources Board, 151–152
Canada
  automobile manufacturers, 107
  company average fuel consumption standards in, 113, 146, 153–154
  feebates program in, 107–126
  fuel consumption in, 115–118
  fuel economy standards in, 145t, 153–154
  light-duty vehicle market in, 111–113
Cap-and-trade programs
  description of, 89, 259
  downstream, 94–99
  studies of, 89–93
  upstream, 93–94
  upstream/downstream hybrid, 99–101
Car(s). *See* Vehicle(s)
Carbon burden
  of automakers, 76–81
  of automobiles, 73–74
  reductions in, 85–86
  trends that affect, 81–85
Carbon capture and sequestration, 244, 249
Carbon content reductions, 100–101
Carbon dioxide emissions
  ACEA agreement, 154–155
  calculation of, 99–100
  climate changes and, 248
  in European Union, 144, 146, 154, 156
  global increases in, 3, 4t
  by Honda vehicles, 79–80
  by light-duty vehicles, 75–76, 100
  by Nissan vehicles, 80
  reduction of, 74, 232f
  tons per year, 75
  by Toyota vehicles, 79

Carbon dioxide emissions *(Continued)*
  transportation-related
    in Asia Pacific region, 255–256
    description of, 4, 73, 255
    increases in, 41
    policy approaches, 41–42, 260
    regulatory approach to, 258
  from unconventional hydrocarbons, 244
  by U.S. automobiles, 73–75
  "wedge and slices" approach to, 29–30
Carbon dioxide flooding, 15
Carbon intensity allocation, 97
Carbon repositories, 244
Carbon trading programs. *See also* Cap-and-trade programs
  description of, 89, 259
  hybrid, 99–103
  incorporating vehicles into, 101, 259
Cellulosic ethanol, 53t, 56t, 64, 232
China
  carbon dioxide emissions in, 4t
  fuel economy standards in, 145t, 147, 157–160
  per capita energy demand in, 33t
  population growth in, 30, 31f
Clean Air Act, 213, 258
Clean development mechanism
  description of, 191–192, 264
  in Latin America, 192–200
Clean Water Act, 213
Climate
  in developing countries, 263
  greenhouse gas emissions' effect on, 2
  warming of, 2
Climate change
  global, 256
  policy agenda for, 265–266
  rapid and irreversible, 245–246
  research of, 256
  transportation policy agenda for, 253–266
  United Nations Framework Convention on Climate Change, 1
Climate Stewardship Act, 101
Coal
  liquefaction of, 16
  transportation uses of, 38
Committee on Nuclear and Alternative Energy Systems, 58
Community planning, 132–133

Company average fuel consumption, 153–154
Congestion. *See* Traffic congestion
Congestion pricing, 183, 194
Connectivity, for traffic congestion reduction, 136–137
Constant dollar rate per liter feebate, 110
Consumers
  changing the behavior of, 209–210, 257–258
  feebate effects on, 118
  outputs of, 229–230
  preexisting preferences, 229–230
  rational, 219–221
  role of, 209–210
Context sensitive street designs, 136–137
Corporate Average Fuel Consumption standards, in Canada, 113, 146
Corporate Average Fuel Economy standards
  average speed, 160
  changes in, 150
  company average fuel economy standard vs., 154
  creation of, 150
  criticism of, 60
  customer interest in, 86
  description of, 19, 45, 50t, 75, 146
  efficacy assessments for, 52t
  entrepreneurial classification of, 60
  flexible-fuel vehicles and, 77
  fuel tax increases and, 61
  implementation of, 55t
  increases in, 69
  for light trucks, 150–151
  light-duty vehicles, 81
  study of, 59
Crash program mitigation, 21–24, 26
Cross-elasticity, 61
"Crossover" vehicles, 81
Cuenca, 193–194, 197t
Curitiba, 193

D

DaimlerChrysler, 78–79
Dar es Salaam project, 181
Decentralization of governance, 195
Decision processes, 222–223
Delaware Valley Regional Planning Commission, 139
Delayed wedge approximation, 18, 18f

Developing countries
  greenhouse gas emissions in, 189–190, 262–263
  infrastructure in, 189–190
  traffic congestion in, 190
Developing economies, 263
Diesel fuel
  carbon intensity of, 42
  European use of, 36
Diesel vehicles
  as light-duty vehicles, 43
  efficiency of, 44
  in European Union, 155
  gasoline-powered vehicles vs., 44
Discourses, 234
Double counting, 95–96, 103
Downsizing, 136
Downstream trading
  accounting issues, 97–98
  double counting, 95–96, 103
  manufacturer allocations, 96–97, 103–104
  overview of, 94–95
Drivetrains, hybrid electric, 14–15, 19, 38

E

E85 fuel economy, 83
Economic rationality, 220–221
Economics, 217
Education, 214
Electricity
  greenhouse gas emissions reduced by using, 38
  liquid fuel shift to, 20
  sources of, 38
Emissions. *See* Carbon dioxide emissions; Greenhouse gas emissions
Emotion, 224
End-use efficiency, 19
Energy conservation
  fuel efficiency, 14–15
  technologies for, 219–220
Energy efficiency, 241–242
Energy Policy and Conservation Act, 84, 150, 241
Energy use
  gross domestic product and, correlations between, 33, 34f
  by transportation
    future increases in demand for, 33, 34f

Energy use *(Continued)*
  percentage of, 165
  policy trends, 35–37
Enhanced oil recovery technology, 15
Environmentalism, 211–213
Environmentalist, 202
Ethanol
  from biomass, 16–17, 19, 37
  cellulosic, 53t, 56t, 64, 232
  efficacy assessments of, 53t
  implementation considerations for, 56t
European Union
  ACEA agreement, 154–155
  carbon dioxide emissions in, 144, 146, 154, 156
  diesel use in, 36, 155
  fuel economy standards in, 145t, 147, 154–156
  fuel taxes in, 144
  per capita energy demand in, 33t
  vehicle standards in, 259
European Union Emissions Trading Scheme, 35

F

Feebates
  benefits of, 62
  Canadian system, 107–126
  constant dollar rate per liter, 110
  consumer effects, 118
  definition of, 107, 260
  design of, 260
  efficacy assessments of, 51, 52t
  fuel consumption affected by, 64, 115–116, 119, 260
  fuel economy improvements from, 108
  greenhouse gas emission reductions and, 259
  implementation of, 56t
  North American. *See also* North American Feebate Analysis Model
    analysis of, 108–109
    model of, 109–111
  obstacles to implementing, 120
  summary of, 126
  types of, 110
  in United States, 121–123
Fiscal incentive programs, 144
Fisher-tropsch process, 16
Fleet averages, 147, 149f, 163

Flexible-fuel vehicles
  alternative fuels for, 98, 104
  credits for, 77–78, 84
  description of, 77
  E85 fuel economy, 83
  by Ford, 78
  fuel economy, 83–84
  future of, 98
  incentives for, 83
Ford
  carbon dioxide emissions, 78
  hybrid electric vehicles, 84
Former Soviet Union
  carbon dioxide emissions in, 4t
  per capita energy demand in, 33t
Fossil fuels
  carbons released from, 89, 255
  energy derived from the burning of, 31
Freeways, 134
Fuel
  alternative, 53t, 56t, 98, 104
  diesel. *See* Diesel fuel
  future pathway for, 36f
Fuel cells
  description of, 44
  hydrogen, 167–169
Fuel consumption
  decreased, 63
  feebate effects on, 64, 115–116, 119, 122, 260
  greenhouse gas emissions and, 161
  by hybrid electric vehicles, 45
  integrated policy approach to, 60–69
  by light-duty vehicles
    average, 46, 47f, 65f, 112, 116
    delayed action scenarios, 48–49
    feebate eligibility, 120
    policy measures to reduce, 50t, 58–69
    projections for, 44–49
    scenarios, 45–47
  policy scenarios for, 64–65
  scenarios regarding, 45–46
  vehicle price elasticity effect on, 124
Fuel economy
  fiscal incentive programs for, 144
  greenhouse gas emissions affected by, 236
  information sources about, 111
  National Research Council studies of, 111, 116

Fuel economy *(Continued)*
  North American Feebate Analysis Model, 123
  of flexible-fuel vehicles, 83–84
  programs to increase, 144
  standards for, 59
  symbolic meaning of, 234
  test driving cycles used to establish, 160–161
Fuel economy market, 58–59
Fuel economy standards
  in Australia, 145t
  in California, 145t, 151–153
  in Canada, 145t
  in China, 145t, 147, 157–160
  comparison of, 147, 148f–149f
  in European Union, 145t, 154–156
  fleet averages, 163
  forms of, 146
  global
    description of, 144
    issues and methodologies involved in comparisons of, 160–163
  in Japan, 145t, 147, 156–157
  regulatory approach to, 161
  in South Korea, 145t
  summary of, 163–164
  in Taiwan, 145t
  in United States, 145t, 150–151
  voluntary approach to, 161
Fuel efficiency
  increases in, 14–15
  promotion measures for, 145t
  vehicle travel increased because of improvements in, 51
Fuel prices, 67
Fuel tax
  Corporate Average Fuel Economy standards and, 61
  description of, 50t
  efficacy assessments for, 52t
  in European Union, 144
  fuel economy standards and, 59
  implementation of, 55t

G

Gasification, of coal, 16
Gasoline costs, 93
Gasoline tax, 62
Gas-to-liquids process, 16
Gdansk project, 175–176
GEF. *See* Global environmental facility

General Motors
  carbon dioxide emissions, 78
  hybrid electric vehicles, 84
GHG emissions. *See* Greenhouse gas emissions
Global environmental facility
  bicycle infrastructure projects, 174–178
  bus rapid transit programs, 178–182
  description of, 165–166, 191, 264
  funding of, 170–172
  Gdansk project, 175–176
  hydrogen fuel cells, 167–169
  in Latin America, 172, 192–200
  Lima project, 176
  Marakina project, 175
  nonmotorized transport projects, 174–178, 184–186
  Operational Program #11, 166–167, 191
  rickshaws, 185–186
  Santiago de Chile project, 176–177
  sustainable transportation projects, 191
  traffic avoidance, 183–184
  traffic demand management, 182–183
  transport priorities of, 169–174
  transportation-focused projects sponsored by, 166
  travel blending, 186–188
  United Nations Development Programme funding of, 170–171
  United Nations Environment Programme, 166, 173
  World Bank funding and influence, 170–171, 173–174
Global warming, 85, 89, 207, 211–212
Governance reform, 199
Greenhouse gas emissions
  in California, 146, 153
  cap-and-trade programs for. *See* Cap-and-trade programs
  climate effects of, 2
  in developing countries, 189–190, 262–263
  equation for calculating, 42
  fleet averages, 147
  fuel consumption and, correlation between, 161
  fuel economy effects on, 236
  global comparison of, 147, 148f–149f
  by light-duty vehicles, 42–44, 73
  modes of, 168t
  public attitudes about, 36

Greenhouse gas emissions *(Continued)*
  reductions in
    Canadian government efforts, 107
    description of, 233
    development projects for, 262
    electricity use for, 38
    governmental support of, 39
    mandatory, 1
    policy measures for, 49–58, 55t–57t
    technologies for, 69, 108, 258
    from traffic demand management, 182–183
    voluntary, 1, 35
  sources of, 2, 168t
  test driving cycles used to establish, 160–161
  transportation-related
    description of, 2, 165, 168t, 253
    in developing countries, 189–190, 262–263
    growth of, 165–166
    increases in, 189, 257
  in United States
    description of, 73
    main sources of, 2
    strategies to reduce, 2
  upstream production of, 255
Gross domestic product, 33, 34f

H

Heavy oil, 15–16
Honda, 79–80
Household, 221
Hybrid electric vehicles
  automakers who produce, 84–85
  carbon dioxide emissions reduced using, 29–30, 85
  consumer use of, 230–231
  description of, 38, 74, 84–85, 224
  drivetrains, 14–15, 19, 38
  energy reduction benefits of, 44
  fuel consumption by, 45
  fuel economy benefits of, 14, 218
  high-efficiency vehicles, 231
  increasing sales of, 81
  light-duty vehicles, 14
  market share of, 85
  medium-duty vehicles, 15
  performance-tuned, 236
  sales, 85
  in symbols, 231
  transactions, 224

Hydrocarbons, 243
Hydrogen
  description of, 17
  from natural gas, 30
  power stations powered by, 30
Hydrogen fuel cells, 167–169
Hyundai, 80

I

Ignorance, 241, 245–246
Imitation, 226
Incremental efficiency, 232
India
  carbon dioxide emissions in, 4t
  per capita energy demand in, 33t
Induced demand phenomena, 131, 132f
InterAmerican Development Bank, 191–192
Intergovernmental Panel on Climate Change, 2
Intermediate technology development group, 185–186
Internal combustion engines, 44, 46
Issue framing, 210

J

Jakarta, Indonesia project, 181–182
Japan
  carbon dioxide emissions in, 4t
  fuel economy standards in, 145t, 147, 156–157
  per capita energy demand in, 33t
Japanese Automobile Manufacturers Association, 155

K

Korean Automobile Manufacturers Association, 155
Kyoto Protocol
  background on, 1
  Canada's agreement to, 107, 154
  clean development mechanism, 191–192
  opposition to, 1
  United States participation in, 206

L

Land use planning, 261
Latin America, 172, 192–200
Life politics, 234

Light-duty vehicles
 in Canada, 111–113
 carbon dioxide emissions by, 75–76, 100, 232f
 corporate average fuel economy standards for, 150–151
 by DaimlerChrysler, 78–79
 definition of, 150
 diesel vehicles as, 43
 by Ford, 78
 fuel consumption by
  average, 46, 47f, 65f, 112, 116
  feebate eligibility, 120
  policy measures to reduce, 50t, 58–69
  projections for, 44–49
  scenarios, 45–47
 fuel economy of, 82–83
 fuel efficiency requirements for, 14
 by General Motors, 78
 greenhouse gas emissions by
  description of, 42–44, 73
  projections for, 44–49
 heavier, 77
 hybrid electric technology for, 14
 increased sales of, 81–83
 market share of, 82
 travel statistics, 65f
 in United States
  description of, 111–113
  increases in, 43
  sales of, 111t
Likelihood thinking, 239
Lima, Peru
 bicycle projects in, 176
 bus rapid transit programs in, 180
Liquefied natural gas, 16
Local air quality, 35
London, 194
Low-emission vehicle program, 152

M

Manufacturers. *See* Automobile manufacturers
Marakina project, 175
McCain-Lieberman Climate Stewardship Act, 101
MDBs. *See* Multilateral development banks
Medium-duty vehicles, 15
Megacities, 29–31
Mexico City, 180
Mitsubishi, 80
Mobility 2030, 85

Multilateral development banks
 clean development mechanism projects, 191–192, 264
 inadequacy in, 196
 sustainable transportation projects, 191
 unmet needs of, 196

N

Narrative identities, 218
Narratives, 235
National Environmental Education and Training Foundation, 204–205
National Research Council, 111, 116
National Science Foundation, 204
Natural gas
 hydrogen derived from, 30
 liquefied, 16
 in stranded locations, 16
New Hampshire Department of Transportation, 140
New Jersey, 135–140, 261
New Jersey Future in Transportation, 135–136, 261
Nissan, 80
Nonmotorized transport
 in developing countries, 190
 Global Environmental Facility promotion of, 174–178, 184–186
North American Feebate Analysis Model
 assumptions, 114–115
 consumer decisionmaking, 113–114
 description of, 109
 fuel consumption effects, 115–116
 fuel economy technologies, 123
 manufacturer decisionmaking, 113–114
 pivot points, 117–120
 results, 115–125
 structure of, 109–111
 U.S.-specific findings, 121–123
 vehicle price elasticity, 124, 125f
NRC. *See* National Research Council

O

Obesity, 130, 131f, 135
Oil
 foreign, U.S. dependence on, 49

Oil *(Continued)*
  formation of, 10
  global demand for, 9, 12, 257
  heavy, 15–16
  nontransportation uses of, 20
  production of, 243
  unconventional types of, 15–16
Oil consumption, 31, 32f
Oil demand. *See* Oil supply and demand
Oil embargoes
  of 1973, 12–13
  of 1979, 12–13
Oil peaking
  ambiguity of, 242–244
  description of, 9–11
  economic effects of, 26
  factors that could affect, 24–25
  mitigation scenarios for
    candidates, 18–20
    crash program mitigation results, 21–24
    delayed wedge approximation used in, 18, 18f
    overview of, 17–18
  mitigation strategies for
    biomass, 16–17
    coal liquefaction, 16
    energy conservation, 14–15
    gas-to-liquids, 16
    heavy oil, 15–16
    hydrogen, 17
    improved oil recovery, 15
    oil sands, 15–16
  modeling of, 20–21
  predictions for, 24–25
  price increases secondary to, 14
  projections of, 12, 12t
  risk management considerations, 24
  world production at, 21
Oil price
  alternate fuel methods and, 26
  increases in, 10, 13, 143
Oil recovery, 15
Oil reserves, 11–12
Oil reservoirs, 10–11, 11f
Oil sands, 15–16
Oil supply and demand
  forecasts for, 21
  global, 20–21
  increases in, 143
  modeling of, 21
  regional imbalances in, 31, 32f

Oil supply disruptions
  description of, 245
  economic effects of, 13, 25
  history of, 13–14
  from terrorist attacks, 245
Options analysis, 250–252
Organization of Economic Cooperation and Development, 13, 69
Outcome thinking, 239

P

Partitioning, of risk, 241–246
PATP, 53t, 56t
Pennsylvania Department of Transportation, 139
Political activism, 229
Population growth
  in China, 30, 31f
  global statistics, 30
Portland Area Comprehensive Transportation Study, 140
Poverty, 172, 185
Price-Anderson Act, 244
Prius, 84
Public awareness, 202–204
Public health
  deterioration of, 130, 135
  vehicles miles traveled and, 130, 135
Public opinion
  information effects on, 207–208
  strength of, 208–209
Public opinion polls, 201
Public support, 214
Public transportation, 192–200. *See also* Transportation
Public understanding, 204–207

R

Rational consumers, 219–221
RD&D, 54t, 57t
Real options analysis, 250–252
Reasoned judgment, 223
Rebound effect, 43, 68f, 229
Rickshaws, 185–186
Risk
  energy efficiency investing and, 241–242
  historical experience and, 240
  partitioning of, 241–246
  time-dependent nature of, 240
Risk sharing, 244

Roadway design practices, 136–137
Russia, 33t. *See also* Former Soviet Union

S

Santiago, Chile
  bicycle transportation projects in, 176–177
  bus rapid transit programs in, 180–181
Scenario analysis, 248
Scenario matrix, 248
Scenario planning
  in action, 248–250
  alternative approaches to, 246
  example of, 248–250, 262
  federal agency use of, 246
  practices involved in, 246–247
  principles of, 247
  real options analysis and, 248
Scenario thinking, 247
Self-identity, 222, 227–229
Self-interest, 227–229
Shale oil, 19
Singapore, 194
Smart consumerism, 225, 258
Social constructionist, 221, 233
Social discourses, 234
Social exchanges
  description of, 225
  imitation, 226
  personal, 226
Social marketing, 186–188
South Korea, 145t
Soviet Union, former
  carbon dioxide emissions in, 4t
  per capita energy demand in, 33t
Sprawl factor, 131–132
Street design and planning, 134, 136–137
"Super giants," 10
Sustainable Mobility Project, 33, 35
Sustainable transportation
  air pollution control and, 263
  description of, 190–191
  donor agencies involved in, 191
  elements of, 263
  establishing of, 263
  Global Environmental Facility participation in, 191
  multilateral development banks in, 191
Symbols, 231, 236

T

Taiwan, 145t
Takebacks, 43, 63
Tax incentives, 54t, 57t
Technologies
  energy conserving, 219–220
  greenhouse gas emissions reductions through, 69, 108, 258
  transportation and, 37–39
Terrorism, 245
Test driving cycles, 160–161
Texas Transportation Institute, 130, 261
Total carbon approach, 96–97
Toyota
  carbon dioxide emissions, 79
  hybrid electric vehicles, 84
Traffic avoidance, 183–184
Traffic calming, 137
Traffic congestion
  community planning effects on, 132–133
  Department of Transportation efforts to reduce, 261
  in developing countries, 190
  increases in, 130
  New Jersey model of, 135–140, 261
  poor land use and, 135
  reasons for growth of, 131–133
  relief approaches for, 135–140
  roadway design practices and, 136–137
  sprawl factor and, 131–132
  state programs to reduce, 135–140
  Texas Transportation Institute findings regarding, 130, 261
  transportation approach to, 129–134
  in United States, 131–133
Traffic demand management, 182–183
Traffic development patterns, 133–134
Traffic modeling, 137
Transportation. *See also* Light-duty vehicles
  air pollution from, 199
  Asilomar Declaration focus on, 5–6
  carbon dioxide emissions from
    in Asia Pacific region, 255–256
    description of, 4
    increases in, 41
    policy approaches, 41–42, 260
    regulatory policies, 258–259
  energy uses

Transportation *(Continued)*
  future increases in demand for, 33, 34f
  percentage of, 165
  policy trends, 35–37
  fuel efficiency
    barriers to improvements in, 49
    increases in, 14–15
    promotion measures for, 145t
    vehicle travel increased because of improvements in, 51
  Global Environmental Facility-sponsored projects, 166
  global increases in, 254–255
  greenhouse gas emissions from
    description of, 2, 165
    in developing countries, 189–190
    growth of, 165–166
    increases in, 189, 257
  land use planning and, 261
  in Latin America, 192–200, 254
  nonmotorized, 174–178, 184–186
  reforms in, 193
  sustainable. *See* Sustainable transportation
  technological changes in, 37–39
  upstream cap-and-trading scheme effect on, 94
Transportation demand management, 51
Transportation master plans, 200
Travel blending, 186–188

U

Uncertainty
  description of, 241
  oil disruption, 245
United Nations Development Programme, 165, 170–171, 191
United Nations Environment Programme, 166, 173, 191
United Nations Framework Convention on Climate Change, 1
United States
  automobiles in
    carbon dioxide emissions, 74–76
    Corporate Average Fuel Economy requirements for, 19
    greenhouse gas emissions by, 73
  carbon dioxide emissions in, 4t, 73
  feebates in, 121–123
  foreign oil dependence by, 49

United States *(Continued)*
  fuel economy standards in, 145t, 150–151
  greenhouse gas emissions in
    description of, 73
    main sources of, 2
    strategies to reduce, 2
    in Kyoto Protocol, 206
  light-duty vehicles in
    description of, 111–113
    increases in, 43
    sales of, 111t
  per capita energy demand in, 33t
  traffic congestion in, 131–133
  traffic development patterns in, 133–134
Upstream emissions, 255
Upstream trading
  description of, 93–94
  hybrid, 99–103
Urban development master plan, 195, 199
Urbanization, 31

V

Vehicle(s)
  carbon burden of, 73–74
  carbon dioxide emissions by, 4
  in carbon trading programs, 101, 259
  classification of, 73
  consumer concerns when buying, 58
  diesel. *See* Diesel vehicles
  efficiency of, 260–261
  flexible-fuel. *See* Flexible-fuel vehicles
  fuel consumption by
    average range for, 42
    delayed action scenarios, 48–49
    oil embargo effects on, 42
    reductions in, 3, 42
    travel reductions and, 48
  global increases in, 254–255
  greenhouse gas emissions by, 73
  hybrid electric. *See* Hybrid electric vehicles
  increased driving per year, 43
  increases in, 29
  light-duty. *See* Light-duty vehicles
  manufacturers of. *See* Automobile manufacturers
  medium-duty, 15
  new

Vehicle(s) *(Continued)*
    fuel consumption by, 42, 45–46
    sales increases in, 45, 48
    older, retirement of, 53t, 57t
    performance features of, 51
    regulation of, 259
    technological changes in, 43
Vehicle kilometers traveled, 43
Vehicle miles traveled
    development projects to reduce, 262
    obesity and, 130, 131f
    public health effects, 130, 135
    reduction in, 129, 135–140
Vehicle price elasticity, 124, 125f

Vehicle travel
    decreased fuel consumption effects on, 63
    fuel efficiency effects on, 51
Vermont Agency of Transportation, 140
Volkswagen, 80

W

World Bank, 170–171, 173–174
World Business Council for Sustainable Development
    description of, 33
    Sustainable Mobility Project, 33, 35